Imperial Borderlands

Mapping the Past

VOLUME 1

The titles published in this series are listed at *brill.com/mtp*

Imperial Borderlands

Maps and Territory-Building in the Northern Indochinese Peninsula (1885–1914)

By

Marie de Rugy

Translated by

Saskia Brown

BRILL

LEIDEN | BOSTON

Library of Congress Cataloging-in-Publication Data

Names: Rugy, Marie de, author. | Brown, Saskia, translator.
Title: Imperial borderlands : maps and territory-building in the northern Indochinese Peninsula (1885-1914) / by Marie de Rugy, translated by Saskia Brown.
Other titles: Aux confins des empires, cartes et constructions territoriales dans le nord de la péninsule indochinoise (1885-1914). English
Description: Leiden ; Boston : Brill, [2022] | Series: Mapping the past, 2589-9945 ; vol. 1 | Includes bibliographical references and index.
Identifiers: LCCN 2021031774 (print) | LCCN 2021031775 (ebook) | ISBN 9789004456211 (hardback) | ISBN 9789004469853 (ebook)
Subjects: LCSH: Cartography—History—18th century. | LCGFT: World atlases. | World maps. | Early maps.
Classification: LCC G1025 .R7 2021 (print) | LCC G1025 (ebook) | DDC 911/.59—dc23
LC record available at https://lccn.loc.gov/2021031774
LC ebook record available at https://lccn.loc.gov/2021031775

Typeface for the Latin, Greek, and Cyrillic scripts: "Brill". See and download: brill.com/brill-typeface.

ISSN 2589-9945
ISBN 978-90-04-45621-1 (hardback)
ISBN 978-90-04-46985-3 (e-book)

This book is printed on acid-free paper and produced in a sustainable manner.

Contents

Foreword

It is an honour to have been granted a preview of Marie de Rugy's new book. *Imperial Borderlands* might seem narrowly restricted, to a very particular portion of the world over just three decades, but de Rugy contributes significantly to a still new direction in the study of mapping and spatial history. To appreciate its innovation and significance, we need to situate her work within the interdisciplinary arena of map history and within the approaches that have been taken to maps and empire.

Right from the start, map history has pivoted about Europe's empires and Western imperialism. When, during the Parisian cholera epidemic of Spring 1832, Alexander von Humboldt engaged in perhaps the first systematic inquiry into early maps – searching for the first maps to depict the new world and then seeking to ascertain their interrelationships – he was motivated by the transformation of Europe's geographical knowledge and so of European intelligence and rationality that the "discovery" of the new world had supposedly wrought. The doubling of creation, Humboldt argued, gave a "huge boost to knowledge" that "gradually changed opinions, laws and political customs. Never before had a purely material discovery in extending the horizon produced a more extraordinary and lasting moral change." The result, in line with the widely held stadial theories of cultural development, was the transformation of medieval, superstitious Europe into the modern, rational West.

Just a few years later, in 1840–42, the second viscount of Santarém, a Portuguese émigré in Paris, used early maps to prove the priority of the Portuguese discovery of the western coast of Africa. Portuguese ships under Gil Eames successfully rounded the treacherous Cape Bajador in 1434 and their pilots then mapped the coast, permitting the Portuguese government to claim the territory of Casamance, in Senegal, against French imperial claims.

As Europe's imperial reach intensified, early maps were used to demonstrate the triumph of the West, to foster sentiment in favour of further imperial aggrandisement, and to debate competing imperialistic claims. Perhaps most important in sustaining the racialised ideology of modern empire, maps were narcissistically held up as mirrors in which Westerners could see and admire their own true, rational self to the exclusion of the mystical and irrational non-European other. Cartography was a marker of inherent Western superiority. Non-Western maps were noticed only to the extent that they could be presented as the timeless works of a different kind of mind, of a non-scientific mentality. (One of my favourite examples is a 1904 essay that presented an "ancient Chinese map" as characteristic of an irrational and childlike spatial understanding that had lasted unchanged for thousands of years. It was in fact a nineteenth-century, printed Korean world map of the relatively recently developed style of *ch'ŏnhado*; the work might not look anything like a Western world map, but both genre and artefact have histories.)

It should therefore not be surprising that, when map scholars finally queried the nature of maps and cartography, there quickly emerged a new concern with the role of maps in creating and sustaining empires. Several factors, not least the "spatial turn" in the humanities and the "cultural turn" in the social sciences, led scholars to realise that maps are not simply statements of geographical fact. They began, especially after 1980, to study maps as cultural documents that are as rich and as complex as any work of art or literature and as social instruments by which their users have organised and modified the world. The effects of this sociocultural reconfiguration were widespread. Entirely new arenas of study have opened up as scholars quickly abandoned the traditional fixation on regional and marine mapping in the early modern West. Map history still featured many studies of early modern maps, especially in their interrelations with art and literature, but it now paid close attention to *all* kinds of maps and plans from antiquity to the present.

In studying maps as social instruments, the initial move was to argue that modern cartography emerged in the Renaissance as a tool of social and political elites, that cartography was integral to the formation of *modern* states and empires. Maps were instruments for creating and preserving unequal power relations. As Brian Harley stated in a 1988 essay, "the ideological arrows have tended to fly largely in one direction, from the powerful to the weaker in society"; he continued by asserting that the "social history of maps ... appears to have few genuinely popular, alternative, or subversive modes of expression. Maps are preeminently a language of power, not of protest." In this context, cartography has been construed as thoroughly, perhaps completely, implicated with Western imperialism, with the construction of imperialistic sentiment, and with the exercise of power within particular empires. Cartography – understood as embracing *all* kinds of mapping and especially the systematic territorial surveys of the modern era – is a central plank of the infrastructure of knowledge that undergirds modern empire.

Detailed studies of mapping and empire were not as monolithically one-sided as this conceptual position has seemed. They have engendered a great deal of new historical insight into the spatiality of modern empires and the necessity of geographical envisioning in early modern and modern imperialism. Most recently, a few studies have explored how geographical knowledge has shaped the actions of imperial agents. I think in particular of works such as

Max Edelson's *The New Map of Empire* (2017) and Ricardo Padrón's *The Indies of the Setting Sun* (2020).

Yet they all have perpetuated statist strictures. Even Thongchai Winichakul's *Siam Mapped* (1994), the first and perhaps the best of the new studies, focused on the Thai state's cartographic response to territorial pressures exerted by its neighbouring French and British empires. The scholarly emphasis has consistently been on specific empires (British India, Dutch East Indies) and specific postcolonial states. There are, of course, sound intellectual and archival reasons for such precise spatial focus, but it does preclude the kind of systemic, connected analysis now advanced by de Rugy, of British and French mapping in the same interior, upland, and contested region of Indochina.

A particular benefit of de Rugy's systemic analysis is that it substantially enlarges the relationship between imperial agents and the local Chinese and Vietnamese political structures from one solely of knowledge extraction (still absolutely important of course, as in James Scott's forceful collection of local maps) to embrace persistent knowledge encounters, negotiations, and circulations. De Rugy identifies a certain commensurability between imperial and local powers in their mutual interest in mapping communication routes. In the difficult terrain of upland Indochina, the systematic mapping of territory – the supposed *sine qua non* of Western spatial rationality – was simply irrelevant; what was spatially important were the linear traces of the roads and rivers by which the inlands were accessed.

De Rugy thus questions the now accepted role of mapping in modern empires, not to dispute its importance but rather to qualify it. Rather than placing maps at the heart of her work, which is of course where map scholars have always placed them, de Rugy decentres them. This intellectual act requires a fundamental change in perspective, to realize that neither Western spatial rationality nor cartography exist and that, even in the West, people have always had multiple spatial conceptions and pursued multiple mapping strategies. Even as modern imperialists deluded themselves that they thought rationally and that their rationally structured maps distinguished them from the (supposedly) lesser peoples they now ruled, they actually thought about the varied places and spaces of states and empires in several different ways that were often commensurate with some of the ways in which the local populace thought about and mapped spaces and places.

Imperial Borderlands deals with the territorial edges of empire, where the frayed natures of imperial systems are more than evident. We would be remiss, however, to think that the multiplicity of mapping strategies and the active interweaving of imperial and local mapping practices are a function solely of the frontier. De Rugy demonstrates that the future direction for studies of mapping and empire, and in map history generally, needs to displace and deprivilege "the map" and to turn instead to how people across cultures engage with maps without the *a priori* assumptions of "cartography."

Matthew H. Edney
University of Southern Maine
University of Wisconsin – Madison

Acknowledgements

This book, published in French in 2018, is a revised version of my doctoral thesis, which was awarded by the University of Paris 1 Panthéon-Sorbonne in 2016. It originated in a trip to Vietnam, the first of several, in 2010. All those who initiated me into the world of Asia, and who accompanied and supported me in my research should rightfully be thanked here. They will forgive me if I cannot mention them individually.

I have a particular debt of gratitude to my thesis supervisors, Hélène Blais, and Hugues Tertrais, as well as to Philippe Papin. This book is the fruit of their guidance. I thank the organisers and regular participants at the "Empires" seminar at the *Ecole normale supérieure*-rue d'Ulm, who helped shape my thinking and fuel my enthusiasm. I thank all the colleagues at the University of Paris 1 with whom I had the good fortune and honour to work, and the many researchers who gave me their time, and opened up new avenues on South-East Asia for me, on the history of colonisation and the history of cartography: they will recognise themselves in this description. During a three-month stay in London for archival research, I was particularly lucky to meet William Gervase Clarence-Smith, who agreed to be one of my thesis examiners, Felix Driver, who was always helpful and enthusiastic about my research, and Natascha Pairaudeau, whose advice accompanied me throughout the years.

I would never have been able to carry out research in several countries had it not been for the financial support of a number of bodies: the *Collège des Écoles doctorales*, the *Labex EHNE*, the *Centre d'histoire de l'Asie contemporaine* and the *École française d'Extrême-Orient*. I am also grateful to those who opened their homes to me during my sometimes lengthy research trips in Ho Chi Minh City, Yangon, Aix-en-Provence, and elsewhere. I thank Nguyễn Thu Hoài and Vũ Thị Mai Anh, François Tainturier, U Htun Yi and Tin Naing Win for deciphering the Vietnamese and Burmese maps for me. Not forgetting the staff at the various archives in which I worked, nor Laura Marguerritte for her patience in producing the maps.

The idea of an English translation of the book came when at Cambridge as a British Academy Newton International Fellow. My thanks go to Sujit Sivasundaram for providing the initial impetus for this project, to Tom Simpson for his sustained support, and to Wolfson College for contributing to funding the translation, and for enabling so many fruitful discussions on maps in South-East Asia. I also thank the translator Saskia Brown, and Brill for taking on the English-language publication as the opening volume in their new "Mapping the Past" Series.

Figures

Acronyms and Transcription Conventions

Acronyms

AIP	*Archives de l'Institut Pasteur* [Archives of the *Institut Pasteur*]
ANOM	*Archives nationales d'Outre-mer*, Aix-en-Provence [French National Overseas Archives]
ANM	National Archives of Myanmar, Yangon
ANV1	National Archives of Vietnam, Hanoi (centre 1)
BEFEO	*Bulletin de l'École française d'Extrême-Orient* [Bulletin of the French School of the Far East]
BL	British Library, London
BNF	*Bibliothèque nationale de France*, Paris [French National Library]
BSG	*Bulletin de la Société de géographie de Paris*, later, in 1900, *La Géographie*. [Bulletin of the Paris Geographical Society]
EFEO	*École française d'Extrême-Orient* [French School of the Far East]
GGI	*Fonds du gouvernement général de l'Indochine* [Records of the Government-General of Indochina, French National Overseas Archives]
IOR	India Office Records
MEP	*Archives de la Société des missions étrangères de Paris* [Archives of the Paris Foreign Missions]
RGS	Royal Geographical Society, London
RST	*Fonds de la Résidence supérieure du Tonkin* [Records of the Office of the Resident Superior of Tonkin, National Archives of Vietnam / French National Overseas Archives]
SGEOCOL	*Service géographique des colonies* [Geographical Office of the Colonies, French National Overseas Archives]
SHD	*Service historique de la Défense*, Vincennes [French Ministry of Defence Archives]

Transcription Conventions

Burmese nouns, in italics in the text, are transcribed using the commonest system in the English-language world, the *Guide to the Romanization of Burmese*, 1971, by John Okell. The tones have not been included.

The names of Chinese towns and provinces are give in *pinyin*, with the present name placed after it in square brackets, if it has changed since the end of the nineteenth century. Anachronistic names (for example, 'Vietnam' before 1945) are sometimes used, for clarity.

Introduction

On 5 June 1866, the French Mekong Exploration Commission, led by Captain Doudard de Lagrée, left Saigon. Its mission was clear: to follow the Mekong River upstream as far as possible, in order to test the river's navigability and to explore its banks. Just north of Vientiane, some local officials reported that English officers were advancing in the other direction, heading south from Luang Prabang:

> This last piece of news was a severe blow to us. We believed we had been outstripped, in the region we wished to explore, by a rival scientific expedition. The interest displayed by the English in the geographical exploration of northern Indochina, and the efforts they had already made in the preceding years, gave this announcement a degree of plausibility which left no room for doubt.[1]

Further reports from local dignitaries that a lavishly appointed British mission comprising a 40-strong contingent of Europeans had been sighted – a *topos* of the period – did nothing to allay French fears. Not only did this mean a loss of political influence and prestige, but it also dashed French scientific hopes of being the first to discover these unknown lands. However, Francis Garnier, the author of these reflections, and second in command of the expedition, sought to reassure his party by pointing out that since the British had left Burma and were already back in the south, they cannot have reconnoitred the sources of the Mekong. The French could therefore still arrive first at the uppermost reaches, where no European had yet set foot, even if the British had beaten them to it in the middle section of the river. This idea worked like magic:

> Rivalry over scientific discoveries is a tremendously powerful spur. The despondency which had gripped us when we first saw that we had been outstripped became a stimulant which roused us to greater fervour and a renewed faith.[2]

This was the frame of mind in which the French party eventually encountered the anxiously awaited British mission. The latter proved to comprise

1 Francis Garnier, *Voyage d'exploration de l'Indo-Chine*, Geneva, Olizane, 2009, p. 236.
2 *Ibid.*, p. 237.

 | DOI:10.1163/9789004469853_002

not the dozens of Europeans and their sumptuous escort, as announced, but one man alone, simply dressed, who greeted them in their own language. For the Dutch-born Mr Duyshart had been employed in the French colony of Cochinchina before being appointed Permanent Geographer to the King of Siam. Both parties were very surprised, and the unwarranted alarm generated was blamed on exaggerated reports by the native population. The relief was palpable, the atmosphere thawed instantly, and the two parties exchanged information on the roads to take and the geographical coordinates of various locations. In retrospect, Garnier considered this unexpected encounter and, thanks to it, the vital information obtained concerning the upper course of the Mekong, to be the highlight of his journey since leaving Saigon.

This episode is interesting on several accounts. The exploratory missions undertaken in the second half of the 19th century by the French and the British in the northern Indochinese Peninsula express the rivalry between them as they rushed to take possession of foreign territories. The image each party had of the other, the fears of being outstripped or of conveying a less admirable image to the local population, generated political tensions, but mixed with a real scientific curiosity, since what both the French and the British wanted most of all was to be the first to discover these as yet unknown lands ("unknown" because untouched by Europeans). Yet when these adversaries happened to come upon one another in the far-flung confines of some Laotian principality, cooperation and the sharing of scientific data seems to have been more common than has been suggested hitherto. This episode also implies that the local population played the role of intermediaries in the European exploration of these spaces, and in the encounters that took place.

In the decades following the French Mekong expedition, throughout the 1880s, the northern Indochinese Peninsula became the locus of collaboration, but also rivalry, between the imperial powers, in an accelerating race for territories that resembled, in form if not in extent, the contemporaneous "Scramble for Africa". The area, inhabited by mountain-dwellers – today called ethnic minorities – was ruled by a number of shifting sovereign powers, variously Burmese, Siamese, Chinese and Vietnamese. After the third Anglo-Burmese War, in 1884–1886, Upper Burma came into British hands, and the French established a Protectorate in Annam-Tonkin. To govern their possessions effectively, the imperial powers had to demarcate, and map, these new territories. What this book aims to show is that the ambiguities of colonial domination appear with particular acuity in the mapping of an empire's borderlands, those liminal zones that attract disproportionate attention during border demarcation, only to be abandoned by the central authorities later.

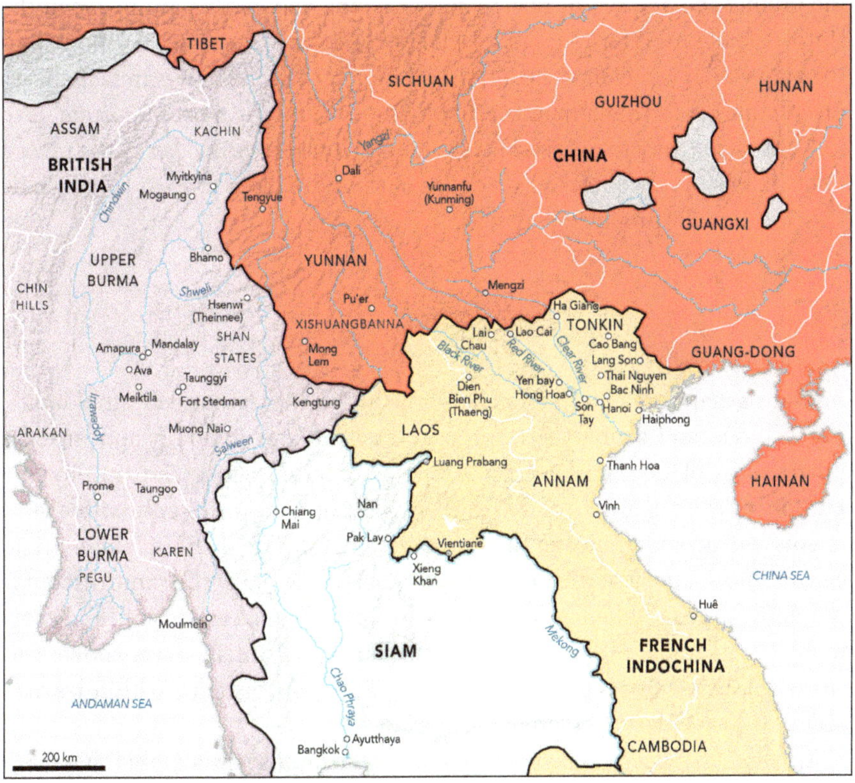

FIGURE 1 General map of the northern Indochinese Peninsula in 1901

For both Britain and France, the territorial gains of the 1880s simply extended an established colonial presence: British India (the Raj), and Cochinchina and Cambodia, respectively. Both countries came into possession of new territories simultaneously, and both tried to shape their resultant expanded lands according to some distinctive form and organisational principle. This is why their territory-building only fully makes sense when analysed side by side, from a joint Franco-British perspective, which includes Siam and China. This brings us to another aim of this book, which is to show that although the colonial territories are mapped at different times by the British and the French, sometimes using different scientific methods, their objectives and practices are largely the same.

Few of the protagonists involved conceived the mapping project in the same way: ministers, diplomats, explorers, missionaries, colonial administrators and surveyors each held different views. Some discourses conjure up the vision of an organised map-making programme, but the complex processes on the ground tell a different tale.[3] There is a long and costly path between the simple sketch used for reconnaissance purposes and the printed topographical map which is meant to be a finished product (although it is in fact always provisional). A further aim of this book, and perhaps the most important, is to show that, contrary to received ideas, maps actually have had very little power, even in a colonial context, if one goes beyond the discourses to look at how they are actually produced and used.

1 A History of Maps and Territories

Insofar as maps are commissioned by the authorities, are used in diplomatic negotiations, and enable orientation in the field, they naturally appear to be instruments of domination.[4] With this assumption about cartography, the Europeans can be understood as exploring distant territories only with a view to conquering and annexing them, thereafter imposing a one-sided and rigid geographical representation on their colonial possessions. The same discourse presents the geographer as a fervent advocate of colonial expansion, and geography as the imperialist science par excellence.[5] Historians have shown this close link between geography and colonisation, for instance within learned societies, particularly Geographical Societies.[6]

Yet were knowledge and power, maps and territory, so inseparable? Research has shown that this is by no means evident. Sometimes the efforts of mapmakers fell on deaf ears, and sometimes mapmakers did not toe the political line.[7] The discrepancies between political and scientific discourses are particularly glaring in this field. Discourses on space should consequently be analysed in

---

3 Discrepancies between the ideas on paper, and the reality, are a standard feature of the colonial situation. See Ann Laura Stoler, *Along the Archival Grain. Epistemic Anxieties and Colonial Common Sense*, Princeton, Princeton University Press, 2009; Frederick Cooper, Ann Laura Stoler (eds), *Tensions of Empire. Colonial Cultures in a bourgeois World*, Berkeley, University of California Press, 1997.

4 John Brian Harley, *The New Nature of Maps. Essays in the History of Cartography*, Baltimore, Johns Hopkins University Press, 2001.

5 Anne Godlewska and Neil Smith, *Geography and Empire*, Oxford, Blackwell Publishing, 1994; Felix Driver, *Geography Militant. Cultures of Exploration and Empire*, Oxford, Blackwell Publishing, 2000.

6 Vincent Berdoulay, *La formation de l'École française de géographie, 1870–1914*, Paris, Bibliothèque nationale, 1981; Dominique Lejeune, *Les sociétés de géographie en France et l'expansion coloniale au XIXe siècle*, Paris, Albin Michel, 1993.

7 Pierre Singaravélou (ed.), *L'empire des géographes. Géographie, exploration et colonisation, XIXe–XXe siècle*, Paris, Belin, 2008.

relation to their contexts of production and the practices of their producers.[8] Just like any other claimed knowledge of reality such as a text or an illustration, maps should be read in relation to where and how they were created, and in terms of their reception, because they mediate "between two mental images, that of their producer and that of their user".[9] Far from being reducible to a type of document of a certain size and containing a key, maps are social and cultural constructs that need to be interpreted. To understand them, it is necessary to take into account the system of references of the societies producing and receiving them, which in turn requires us to reflect on our own categories and concepts.

Observation and measurement were common practices in mapmaking, but they proved unsuitable for difficult terrain. Recent research has shed light on the contacts between European and local populations,[10] and on all aspects of their interaction (what is said and written, gestural language, and so forth).[11] Material in this area yields precious insights into how geographical knowledge was used by local populations, but there is little of it in the archives, although encounters with the local population are described in accounts of explorations. The sources mostly concern Africa. We shall ask what might limit or modify their application to the case of South-East Asia, and also whether these analyses can be taken further, especially due to the existence of a large number of maps.

I shall refer essentially to maps made by European explorers, by military personnel and by surveyors from geographical agencies, some of which also show infrastructures and population centres. These printed topographical relief maps can sometimes be analysed in relation to the available preparatory documents, in order to show how the finished product came into being. Large-scale maps and plans, such as land registry documents or town plans, are not included in this study. I pay particular attention to locally-produced maps, with a view to elucidating how they were used by the Europeans, for example, the maps made by a region's inhabitants, by Vietnamese officials, by merchants

---

8 See Marie-Noëlle Bourguet, Bernard Lepetit, Daniel Nordman, Maroula Sinarellis (eds), *L'invention scientifique de la Méditerranée. Égypte, Morée, Algérie*, Paris, EHESS, 1998; Hélène Blais, *Voyages au grand océan. Géographies du Pacifique et colonisation, 1815–1845*, Paris, CTHS, 2005; Isabelle Surun, *Dévoiler l'Afrique ? Lieux et pratiques de l'exploration (Afrique occidentale, 1780–1880)*, Paris, Editions de la Sorbonne, 2018.

9 Christian Jacob, *L'empire des cartes. Approche théorique de la cartographie à travers l'histoire*, Paris, Albin Michel, 1992, p. 137.

10 Felix Driver, Lowri Jones, *Hidden Histories of Exploration: Researching the RGS-IBG Collections*, London, Royal Holloway / University of London, 2009.

11 Camille Lefebvre, "Itinéraires de sable. Parole, geste et écrit au Soudan central au XIXe siècle", *Annales. Histoire, Sciences Sociales*, Vol. 4, 2009, p. 797–824.

from Yunnan, in the different courts, or equally in the field. Alongside this analysis of maps, I draw material from military archives and narratives of expeditions to explore who these maps' authors were, how they proceeded, and what knowledge they helped produce. I also use administrative reports, to assess the link between the lines on the map and the realities on the ground.

By examining these different interfaces, articulating distinct scales of analysis and taking into account the multiplicity of actors involved in constructing knowledge of these confines, I seek to understand to what extent maps produced territories in the northern Indochinese Peninsula, in the eyes of the colonisers and the local people. Taking map-making as the prism through which to examine these confines, how they were demarcated and the populations which inhabited them, permanently or provisionally, I hope to circumscribe relations to a territory which are not defined by national boundaries, in spaces situated between the zone and the line.

Hence this book seeks to go beyond an analysis of representations and the already familiar diplomatic history of imperial territorial expansion and border treaties, in order to examine the correlations between maps and territories, to see how maps are really made, how territories are gradually constructed, and the extent to which maps actually produce the territories they designate. Is it enough to draw a line on a map for a frontier to exist? To what extent are maps considered to be useful instruments of government in these outlying areas? I shall examine the different conceptions of space underlying political choices, for instance the decision to annex one territory rather than another, as well as the effects of these decisions on the ground in the process of territory-production, the problems encountered, especially in the form of the local population's reactions, and the solutions devised by the colonial powers.

2 Confines, Margins, Frontiers: Space as an Object

Space is not represented in the same way by European colonial civil servants, Chinese or Siamese soldiers, or local inhabitants; their perspective will depend on whether they are living in the territory, travelling through it, planning to govern it, or demarcating its borders. Analysing these varied coexisting logics of space in this colonial situation, the practices attached to them, and the actual territories they engender, obliges us to go beyond historical discursive representations and also beyond a binary opposition between European and indigenous conceptions, which are often intertwined in these borderlands. Following on from research which has taken on board issues of spatiality when writing the history of certain colonial territories, this book can be described as a spatial history of the northern Indochinese Peninsula.[12] I maintain that this spatially marginal territory, characterised by a "geography of ignorance"[13] – since the population of the plains knows little about it, and the Europeans even less, such that for a long time it was a blank space on the maps – is nonetheless heuristically central. That is, the transitional geographical zone (mountains), and the transitional political zone (frontiers), which characterise the northern Indochinese Peninsula, do not simply define the limits of our inquiry, but constitute its very subject matter.

Without going as far as James C. Scott, who artificially homogenises the frontier populations in saying that they escape the grip of any political authority, I agree with him on the importance of the South-East Asian uplands (although I would argue that the relevant area is smaller), and on frontiers as the analytical focus.[14] The vantage point from which the geographical policies of imperial power centres will be observed is therefore the frontier zone or confines. The process of mapping these confines reveals how territories are produced, and how States and colonial entities come to be defined. A frontier can have the sense of a political line between two States, like a "border" or "boundary"; or else it can be conceived as a pioneer front, as in the "American frontier".[15] We generally maintain this distinction in this book. The limit between China and Vietnam, for example, can be called a "boundary", whereas the rest of the peninsula consisted rather of "frontiers" until the introduction of border

12 Hélène Blais, Florence Deprest, Pierre Singaravélou (eds), *Territoires impériaux. Une histoire spatiale du fait colonial*, Paris, Publications de la Sorbonne, 2011. On the annexation of Australian territories by the British, see Paul Carter, *The Road to Botany Bay: An Exploration of Landscape and History*, London, Faber & Faber, 1987; for Egypt, see Timothy Mitchell, *Colonizing Egypt*, Cambridge, Cambridge University Press, 1988; and for a historiographical note on the question, see Daniel Clayton, "Le passé colonial/impérial et l'approche postcoloniale de la géographie anglophone", in Pierre Singaravélou (ed.), *L'Empire des géographes. Géographie, exploration et colonisation. XIXe–XXe siècle*, Paris, Belin, 2008, p. 219–234.

13 Willem van Schendel, "Geographies of knowing, geographies of ignorance: jumping scale in Southeast Asia", *Environment and Planning D: Society and Space*, 20–6, 2002, p. 647–668.

14 Scott, James C. *The Art of Not Being Governed: An Anarchist History of Upland Southeast Asia*. Yale: Yale University Press, 2009.

15 Michel Foucher, *Fronts et frontières. Un tour du monde géopolitique*, Paris, Fayard, 1988. The notion of "frontier" was first theorised in 1893 by J. F. Turner with reference to the conquest of the American West and the model of territory-building developed in the United States.

demarcation, which tended to stipulate a line.[16] However, some recent works have nuanced the idea of the boundary between Vietnam and China, particularly in the light of robber gangs in the area.[17]

What has emerged from research – principally focused on Africa – on the processes involved in establishing frontiers, is the large number of interactions involved in boundary delimitation. This work gave the lie to the prevailing ideological assumptions concerning frontiers under colonialism.[18] African territorial boundaries, long considered artificial and imposed arbitrarily by the colonisers, were criticised for ignoring existing geographical and human realities. They were also portrayed as impassable barriers maintained by the imperial powers' use of force. However, the combined roles of diplomats, surveyors, and local populations in the process of determining the boundary reveal a quite different reality. These are some of the issues I redeploy in this book, for the case of South-East Asia.

The conclusions of my research tally with this scholarship concerning another continent and an earlier period. I demonstrate namely that the local populations have a certain agency and room for manoeuvre in their negotiations on territorial issues and on the route taken by the frontier.[19] Building on the work of Eric Vanden Bussche on the demarcation of the China-Burma border, I show that the game played by the Chinese in relation to the British can be understood in terms of their relation to the French.[20] Although I do not place as much emphasis on diplomatic wranglings as does Venden Busche, I draw on his conclusions, while expanding their scope through additional perspectives, including those of the French scientific and political authorities.

16 Pierre-Bernard Lafont (ed.), *Les frontières du Vietnam*, Paris, L'Harmattan, 1989. Even today, the Vietnamese border has different signs for the "frontier area" [*khu vực biên giới*], and the "borderbelt" [*vành đai biên giới*].

17 Bradley Camp Davis, *Imperial Bandits. Outlaws and Rebels in the China-Vietnam Borderlands*, Seattle, University of Washington Press, 2017.

18 Hélène Blais, "Frontière coloniale", in Sophie Dulucq, Jean-François Klein and Benjamin Stora (eds), *Les mots de la colonisation*, Toulouse, Presses universitaires du Mirail, 2008; Camille Lefebvre, *Frontières de sable, frontières de papier. Histoire de territoires et de frontières, du jihad de Sokoto à la colonisation française du Niger, XIXe–XXe siècle*, Paris, Publications de la Sorbonne, 2015.

19 Erbig, Jeffrey Alan. *Where Caciques and Mapmakers Met. Border Making in Eighteenth-Century South America*. Chapel Hill: University of North Carolina Press, 2020.

20 Vanden Bussche Eric, *Contested Realms. Colonial Rivalry, Border Demarcation, and State Building in Southwest China, 1886–1960*, PhD Dissertation, Stanford University, 2014.

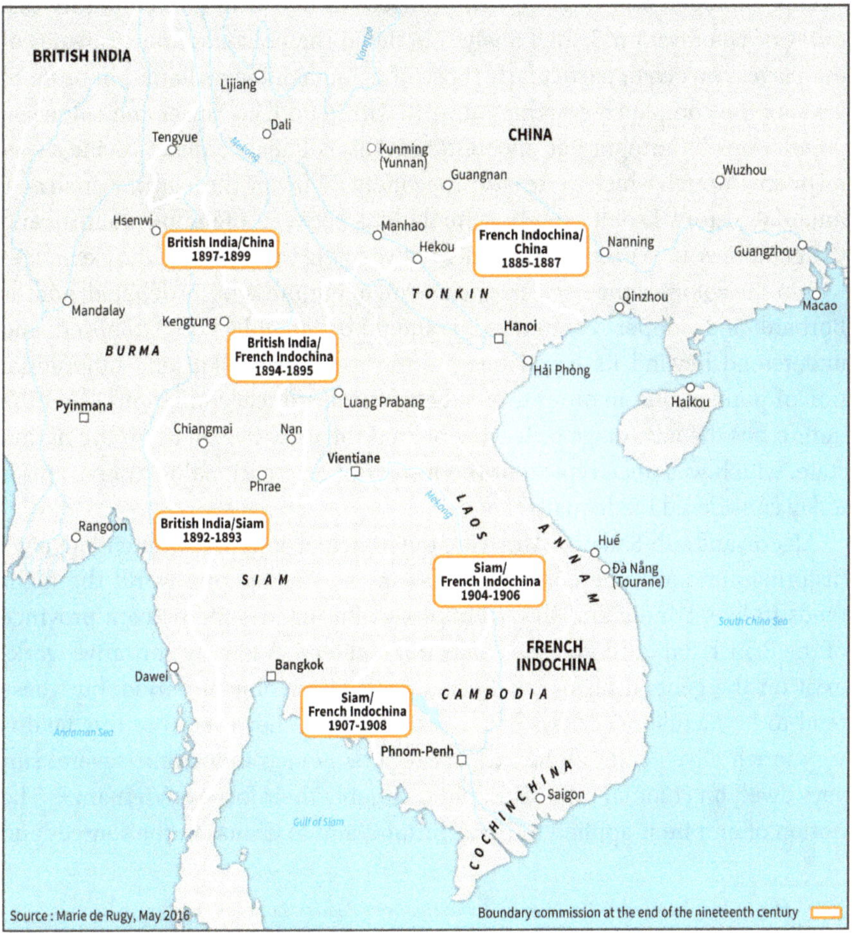

FIGURE 2 Border Commissions in the Indochinese peninsula, end nineteenth – early twentieth century

3 Connecting Empires

Another focus of this book is to track the circulation of knowledge and practices between different imperial systems, and to analyse how each system perceived the other. Moving between the productions of the European metropoles, Asian capitals, and the confines of the northern Indochinese Peninsula, I attempt to understand how imperial territories gradually took shape in maps, and in reality. The different levels explored give access to different aspects of the issue – discourses, practices and concrete effects on the ground.

Our approach will thus be two-pronged. In line with contemporary histories of colonisation, I shall analyse in detail the practices and strategies of the players involved, particularly those of colonial intermediaries, in order to lay bare the complex hierarchies at work within both coloniser and colonised populations.[21] Cartographic encounters in this colonial context provide a useful prism through which to view this. Secondly, following the lead of British new imperial history, I shall examine circulations between the home country and their colonies, the complex dynamics between the two, and also the exchanges within the colonies themselves. The notion of empire is hotly debated, and, as Burbank and Cooper recommend, it should be carefully contextualised, and understood beyond its broad lines of territorial expansion and differentiation of populations, in order to avoid aberrant historical conflations.[22] Yet the notion has the advantage of leaving behind the exclusive focus on the nation state, which was once typical of the history of international relations, and is today considered inadequate.[23]

Algeria and sub-Saharan Africa are the principal objects of French-language historiography of imperialism, whereas in the Anglophone world the focus tends to be on India and the "Dominions". Burma, despite being a province of the British Raj until 1937, is usually not included. A few authoritative works exist on the general history of Burma during the colonial period, but these tend to be factual overviews.[24] It is worth noting, however, that this field is very much alive at the moment.[25] As regards French Indochina, several surveys dwell on colonial interfaces, but invariably their focus is Vietnam.[26] The notion of empire is applied to Great Britain (and to China) in the sources and

21 Howe, Stephen (ed.), *The New Imperial Histories Reader*. London, Routledge, 2010.

22 Jane Burbank and Frederick Cooper, *Empires in World History. Power and the Politics of Difference*, Princeton, Princeton University Press, 2010.

23 Pierre Grosser, "Comment écrire l'histoire des relations internationales aujourd'hui? Quelques réflexions à partir de l'Empire britannique", Histoire@Politique, 10, January 2010; Robert Frank, *Pour l'histoire des relations internationales*, Paris, PUF, 2012.

24 Dorothy Woodman, *The making of Burma*, London, Cresset Press, 1962; Thant Myint-U, *The Making of Modern Burma*, Cambridge, Cambridge University Press, 2001.

25 Mandy Sadan, *Being and becoming Kachin. Histories beyond the State in the Borderworlds of Burma*, Oxford, Oxford University press, 2013; Alicia Turner, *Saving Buddhism. The Impermanence of Religion in Colonial Burma*, Honolulu, University of Hawaii Press, 2014.

26 See the as yet unrivalled *Indochine. La colonisation ambigüe (1858–1954)*, by Pierre Brocheux and Daniel Hémery (Paris, La Découverte, 1995). On Vietnam, the works of Philippe Devillers, Charles Fourniau and Nguyễn Thế Anh, and, more recently: Christopher E. Goscha, *Vietnam: A new History*, New York, Basic Books, 2016; on Laos, see Søren Ivarsson, *Creating Laos: the making of a Lao space between Indochina and Siam, 1860–1945*, Copenhagen, NIAS, 2008; Yves Goudineau and Michel Lorrillard (eds), *Recherches nouvelles sur le Laos*, Paris, École française d'Extrême-Orient, 2008.

in academic literature, and the British world-system is treated as a definable and organised totality of colonial possessions, within which each colony plays a specific role. However, this notion is contested in the case of Vietnam (called both the Empire and the Kingdom of Annam), and France is only rarely treated as an imperial system. In this context, it is particularly fruitful to attempt a connected history of these colonial territories, despite the difficulties raised by the historiographical imbalance in favour of studies on British imperialism.[27]

In the wake of decolonisation, academic research tended to broach cartography in the framework of the nation state, and to focus on the country's former colonies – Siam, India or Algeria[28] – instead of analysing imperial systems in their inter-relation.[29] The advantage of a comparative or connected approach is that it can mobilise scales of analysis other than those of the nation. A magnificent early example of this, in French research, are the works of Denys Lombard, including his "Javanese Crossroads", which presents the outlines of a "global history",[30] Romain Bertrand's later "history in equal shares" [*"l'histoire à parts égales"*],[31] and, more recently, Margolin and Markovits's survey of European-Indian relations from the 15th to the 21st century.[32] These works aim to harmonise global with imperial perspectives by retaining from each the most relevant paradigms, such as transfers and exchanges across borders, imperial practices that are similar or divergent, and so forth.[33]

27 John Darwin, *The Empire Project: the Rise and Fall of the British World-system, 1830–1970*, Cambridge, Cambridge University Press, 2009.

28 Thongchai Winichakul, *Siam mapped: a history of the geo-body of a nation*, Honolulu, University of Hawaii Press, 1994; Matthew H. Edney, *Mapping an empire: the geographical construction of British India, 1765–1843*, Chicago, University of Chicago Press, 1997; Hélène Blais, *Mirages de la carte. L'invention de l'Algérie coloniale*, Paris, Fayard, 2014.

29 With some exceptions, such as Véronique Dimier, *Le gouvernement des colonies. Regards croisés franco-britanniques*, Brussels, Université de Bruxelles, 2004. Comparative work has also been carried out on African imperial frontiers, for example, William F. S. Miles, *Hausaland Divided. Colonialism and Independence in Nigeria and Niger*, Ithaca, Cornell University Press, 1996.

30 Denys Lombard, *Le carrefour javanais. Essai d'histoire globale. Vol. 1: Les limites de l'occidentalisation; Vol. 2: Les réseaux asiatiques; Vol. 3: L'héritage des royaumes concentriques*, Paris, Éditions de l'EHESS, 1990.

31 Romain Bertrand, *L'histoire à parts égales. Récits d'une rencontre Orient-Occident (XVI*e*–XVII*e *siècle)*, Paris, Seuil, 2011.

32 Jean-Louis Margolin and Claude François Markovits, *Les Indes et l'Europe. Histoires connectées, XV*e*–XXI*e *siècle*, Paris, Gallimard, 2015.

33 The concept of "cultural transfer", which was coined with reference to Franco-German relations between 1750 and 1914, includes the notion of a culture's influence on another culture, as well as the gradual assimilation of initially foreign elements. It was devised originally to compensate for the shortcomings of comparative analyses which highlighted differences at the expense of interactions. This resulted in territories being studied

Our spatial study of the northern Indochinese Peninsula starts with a summary of the geographical knowledge available to a European in 1885, and compares the scientific exploration of the area carried out by the British and the French. Chapter 2 leads on from this to discuss the institutional history of the geographical agencies, and the beginnings of the systematic mapping of territories. The many collaborations and rivalries this generates are studied in the ensuing chapter. The conditions of map production at the time are shown by examining how mapmakers deal with the terrain once in the field, using the prevailing geographical methods. Since observation and measurement are not always possible, the Europeans also have recourse to gathering information from the local populations. This information-gathering, as well as the ways in which Europeans used Asian maps, will be discussed in chapters 5 and 6. I shall go on to explore colonial transport routes, which themselves generate territories and have a prominent place on the maps. Chapter 8 focuses on the issue of frontiers, demarcated and also disregarded, despite border controls. Lastly, I shall address the various administrative logics deployed in the region, and the persistence of territorial anomalies. It is only at the end of this study in connected history, which describes the multifactorial processes of territorial construction in the northern Indochinese Peninsula, that this region where spaces invented by imperial powers come into contact can fully make sense.

without their "frontier zones" and their "crossing points". The notions of "importing" and of "unilateral influence" gave way to those of "hybridisation" and "cross-fertilisation", showing that the exchanges were not a one-way traffic. Michel Espagne and Michael Werner (eds), *Transferts. Les relations interculturelles dans l'espace franco-allemand, XVIIIe et XIXe siècle*, Paris, Éd. Recherche sur les civilisations, 1988.

CHAPTER 1

Connected Histories of Exploration

In 1884–1885, as the Berlin Conference was drawing up the rules for the "Scramble for Africa", the French and the British were competing for new colonies in the Indochinese Peninsula. Following the annexation of Upper Burma at the conclusion of the third Anglo-Burmese war (1885), and the annexation of Annam-Tonkin as a French Protectorate the same year, the two European powers' first priority was to integrate these new territories into their pre-existing possessions. They also had to contend with the expansionist designs of Siam, an independent monarchy, and of Imperial China in Yunnan, which set limits on their own territorial ambitions. In order to understand how these territorial configurations developed, we shall examine the British and French exploratory missions that preceded and accompanied these conquests.

What was the extent of European knowledge of the Indochinese Peninsula, especially of its frontier areas, at the time of the 1885 conquests? An overview of the knowledge available will highlight what the European powers did not know, and what they avoided or concealed. Rather than aim at an exhaustive inventory (an impossible and fruitless task), we shall identify the types of knowledge available to the British and French at the time, and ask whether this knowledge was used or not; and if not, whether this was on purpose, or through oversight. That is, taking the documents quoted by British and French sources in the 1880s, we shall describe what Valentin Mudimbe has called a "colonial library" – all the texts and oral knowledge at the disposal of the Europeans, which contributed to building up a representation of their Empire.[1] The issue is not only the knowledge itself, but how it circulated between the British and the French, in order to assess whether we can really speak of a *European* knowledge of the Indochinese Peninsula, and hence of a representation of these imperial territories common to the two powers.

The representations of the Peninsula were fashioned by specific categories of thought and analysis. What was seen, and what were the blindspots? First, we should be aware that explorers, who produced geographical knowledge, not only occupied a particular place in their home societies, but also participated in a variety of networks, large and small, in the metropole, the colonies, and elsewhere abroad, in the case of members of Geographical Societies. Second,

1 Valentin Mudimbe, *The Invention of Africa: Gnosis, Philosophy, and the Order of Knowledge*, Bloomington, Indiana University Press, 1988.

DOI:10.1163/9789004469853_003

the political and scientific instructions issued to explorers raise the issue of the links between knowledge and power: did every exploratory mission obey a covert political goal, and did they correspond overall to a precise plan? Exploration started before territorial conquest, but it also took place alongside it, thus raising the question of whether the expeditions prepared for, and were directly linked to military campaigns; and whether the military presence affected the type of scientific inquiry carried out. English and French explorers, in the footsteps of previous decades of exploration by their predecessors, and steeped in the knowledge already accumulated, were part of the same movement. These Europeans sometimes explored the same territory, they met each other, read each other's books, and used their research to increase geographical knowledge of the marginal spaces about which they knew so little. So what new geographical knowledge did the British and French contribute? What had they discovered about the Indochinese Peninsula, and its imperial territories, by the early 20th century?

1 A Well-Stocked Colonial Library with a Long History

In his *Précis de géographie universelle* of 1810, Conrad Malte-Brun described the paltry amount of information the Europeans had on the Indochinese Peninsula: "The vast regions which, in the shape of a double peninsula, extend between the Bay of Bengal and the China Sea, are almost unknown, except along their coasts, and as for the interior, we have only a set of useless and complicated conjectures."[2] The sources of the existing knowledge were initially ancient maps, such as Ptolemy's, which was still an authoritative reference for the French when they mapped Algeria in the 19th century.[3] However, Ptolemy is never mentioned by the British or the French for the Indochinese Peninsula, although he certainly made a map, which was used by Jean-Baptiste d'Anville in 1763 for his *Mémoire sur les limites du monde connu des anciens au-delà du Gange* [Study on the limits of the world known to the Ancients beyond the Ganges] [Figure 3]. Such 18th-century documents appeared on rare occasions in the key of later maps, for example on the British Officer Henry Yule's

2 Conrad Malte-Brun, *Précis de la géographie universelle, ou Description de toutes les parties du monde…*, Paris, F. Buisson, 1810–1829, 8 vol. (p. 145 for the 5th edition of 1845).

3 On the use of Ptolemaic geography by the French in Algeria, see H. Blais, *Mirages de la carte. L'invention de l'Algérie coloniale, op. cit.*

FIGURE 3 *Pour le mémoire sur les limites du monde connu des anciens au-delà du Gange* (detail), Jean-Baptiste d'Anville, 1763
SOURCE: BNF [http://gallica.bnf.fr/ark:/12148/btv1b84692911]

1857 map of Burma.[4] Subsequently, these distant lands were known through the sparse records of missionaries, diplomatic envoys and solitary adventurers. Given this dirth of documentation, the Europeans arriving in the Peninsula tended to consult local geographical information held at the royal or princely

4 Henry Yule, *The New Map of Burma and the Regions Adjacent*, 1857 (MEP, K MYA 3).

courts. Did the British and the French go about this assembly of a "colonial library" in the same way?

1.1 *Maps by Missionaries, Necessary but Insufficient*

The earliest European geographical descriptions of the Indochinese Peninsula were the work of French missionaries who settled in the 17th century, whereas their British and American counterparts only arrived with the colonisation of Burma some two centuries later. One of the oldest maps, dating from 1653, by the French Jesuit Alexandre de Rhodes,[5] shows "Cocincina", Tvnkin", the China Sea [*Mare sinicvm*], and the "solitude" [*Solitudo*] of Laotian spaces beyond the mountains of the Annamite Range, which were unknown to the Europeans and hence left blank on the map [Figure 4]. Another map, which became famous among French cartographers, is the *Tabula Geographica Imperii Anamitici* by Mgr Jean-Louis Taberd, a priest with the Paris Foreign Missions Society, who was appointed apostolic vicar for Cochinchina in 1827.[6] This map, published in 1838, was reprinted in 1862 at the request of Chasseloup-Laubat, the Minister for the Navy and the Colonies – a sign of his interest in the French presence in the early years of colonisation [Figure 5]. Taberd's name was often found in the keys to maps, and his work was frequently re-used. The same goes for the geographical surveys carried out by the Jesuits in 18th-century China, at the Emperor's request. Their general map of China was the basis for French maps drawn a century later, especially for the path of the Tonkin – China border. Similarly, Francis Garnier's map of 1873 used Jesuit maps of China, and he "carefully compared" his own map "with the original maps from 1735". Jesuit missions in Tibet also used their predecessors' cartography, and Mgr Taberd's work was used for mapping Cambodia and Cochinchina.[7] The role of these cartographic forbears was particularly decisive for cartographic norms

5 Alexandre de Rhodes (1591–1660) landed in Cochinchina, along with five other Jesuits, in 1624. He was one of the first Frenchmen to set foot there. He worked on transcribing Vietnamese into Latin script, creating an alphabet called the *quốc ngữ*, which is still in use today. It is significant that his *Voyages et missions du Père A. de Rhodes ... en la Chine et autres royaumes de l'Orient, avec son retour en Europe par la Perse et l'Arménie*, first published in 1653, was reprinted in 1884 (Lille, Desclée de Brouwer).

6 The *Société des missions étrangères de Paris*, located in the rue du Bac, Paris, was founded in 1664 after Alexandre de Rhodes's journey, and on the instigation of King Louis XIV, to train missionaries to evangelise Asia. The first apostolic vicars to be appointed for Indochina were François Pallu and Pierre Lambert de La Motte. See the entry on Jean-Louis Taberd: http://archives.mepasie.org/notices/notices-necrologiques/taberd-1794-1840 [consulted on 10 May 2020].

7 Ernest Doudart de Lagrée, Francis Garnier, Louis Delaporte, *Voyage d'exploration en Indo-Chine*, Paris, Hachette, 1873, p. 10.

FIGURE 4 Map of the Kingdom of Annam by Alexandre de Rhodes, 1653
SOURCE: *Voyages et missions du Père A. de Rhodes …* (uncoloured at BNF [http://gallica.bnf.fr/ark:/12148/btv1b8607026z])

FIGURE 5 Map of the Kingdom of Annam by Monseigneur Taberd, 1838 (detail)
SOURCE: BNF [http://gallica.bnf.fr/ark:/12148/btv1b530936o5j]

regarding the spelling of names. Francis Garnier, for example, explained that he preserved the missionaries' spelling of Chinese and Vietnamese names. The British do not seem to have consulted these documents, because they did not concern their own colonies. They had no prior information from missionaries to guide them when they annexed Burma.

With colonisation, missionaries were increasingly consulted for their long experience in the field. They were "irreplaceable as information sources and advisors, since, due to their long immersion in Indochinese societies, they were the only Europeans to possess first-hand knowledge."[8] By their regular journeys up and down the country on pastoral missions, they gained good knowledge of the territory, and so their itineraries were used at the moment of conquest. "The Reverend Mr Cushing's travels from Toungoo to Zimmay via Takkong Ferry and Kiangtung" and "The Reverend Mr Rose's journey to Mandalay, Theinnee, Monai and Poayhla, were used to draw up the *The Map of the Teak localities in British Burma, Burma Proper and the Shan and Karennee States*, in 1871.[9] Josiah Cushing, who had been sent by the American Baptist Mission in the late 1860s first into the Shan territories and later to the Kachin areas, is better known for his works on linguistics, particularly his English-Shan dictionary.[10] Indeed,

8 P. Brocheux, D. Hémery, *Indochine. La colonisation ambiguë (1858–1954)*, *op. cit.*, p. 29.

9 These two accounts are mentioned in the key to the *Map of the teak localities ...*, RGS, mr Burma G.13.

10 Josiah Nelson Cushing, *A Shan and English Dictionary*, Rangoon, C. Bennett, American Mission Press, 1881; for further information, see Wallace St. John, *Josiah Nelson Cushing, Missionary and Scholar, Burma*, Rangoon, American Baptist Mission Press, 1912.

map-making was far from being the principal activity of the missionaries, and it depended on the geographical curiosity and the particular talents of the individual. The same was true for the French. A map produced in 1885–1886 of the central provinces of Quảng Bình and Nghệ An was based on the one published by Père Mourouziès in the *Missions Catholiques* in 1876; the map of the Province of Hué was based on the itinerary of Father Lesserteur.[11] Not all the missionaries drew maps, and those who did tended to make roughly-drawn location sketches, which proved difficult to use.[12] Francis Garnier, using a sketch map passed to him by the *Dépôt des cartes et plans de la marine*, and drawn by Père Arnoux, a French missionary in Cochinchina, complained that it contained "mistakes in the parts overlapping with the Commission's routes which are too serious to inspire much confidence in the rest."[13] So although late 19th-century European cartographers needed these sorts of documents due to the lack of maps, they considered them very imperfect.

Missionaries sometimes played a political role, to different degrees, and were consulted on delicate issues.[14] Mgr Puginier is a case in point, a priest with the Paris Foreign Missions Society, and apostolic vicar of Western Tonkin. In response to an inquiry by the Resident Superior concerning the line of the border between Tonkin and China, he wrote, on 28 January 1888:

> I have the honour of conveying to you some notes I made to help expose the truth about the frontiers of Tonkin and Siam. I wish these notes to remain secret. You may use them as you see fit in the decision you will have to take on border delimitation, but I beg you not to let it be known outside that you have this information.[15]

Here the bishop agrees to share his knowledge of the northern territories, but with great caution. Analysing the Western provinces of Tonkin, he warns the French against the British, whom he considers to have designs on some of the Laotian territories in order to reach Yunnan, while forming an alliance with

11 SHD, 10H2, Dossier 5.

12 On the different types of missionary cartography and especially the scales envisaged, for Africa, see Jean-Michel Vasquez, *La cartographie missionnaire en Afrique. Science, religion et conquête (1870–1930)*, Paris, Karthala, 2011.

13 F. Garnier, *Voyage d'exploration de l'Indo-Chine, op. cit.*, p. 11.

14 James Patrick Daughton, *An Empire Divided: Religion, Republicanism, and the Making of French Colonialism, 1880–1914*, Oxford, Oxford University Press, 2006; Charles Keith, *Catholic Vietnam: a Church from Empire to Nation*, Berkeley, University of California Press, 2012. There are to date no equivalent works on Burma.

15 ANOM, GGI 14456, Renseignements sur les Chaus de Hung Hoa et Thanh Hoa fournis par Mgr Puginier, January 1888.

Siam. The discourse of rivalry between the British and the French is nothing new, but it is, interestingly, couched in cartographic terms in this document:

> The map made in Siam is wrong, it places as a Siamese Laotian enclave the *huyens* [districts] of Trinh-Co, Man-Tay, Can-Nua, and Xam-to, which Annam can rightly claim as its own, or at least as its tributary states, as I mentioned in my earlier notes. It is certain that they have been paying it tribute for a long time.[16]

As well as accusing Siam of unilaterally attributing territories to itself, he notes that the same map makes territories which are fully Vietnamese into mere tribute-paying states, concluding that all these mistakes are intentional. The patriotic dimension is less important here than the cartographic analysis, based on historical rights. Mgr Puginier is consulted because of his good knowledge of the situation on the ground, and internationally. Missionaries in the field were thus precious sources of advice and information, even if the information was not used later.

1.2 *The Importance of Hydrography and Marine Maps*

From their arrival in the Peninsula and until the early 20th century, the Europeans also used another type of map: the hydrographic survey. The profession of hydrographic engineer was officially created in France in 1814, but it had existed as an activity since the 18th century, and became extremely important under the influence of the hydrographer and cartographer Beautemps-Beaupré (1766–1845).[17] These engineers were trained at the *École Polytechnique* of Paris where they learnt to make marine maps, and studied the tides, the currents and the weather, in order to be able to carry out surveys in the field.[18] Hydrographic maps provide information about seas, oceans and rivers, and they were used principally for navigation, to show the path, depth and width of watercourses, in order to inform decisions about which kind of boat, if any, could travel there (depending on its draught). The maps were published by the *Dépôt des cartes et plans de la marine*, a prestigious institution founded in 1720, and absorbed in part by the *Service hydrographique et océanographique de la marine* in 1887. It was a hub of scientific research and publication, and it also kept archives on

16 *Ibid.*

17 Olivier Chapuis, *À la mer comme au ciel. Beautemps-Beaupré & la naissance de l'hydrographie moderne, 1700–1850*, Paris, Presses de l'université Paris-Sorbonne, 1999.

18 Maurice Rollet de l'Isle, *Étude historique sur les ingénieurs hydrographes et le service hydrographique de la marine (1814–1914)*, Paris, Imprimerie nationale, 1951.

marine cartography, marine exploration, and hydrography. In Bruno Latour's terms, it can be called a "centre of calculation", in the sense of a central and pivotal space of scientific exchange and circulation.[19] Explorers, for instance, could consult documents at the *Dépot* to prepare their expeditions, and send in their own maps afterwards, or spend a few months there preparing their research for publication.[20] As a unique documentation centre and a site of map production, the institution played a vital role for explorers, both before and after they went into the field.

The British Hydrographical Office, founded in 1795, had many similarities with its French counterpart. Its first Director was Alexander Dalrymple, who collected and catalogued hydrographic maps. Under the leadership of Captain Thomas Hurd, appointed in 1808, this Office flourished, after obtaining the authorisation to sell maps to the public, and, above all, when he created a separate professional body of hydrographers, to provide specialised staff.[21] Captain Edward Parry succeeded him in 1823, and formulated stringent guidelines for hydrographers and the burgeoning discipline of hydrography. In 1828, he brought together French and Spanish Offices on a scientific expedition he organised in the South Atlantic. He was succeeded as Head of this Office by Sir Francis Beaufort, known for his wind force scale. Large numbers of hydrographic maps were produced in both France and Great Britain during the 19th century, for use by their respective navies. The relations between cartography, hydrography and colonial expansion intensified, and different zones of influence began to emerge. Early 19th-century circumnavigation in the Pacific improved existing knowledge of these spaces' geography and hydrography. Some of these areas later became colonial territories, but there is no consistent correlation between the recommendations of the hydrographic engineers and the political decisions taken.[22]

Hydrographic research in the Indochinese Peninsula focused on improving knowledge of the coasts. The first maps of the Vietnamese shores used by the French at the end of the 19th century were made by Jean-Marie Dayot and his brother, between 1790 and 1795, when Dayot, an officer in the French Navy, was

19 Bruno Latour, *La science en action*, Paris, La Découverte, 1989.

20 Jules-Léon Dutreuil de Rhins spent two years there working on his map of Indochina. He was one of a series of French explorers and officers who were temporarily integrated into the *Service hydrographique* on their return from the field. Others include Bigrel, Garnier, Delaporte and Gouin.

21 Adrian Webb, *The Expansion of British Naval Hydrographic Administration, 1808–1829*, Doctoral Thesis at the University of Exeter, 2010. See particularly chapter 4 on the scientific training required, and the techniques used, p. 143–192.

22 H. Blais, *Voyages au grand océan. Géographies du Pacifique et colonisation, 1815–1845*, *op. cit.*

working for Nguyễn Ánh, the future king Gia Long (1762–1820). The maps, published in 1807 at the *Dépôt de la marine*, show the coasts from the Southern tip of Cochinchina to the Gulf of Tonkin.[23] British hydrographers also produced maps of the Indochinese Peninsula, some of which were translated into French, such as the *Map of the Gulf of Siam*, drawn at a scale of 1:100,000 and 1:292,000 by Captain John Richards from his surveys aboard *H.M.S. Saracen* between 1856 and 1859.[24] They were published by the British Hydrographic Office in 1858, and by the *Dépôt des cartes et plans de la marine* in 1862, after translation into French.[25] The French explorer Francis Garnier used and quoted them in 1866, as part of his overview of the geographical knowledge available on the Indochinese Peninsula.[26]

The role of hydrographic research in defining these territories did not stop at the form of the Peninsula's coastlines, but included watercourses, which were fundamental structuring elements in Vietnamese and also in European cartography. The section of the Mekong between Tonle Sap in Cambodia and the China Sea was a prime focus of maps drawn in the 1860s, invariably placed centrally, and showing the river's tributaries. The course of the Mekong was explored by the Navy hydrographic engineers Manen, Héraud, and Vidalin soon after the French arrived in Cochinchina. Although the maps from 1861 and 1862, which were most probably preparatory sketches, have unfortunately disappeared from the archives, we nonetheless have access to the map published in 1863 by the *Dépôt de la marine*,[27] which is probably therefore the definitive version, with the Mekong occupying a central position. Subsequent works, for example by Francis Garnier ten years later, Théophile Bigrel, and Jules-Léon Dutreuil de Rhins, seem to have referred less to this map than to raw data and

23 Edouard Caspari, "Les cartes marines de l'Indo-Chine", *Annales de géographie*, 15/80, 1906, p. 172–173. The map can be consulted on gallica.bnf.fr.

24 John Richards, *The Gulf of Siam Pilot. From the Survey made in H.M.S. Saracen between the years 1856–58*, London, Hydrographic Office, 1863, second edition. The sheets can be consulted separately in the British Library: Maps SEC.13.(2414.), Maps SEC.13.(2719.), and Maps SEC.13.(2720).

25 BL, Maps H.F.SEC.17(1972), *Carte du Golfe de S., dressée d'après les reconnaissances faites en 1856, 1857 et 1859 par M.J. Richards, Master, et plusieurs autres de la Marine Royale Anglaise*, [Paris], Dépôt des cartes et plans de la marine, 1862. There were several new editions of the map between 1866 and 1874.

26 E. Doudart de Lagrée, F. Garnier and L. Delaporte, *Voyage d'exploration en Indo-Chine, op. cit.*, p. 10.

27 BNF, Cartes et Plans, Service hydrographique de la marine, Portefeuille 180, division 4. These maps may be held elsewhere, but we have not yet been able to locate them.

measurements. Overall, it seems clear that hydrography is the most important discipline for the mapping of Indochina, with a primarily practical function of ascertaining the navigability of the Mekong, for political, military and commercial purposes.

For the British, on the other hand, it was not the Mekong which monopolised their attention, but the course of the Irrawaddy. A hydrographic map published in 1825 charts the mouth of the river and its course as far as Yandabo, in the centre of Burma. The hydrographic surveys were carried out during the First Anglo-Burmese War, which ended with the Treaty of Yandabo of 24 February 1826: Lieutenant Windsor, who piloted the steamship *Diana*, had been commissioned by the Admiralty to use this opportunity to produce a 1:200,000 scale map.[28] Further work took place, more intensively, in the 1850s, giving rise to several published maps made by navy officers exploring the river in the context of the Second Anglo-Burmese War.[29]

For the mapping of Tonkin, Charles Maunoir has shown a similar dominance of the river. Before 1885, maps were little more than the "routes drawn by our hydrographer engineers of the sinuous side-arms which cover the delta."[30] The hydrographic explorations carried out by engineers and navy officers were invariably treated as a vital, if not the only, source of information for mapping the territory. Thus on maps of Tonkin from the late 1870s and early 1880s, only the course of the Red River and the triangle of the Delta appear. The Red River was represented as the backbone of an otherwise little known and little explored Tonkin, since only the river itself had been explored and measured with precision, and knowledge of the province was thus reduced primarily to the knowledge accumulated about the Red River.

So when in 1885 the European powers pursued the expansion of their empires, they had several sources at their disposal for understanding the geography of these spaces they had barely visited, namely maps made by missionaries, and hydrographic surveys. Yet were these documents really used? We know how many books in a library can lie slumbering unopened on the shelves. We have only a few conclusive signs of whether a "colonial library" has really been consulted: an annotation or translation of a document, or a mention of it in a map's key, on record cards, or in travel narratives.

---

28 BL, Maps SEC.12(2135) and (2136), *Sketch of the Irawady*, 1, *From the Sea to Rangoun and Prome*; 2, *from Prome to Yeandabou*, by G. Winsor, 1825, London, 1852.

29 Henri Cordier, *Bibliotheca indosinica. Dictionnaire bibliographique des ouvrages relatifs à la péninsule Indochinoise*, Paris, Imprimerie nationale, 1912, p. 61–62.

30 BSG, septième série, VI, 1885, p. 182.

2 Geographies of Exploration and Travel Narratives before 1885

Reports of expeditions carried out before 1885 were another important information source for later explorers. The territories annexed by the British and the French in the 1880s and 1890s had mostly been explored prior to these conquests, aided by the presence of colonial footholds in the region. However, exploratory expeditions tended to take place in spaces already conquered, as suggested by the monthly journal *Excursions et reconnaissances*, published by the colonial government in Saigon between 1879 and 1890, which featured the latest French discoveries in Cochinchina; or, for the British, the first Burmese *Gazette*, published in 1879, which only covered Lower Burma (annexed by the British in 1852).[31] Yet, as mentioned earlier, the Europeans were also attracted to adjacent and as yet unconquered territories, out of scientific curiosity, but above all for economic and strategic reasons. We shall examine how their expeditions were organised, and how bodies of knowledge were constituted about these territories that until then had hardly ever been observed and measured according to the standards of scientificity of the French and the British. What, then, characterised these expeditions, which took place before any colonial conquest, but which were conducted with an eye to future political moves? The territories recommended by the explorers were not always annexed finally, suggesting that geographical arguments did not pull much weight compared to political ambitions.[32] The question arises more generally of the reception of these travel narratives. Some of them became authoritative sources, that is, they were read, and the geographical information they contained became a yardstick to be confirmed or corrected by future expeditions. So how was this information re-used, when, and by whom?

2.1 *Two Leading Institutions: The* Société de Géographie de Paris *and the Royal Geographical Society*

Geographical Societies incarnated the spirit of scientific curiosity which drove many Europeans to go out and discover the world. They provided a framework, encouraging, commissioning and even sometimes financing expeditions, and they enabled the results obtained in the field to circulate in the home country.

---

31 Horace Ralph Spearman, *The British Burma Gazetteer*, Rangoon, Government Press, 1879.

32 This was the case for the Pacific, as shown by H. Blais, *Voyages au grand océan. Géographies du Pacifique et colonisation, 1815–1845*, *op. cit.* It also applies to expeditions in Africa before the period of colonial conquest: I. Surun, *Dévoiler l'Afrique ?*, *op. cit.*

The first Geographical Society to see the light of day, the Paris Geographical Society, was founded in 1821, with Conrad Malte-Brun as its General Secretary. The Royal British Geographical Society was founded a few years later, in London, in 1830. Both institutions aimed at supporting advances in geographical knowledge internationally, and at disseminating information about European discoveries to their members.[33] They were part of the world of learned societies, bringing together dignitaries and aristocrats in networks of sociability and belonging, which they created and developed. With the gradual institutionalisation of the geographical sciences, Geographical Societies became places of exchange and discussion for an increasingly wide range of people.[34] One can find, for instance, articles written by navy officers, merchants, colonial civil servants, architects, engineers and missionaries. After a quiet couple of decades, both Societies became more active, in the middle of the 19th century, as questions of empire intensified in their respective countries. Without wishing to suggest that an "imperial turn" took place at this point, it is important to note the close links between certain eminent members of these Societies and the pro-colonisation politicians (and they were sometimes one and the same person).

The Royal Geographical Society, for one, was clearly attracted to imperialism, an orientation shared by its members. At the time, Darwinism was taking root in Europe, the British educational system was being exported, and exploration and mapping of "virgin" territories continued.[35] There were close links between the Royal Geographical Society and the Admiralty, in the Society's early years, as embodied by Francis Beaufort, a member of the Society's Board and hydrographer in the Navy, who could thus play the role of intermediary between the colonial government and the nascent Society.[36] Roderick Murchison (1792–1871), a geologist, was one of the Society's founders, as well as serving as its President for 30 years, from 1843 to his death, with some interruptions.[37] Under his leadership, the Society grew considerably, and

---

33 Alfred Fierro, *La Société de géographie, 1821–1946*, Geneva, Droz, 1983; D. Lejeune, *Les sociétés de géographie en France et l'expansion coloniale au XIXe siècle, op. cit.*

34 On the institutional emergence of the human sciences and of geography in particular, see Hélène Blais, Isabelle Laboulais (eds), *Géographies plurielles. Les sciences géographiques au moment de l'émergence des sciences humaines, 1750–1850*, *op. cit.*; Jean-Marc Besse, Hélène Blais, Isabelle Surun (eds), *Naissances de la géographie moderne, 1760–1860: lieux, pratiques et formation des savoirs de l'espace*, Lyon, ENS Éditions, 2010.

35 Max Jones, "Measuring the World: Exploration, Empire and the Reform of the Royal Geographical Society, c. 1874–93", in Martin Daunton (ed.), *The Organisation of Knowledge in Victorian Britain*, Oxford, Oxford University Press, 2005, p. 313–336.

36 F. Driver, *Geography Militant …*, *op. cit.*, p.34.

37 More precisely, he served in the years 1843–1845, 1851–1853, 1856–1859 and 1862–1871.

became a real scientific institution. The "imperial turn" manifested itself in the role played by many of its future Presidents in the colonies.

On the French side, the arrival of Jules Duval (1813–1870), of the marquis de Chasseloup-Laubat (1805–1873) and of Charles Maunoir (1830–1901) in the 1860s brought about a major change. Duval passionately defended developing the colony of Algeria. After a career as a lawyer in the Aveyron region of France, where he was influenced by Fourierist ideas, he left for the province of Oran to head the *Union agricole d'Afrique* [African Agricultural Union]. He became Editor in Chief of *L'Echo d'Oran*, in 1852, then member of the Council of the province in 1858. Duval was convinced of the need to develop trade and communication networks in Algeria, and he was extremely influential in the Paris *Société de géographie*, where he was President of the Central Commission from 1868 onwards. As for Prosper de Chasseloup-Laubat, he had been Minister for the Navy and the Colonies for four years (and would serve in this post for another five) when he became President of the *Société* in 1864, until his death in 1873, a period during which he incarnated its "imperial turn". Lastly, Charles Maunoir had a brilliant career in the *Dépôt des cartes et plans du ministère de la Guerre* while also heading the journal *L'Année géographique* between 1862 and 1878, becoming General Secretary of the *Société de géographie* in 1867, a post he held until 1896. The intertwining of political and scientific roles highlights the impetus colonialism gave to geographical expeditions. Many Presidents' first-hand experience of the colonies led them to encourage exploration of these imperial possessions, considered by the Europeans as lands yet to be discovered. However, although enthusiastically pursued, these explorations were not the only ones – there was also the Antarctic, for example. These other expeditions were motivated by a desire to discover new territories, and by a genuine geographical curiosity, such that even if political annexation may have been on the agenda, no direct correlation can be made between the two.

Geographical Societies also played the very practical role of providing advice for explorers, and constituted a forum for shared experience. They thus indirectly helped institutionalise the geographical sciences. Anatole Petiton (1836–1911), a former student at the *École des Mines*, included the following explanation in an article he published in the *Bulletin de la Société de géographie* on his explorations in Cochinchina:

> I trust I am performing a useful task in providing some practical details concerning the precautions to be taken when carrying out explorations in Indochina, to ensure their success in good conditions. May the fruit of my experience, which I have paid for so dearly in my long and arduous

> explorations, prevent those who come after me encountering the same problems and dangers.[38]

There follows a long description of the items to take with one, particularly food and medicines. The idea seems to be to give an almost exhaustive list of how expeditions can be improved. The British Royal Geographical Society was more scientific in its approach. Its advice tended to be technical, to help explorers carry out topographical surveys.[39] In 1854, it started publishing its *Hints for Travellers*, which contained advice for the solitary traveller rather than official instructions for the colonial scholar.[40] The Royal Geographical Society's mapmaker found it "immensely difficult to harmonise the routes taken by inexperienced travellers, and there is a clear and pressing need for more precise methods".[41] Travellers were thus primed in advance in the hope that the outcome would be higher-quality maps, and so that the work carried out in the field should be more scientific. The Royal Geographical Society also offered courses in astronomy, topography and cartography, as from 1879, disseminating good practice, and enabling explorers to put into practice the advice given in the *Hints*.

Apart from their role in supporting, advising and financing expeditions, Geographical Societies also contributed to developing and publishing maps.[42] The career and position of the official cartographer of the *Société de géographie*, Jules Hansen, much like those of his counterpart in the Royal Geographical Society, are revealing:[43] after meeting Eugène Cortambert, the Director of the *Dépôt*, he approached the *Société de géographie*, created the maps for its *Bulletin* as from 1872, and became the *Société*'s official cartographer in February 1876. He drew its explorers' itineraries, which were published as part of the *Société*'s annual activities report by Charles Maunoir in the *Rapport sur les*

38 BSG, Septième série, tome v, 1884, p. 370–371.

39 Peter Collier and Rob Inkpen, "The Royal Geographical Society and the development of surveying 1870–1914", *Journal of Historical Geography*, 29–1, January 2003, p. 93–108.

40 F. Driver, *Geography Militant …*, *op. cit.*, p. 58–59.

41 Hugh Robert Mill, *The record of the Royal Geographical Society, 1830–1930*, London, The Royal Geographical Society, 1930, p. 385.

42 For the British case, see Dorothy Middleton, "Guide to the Publications of the Royal Geographical Society 1830–1892", *The Geographical Journal*, 144–1, March 1978, p. 99–116.

43 See the article on him and his role in mapping Africa. There are many similarities with the mapping of French Indochina, for instance the way documentation was treated, the relation to explorers, and the work with engravers and publishers. Olivier Loiseaux, "Jules Hansen (1849–1931): dessinateur-géographe", *Revue du Comité français de Cartographie*, 210, December 2011, p. 137–148.

travaux de la Société de Géographie et sur les progrès des sciences géographiques. These drawings were not precisely measured routes, but simplified sketches comprehensible to a readership unfamiliar with the terrain. Hansen thus held a singular position between the field and the office. He reviewed the work done by explorers in the colonies, interpreted their notes and reports, and then sketched their itineraries, compiling the whole into a map. He had good relations with the explorers, and also worked closely with several eminent Parisian printers and engravers, among them Erhard, Delagrave, Barrère and Challamel. He thus kept control of the process of map-making, in his pivotal role as geographer and draughtsman, between explorers in the field and the scholarly community in the metropole.

Lastly, Geographical Societies constituted networks of exchange and sociability in which geographical discoveries and maps circulated, between members of a Society, between Societies – which, for example, sent each other their publications – and between Geographical Societies and other institutions. The Royal Geographical Society's earliest collections of maps thus came partly from the British Hydrographic Office and partly from the French *Dépôt des cartes et plans de la marine*, which contributed its maps of the Chinese coasts.[44] This scientific collaboration demonstrates the high regard in which the Geographical Societies were held. Too often reduced to the role of staunch defenders of colonialism due to their support for exploring and studying the colonies, amongst other distant lands, Geographical Societies were above all concerned with the scientific quality of their work. Thus, with the development of atlases for schools at the turn of the 19th–20th century, the Royal Geographical Society, conscious of the guarantee of quality it represented, tried to stay relatively neutral and not be instrumentalised by publishers. It refused to lend its name to the publishing house Johnston, notwithstanding the debates between those Society members keen to exploit this economic opportunity, and those who advocated a scholarly approach which ruled out recommending one publishing house over another.[45] The usefulness of geography for colonial expansion was not antithetical to the ideal of scientificity, however. Both cohabited in the Geographical Societies, where the mix of scholarly satisfaction and usefulness, as well as the curiosity for the as yet undiscovered, was a strong incentive for explorers.

44 *The Journal of the Royal Geographical Society*, 1843, vol. 13, Introduction.

45 Julie McDougall-Waters, "British School Atlases, 1880–1930: Questions of Relevance, Credibility and Authorship in the Production of Geographical Knowledge", *Imago Mundi*, 66–1, 2 January 2014, p. 82–94.

2.2 *The Primary Objective: Travelling Upstream to a River's Source*

The Indochinese Peninsula was explored first by following the major watercourses inland, into the interior, as was the case with Africa. For the British, the focus was the Irrawaddy and the Salween rivers, for the French, the Mekong and the Red River. These were the main thoroughfares around which the representations of the Peninsula began to take shape. The primary aim of these explorations was naturally to test the navigability of the river, but also to gather information about the territories passed through, which were almost unknown to the Europeans. A spirit of rivalry developed between the British and the French over their progress upstream towards the Chinese border, as we shall see in examining how the territories were represented, in both camps.

Among the many exploratory expeditions launched in the 1860s and 1870s, one stands out – for its size, its length, the authorities that determined its objectives, the quality of its results, and the reception and posterity of the geographical work it produced. This was the French Mekong Exploration Commission. It occupies an exceptional position as the first "scientific expedition" to be undertaken in the Indochinese Peninsula. It was "launched under the aegis of the French State, supported by official institutions (the *Institut de France* and the *Académies*), and it was headed by a scientific unit specially recruited for the expedition, which carried out its work in the presence of, and sometimes with the direct collaboration of, the army."[46] It was first led by Commander Ernest Doudart de Lagrée, and later by Francis Garnier after Doudart de Lagrée's death on 12 March 1868 in China.[47] The official instructions from the Governor of Cochinchina, Admiral La Grandière, were unambiguous:

> Determine the course of the river geographically by a quick reconnaissance which should advance as far as possible; on the way, study the resources of the regions crossed, and find the most efficient ways to join the upper valley of the Mekong to Kambodj and Cochinchina for trade. These are, in brief, the essential objectives you should always bear in mind.[48]

46 M.-N. Bourguet, B. Lepetit, D. Nordman and M. Sinarellis (eds), *L'invention scientifique de la Méditerranée …*, *op. cit.*, p. 15.

47 In addition to Ernest Doudart de Lagrée and Francis Garnier (hydrography, meteorology and cartography), the Mekong Exploration Commission was composed of Louis Delaporte (a draughtsman and ensign in the Navy), Louis de Carné (the son of the Foreign Affairs Minister, responsible for descriptions and information on trade, a civilian), Lucien Joubert (an anthropologist and Navy Doctor, 2e *classe*), Clovis Thorel (a botanist and Navy Doctor, 3e *classe*) and Emile Gsell (a photographer).

48 BSG, Sixième série, tome v, January–June 1873, p. 297.

Economic concerns were interwoven with political issues, and a spirit of scientific inquiry. Regarding the latter, the survey of the Mekong produced a number of geographical breakthroughs, of which not the least was the river's course, which until then had been represented as straight.[49]

The expedition's results were published in 1873, in four volumes. The first volume, entitled *Partie descriptive, historique et politique du voyage* [Descriptive, historical and political part of the journey], presented the diary of events from 5 June 1866 to 29 June 1868. Francis Garnier regretted in the Preface that he "had not managed to reconcile the interest of the narrative with scientific requirements, which are the essence of the present publication."[50] In other words, his primary concern was the scientific content, even if this resulted from observation only. The material gathered comprised the second volume, devoted to "scientific observations, and special studies by the Exploration Commission". The fields covered were astronomy, meteorology, a note on cartography, geology and mineralogy, anthropology, agriculture and horticulture, and vocabularies. Then came the atlas of the expedition, in two parts, one of which was composed of maps and architectural plans, and the other, entitled *Album pittoresque*, of drawings by Louis Delaporte. In the atlas, the primary focus was the itinerary taken, and the river itself, regardless of the scale chosen [Figure 6].

The work's reception was complex. The first map was published as early as 1867, at the request of La Grandière, the Governor of Cochinchina, after M. Duyshart happened to meet some members of the expedition not far from Pak Lay, on 16 April 1867, and sent the French Consul the materials produced thus far. The map, which was more of a sketch, was published in the *Revue maritime et coloniale* in November 1867 and reprinted shortly afterwards in *Petermanns Mitteilungen*, a journal founded in 1855 at Gotha, which made the first geographical findings of the expedition known beyond the borders of France, and across Europe.[51] In France, the results were greeted with general indifference, partly for political reasons, partly because of personal antagonisms, and partly because the Franco-Prussian War slowed the work down. The *Société de géographie* awarded its *grande médaille d'or* [Prime Gold Medal] posthumously to Garnier and Doudart de Lagrée in 1869, and in the following year, the British Royal Geographical Society awarded its Gold Medal

49 Jean-Pierre Gomane, *L'exploration du Mékong. La mission Ernest Doudart de Lagrée-Francis Garnier, 1866–1868*, Paris, Harmattan, 1994, p. 162.

50 E. Doudart de Lagrée, F. Garnier and L. Delaporte, *Voyage d'exploration en Indo-Chine…*, *op. cit.*, Preface.

51 "Die Französische Expedition auf dem Mekhong, 1866 und 1867", *Petermanns Mitteilungen*, November 1867, p. 10.

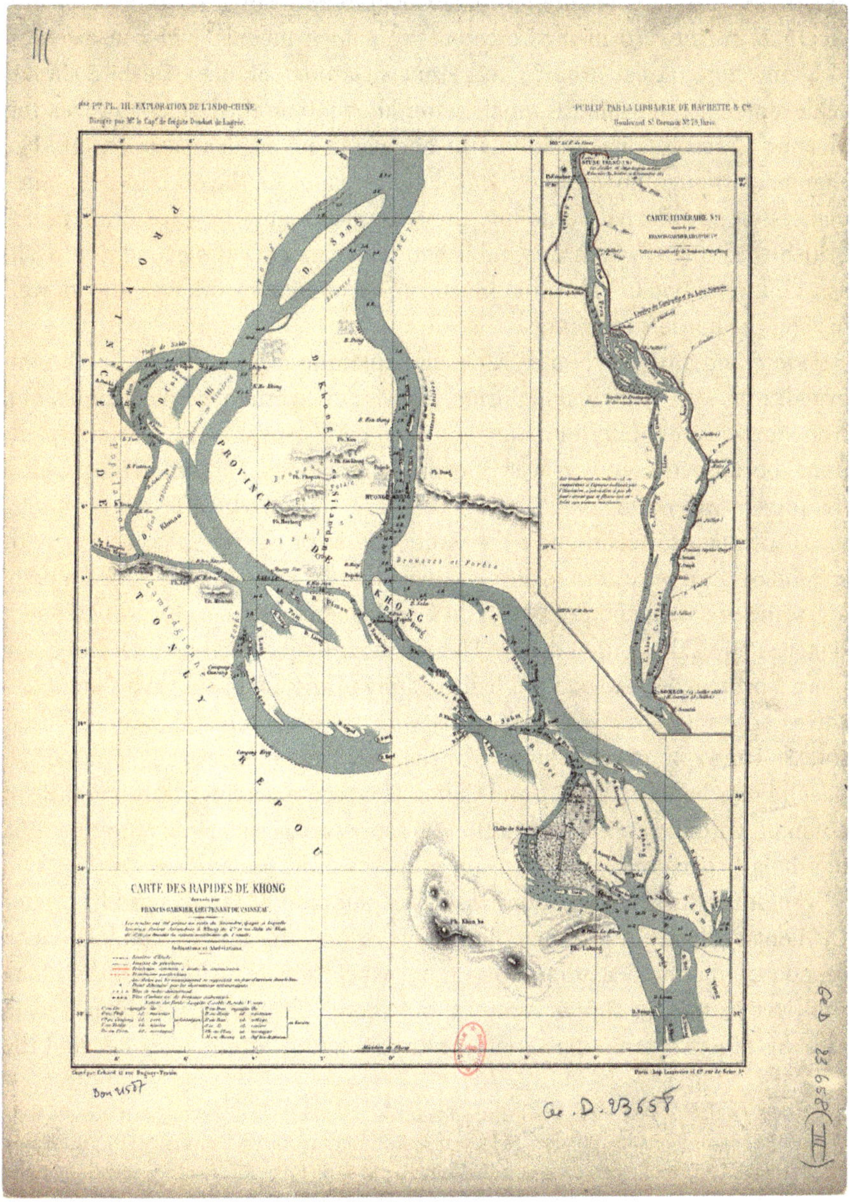

FIGURE 6 Itinerary no. 1, Mekong Exploration Commission by Francis Garnier, 1873
SOURCE: BNF, Cartes et Plans, GE DD-1111 [http://gallica.bnf.fr/ark:/12148/btv1b530704539/f1.item]

to Francis Garnier. The International Geographical Congress, at its very first meeting, in Antwerp in 1871, created two golden medals, which it awarded to Livingstone (Great Britain) and Francis Garnier (France). Despite similar achievements, the former is much better known than the latter. However, the Mekong Commission's exploits also circulated in English through another channel: although no complete translation was made, the account was summarised in a work that contained the narrative of another expedition also – British this time – which was published ten years after the expedition itself, as a children's book.[52] Thus the Mekong Commission's work was known well beyond the borders of France.

Before and after the French Mekong Exploration Commission, two British expeditions were launched in Burma, likewise to ascertain the navigability of a river, but with very inferior means. One can hardly speak of "scientific expeditions", because its members were mostly army personnel posted in Burma, and the project was quite small. The Salween expedition, which lasted from 1863 to 1865, sought to reconnoitre the upper course of the river. In keeping with a standard form of exploration narrative, the two accounts contained a daily log, some practical and political issues arising, meteorological readings, and a sketch-map of the route taken.[53] The resulting map, which had neither a key nor a North arrow, showed the hydrography of the area, and gave some indication of the terrain's relief and crops, with hand-written comments on their state.[54] The explorers concluded that the river's turbulence made it unnavigable for boats between Burma and China. This reconnaissance thus had little in common with the French expedition up the Mekong, since it had no scientific or scholarly personnel, and it produced no scientific findings.

The other British expedition was the exploration of the Irrawaddy. Doctor Clement Williams, a former military doctor and the first Political Agent in Mandalay, mounted a solo exploratory expedition of a few months along the river in 1863. A similar venture, but more substantial, was launched in 1868 by Albert Fytche, the Chief Commissioner of Burma, who selected the

52 *The French in Indo-China. With a Narrative of Garnier's Explorations in Cochin-China, Annam, and Tonquin*, London, Nelson, 1884. The copy in Cornell University Library (Wason DS534 F87) was given to a certain Annie Gray, First Prize for Reading and Geography, 29 June 1893.

53 G. Colquhoun Sconce, *Journal of the Salween Surveying Expedition*, Calcutta, Military Orphan Press, 1865; C. E. Watson, Francis Fedden, *Journal of the Salween Surveying Expedition During the Season 1864–65. With Route Map*, Calcutta, Military Orphan Press, 1865.

54 "Sketch Map. Route of Lieutenants Sconce & Watson During the Salween Expedition From December 1863 to April 1864. Drawn by R. H. Dopoy Draftsman, Executive Engineers Office, Department P. W. Moulmein Division", in G. C. Sconce, *Journal of the Salween Surveying Expedition, op. cit.*

expedition's members, and ensured their protection. Williams, who joined this expedition, reported that the project had received a large amount of money, but he also complained that Queen Victoria did not herself contribute, a sign that the Irrawaddy expedition attracted little interest compared to its counterpart along the Mekong, which had gained significant political support.[55] Its goals were primarily commercial: to reopen the former trade routes between China and Burma. For this, it was necessary to test whether a steamboat could pass beyond the capital, Mandalay, since at the time, none had managed to get through from Bhamo.[56] Led by Capitain Sladen, a British Political Agent in Mandalay, the expedition came to the conclusion that the Irrawaddy was navigable from Rangoon to Bhamo in all seasons.[57] In his report, John Anderson, a doctor and naturalist, provided a 15-page list of the names of all the people who had tried to reach the Irrawaddy and its source before them, or had attempted to depict these.[58] The report also contained the trade routes in Upper Burma connecting to China, scientific notes, and two maps, one of the country, the other of the sources of the rivers. Although, due to its large size, and to the presence of a doctor and naturalist on board, the expedition resembled a scientific project, its scientific findings were too scanty for it to be classed in this category, and it never achieved the renown of the Mekong expedition. It was mentioned in later, including French, works, but never in detail, suggesting that the Report was in fact little read. In the ensuing years, there were other explorations of the Irrawaddy, including James Wheeler's much-cited account of his trip at the end of 1870, when he travelled up the river for a few weeks on a visit to the king of Ava.[59]

---

55 Clement Williams, *Through Burmah to Western China. Being Notes of a Journey in 1863 to establish the Practicability of a Trade-route between the Irawaddi and the Yang-Tse-Kiang*, Edinburgh / London, W. Blackwood and Sons, 1868.

56 Edward Bosc Sladen, "Burma: Exploration via the Irrawaddy and Bhamo to South-Western China", *Proceedings of the Royal Geographical Society of London*, 15/5, 1870, p. 343–364.

57 The members of the expedition, in addition to Captain Sladen, were Captain J. M. Williams (an engineer), Dr John Anderson (a doctor and naturalist from the Calcutta Museum, which came into possession of the collections after the expedition), Theodore Stewart (a representative for trade), Alexander Bowers (another representative for trade, from the Royal Naval Reserve), F. N. Burn (an interpreter), Jemadar Mahomed Ali Khan, and 50 policemen from the Rangoon and Prome Police Force. See Alexander Bowers, *Bhamo Expedition. Report on the practicability of re-opening the trade route, between Burma and Western China*, Rangoon, American Mission Press, 1869; John Anderson, *A report on the expedition to western Yunan via Bhamo*, Calcutta, Office of the Superintendent of Government Printing, 1871.

58 J. Anderson, *A report on the expedition to western Yunan via Bhamo, op. cit.*, p. 178–193.

59 James Talboys Wheeler, *Journal of a Voyage up the Irrawaddy to Mandalay and Bhamo*, Rangoon, J. W. Baynes, 1871.

The surveying of the Mekong lay behind many scientific advances, but it also brought with it some disappointments, namely the conclusion – as Doudart de Lagrée had reported from the Mekong Exploration Commission – that the river was, after all, not suitable for trade. The French thus turned to the Red River, in the light of the recommendations made by Francis Garnier in 1872:

> The Chinese goods travelling downstream to Saigon could make use of the coastal navigation already established between the river's mouth and Saigon, and you have only to glance at the map you have in your hands, Sirs, to see how much shorter and politically simpler this would be than taking the winding route through the Cambodian River Valley, which passes through several sovereign territories.[60]

China had anyway abandoned imperialistic designs on the area as early as the 10th century, since improvements in navigation had enabled Tonkin to be circumvented, and trade from the Chinese province of Yunnan no longer passed through there. Influenced by explorers such as Jean Dupuis, France had "in desperation settled along the coast of Annam, despite its peripheral position regarding the major trade routes".[61] Jean Dupuis, an arms dealer, left Hanoi in November 1872, and reached Yunnan in March 1873, hoping to do trade with China.[62] The Red River became the focus of all the hopes which had been dashed by the Mekong expedition. Jean Dupuis's reconnaissance trip was positive, but on his return he was apprehended by the Vietnamese authorities, and a French force led from Cochinchina by Francis Garnier had to be sent to rescue him. The Citadel of Hanoi was captured, but Garnier was killed in the fighting. The French authorities in Paris did not accept this colonial annexation, since it occurred during the "Pause" (1867–1878) in France.[63] On 15 March 1874, a treaty was signed obliging Annam to open the Red River to international trade. Although the river's exploration did not contribute to the expansion of the French Empire, despite pressures from, especially, Dupuis and Garnier, it did succeed in making the Red River, which itself had been navigated and measured precisely, into the backbone of Tonkin as a whole, which was little known and little explored at the time. Jean Dupuis's report did not

60 BSG, Sixième série, tome III, January–June 1872, p. 155–156.

61 Frédéric Thomas, *Histoire du régime et des services forestiers français en Indochine de 1862 à 1945*, Hanoi Thế Giới, 1999, p. 13–14.

62 Jean Dupuis, *L'ouverture du fleuve rouge au commerce et les événements du Tong-Kin, 1872–1873. Journal de voyage et d'expédition*, Paris, Challamel aîné, 1879.

63 After France's defeat by Prussia in 1870–1871, and the fall of the Second Empire, the nascent Third Republic curbed colonial expansion until the 1880s.

only contain an account of the exploration, but also a wealth of information on the river itself.[64] Exploration and colonisation were less tightly interwoven during this period, especially in the 1870s, when expansion was checked, despite the fact that the explorers themselves – Dupuis, Garnier and Harmand, for instance – recommended annexation. At this historical moment when the respective interests of the British and the French were crystallising in the area, the Indochinese Peninsula was largely represented in terms of its rivers, a vision which persisted in subsequent geographical works because these expedition reports circulated quite widely, and were well known. They contributed to founding a particular representation of the territory.

2.3 *The Ongoing Revision of Accrued Knowledge*

It is clear from the texts produced during the colonial expansion of the 1880s, and from the keys on the maps relating to this, that their authors invariably drew on the work of their predecessors. The discoveries and writings of the previous decades were used to confirm or invalidate new findings, sometimes by introducing a fresh perspective. Thus expeditions tended to start from where previous explorers had left off, using the conclusions of their predecessors to prepare their own journeys. We shall ask how these prior works were received, how they were used in practice in the field, and whether, in fact, they were always consulted. Although it is difficult to quantify these issues, it is fairly clear that previous itineraries were followed.

In the sources to the map from a new expedition, the names of the previous mapmakers were often mentioned. Francis Garnier's map of 1873, for instance, refers to Captain Yule's map of Burma, and to the maps of MacLeod, Richardson and Parkes for Laos and Siam. Garnier's acknowledgment of English material may suggest that he was emphasising the current territorial domination of the French, or, on the contrary, it may demonstrate that scientific endeavour transcended colonial rivalries, and that acquiring authoritative knowledge of the local geography from predecessors took precedence over all else. Travel narratives, and articles published by Geographical Societies, were well-known information sources for explorers. Although this material was rarely translated in full, into French or English, it was cited and summarised, and the findings were also disseminated by learned societies to reach a broader public.[65]

---

64 *Le Tong-Kin et le bassin du fleuve Rouge au point de vue politique et commercial* by J. Dupuis, drawn by Ch. Meyniard, Paris, Challamel aîné, 1879. Bnf, Cartes et Plans, GE C-23937.

65 Jules Harmand's account of his scientific mission in Laos, from 1877, was translated into German: "Im Innern von Hinterindien (nach dem Französischen des Dr. Harmand)", *Globus* 38/14, 1880, p. 212.

In the case of the Mekong Exploration Commission, hand-written annotations on older maps and texts illustrate the process of updating the available knowledge.[66] For instance, a few years after the expedition, in 1871, Francis Garnier annotated Wusthof's *Voyage au Laos*, the account of a seventeenth-century Dutch explorer, who visited Vientiane in 1641–1642. By adding marginal comments concerning the latest discoveries, and correcting false or contradictory information, he effectively updated this work from three centuries previously, for future use.

This and other material from the Mekong expedition was re-used, and it constituted a major reference point for both British and French authors on the area. The itineraries, distances and place names were retained, refined, and sometimes contested. When commenting on the Chinese maps he translated, Devéria used the place names provided by Garnier, which were well known. This shows the importance of taking into account how knowledge circulated and was popularised, when considering the reception of geographical research.

It is worth noting also that explorers' hypotheses, and the questions they left unresolved, determined the itineraries of subsequent expeditions, especially as the individuals knew each other, and corresponded. Garnier's correspondence contains seven letters from Henry Yule, and one from Anne Mouhot, and this contact played a key role in preparations for an expedition.[67] Before leaving, it was important to know what had already been done, the validated knowledge available, the leads suggested by predecessors, and the as yet unexplored routes. What remained paramount was to accrue further knowledge, and to rectify errors progressively, by connecting back to previous research – a feature of all the explorations after 1885. Colonisation of the territories did not change or interrupt this scientific practice, but the focus shifted to marginal territories which the Europeans had not yet explored.

3 Colonial Knowledge and Empire-Building after 1885

Reconnaissance missions in the northern Indochinese Peninsula continued in the 1880s, building on their predecessors' work. However, after the establishment of the French Protectorate of Annam-Tonkin, and British annexation of

66 The absence of sources regrettably makes it impossible to take this point further. Overall, little research has been carried out to date on the revision of existing knowledge, as has been done for Coquebert de Montbret's annotated maps by Isabelle Laboulais, "Les cartes géographiques du fonds Montbret", *Annales historiques de la Révolution française*, 343, March 2006, p. 175.

67 BNF, SG colis 200 (5232), correspondence addressed to Francis Garnier, 1857–1874.

Upper Burma, the status of these territories underwent a radical change, which was signalled on the ground by the appearance of a colonial administration, and European military personnel. From a practical, logistical point of view, the presence of the administration implied points of contact for Europeans in spaces where previously there had been very few – a Christian mission, a consulate – or none at all. In this respect, post-1885 explorations of the imperial confines had an easier task. But did the presence of colonial civil servants and the army alter the nature or objectives of the exploration of these areas?

As in the previous period, the size of expeditions ranged from the individual traveller to the full "scientific" exploration. Moreover, the French embarked on a venture that had no parallel on the British side: the Pavie Mission. It was contemporary with the military conquest of the area, and it raises the question of the closer links between knowledge and power during the period 1885–1914. Did Pavie's twofold role as a tireless explorer and a shrewd diplomat introduce a certain bias into the objectives and results of the project? Other, shorter, expeditions also took place: the British, and other Europeans, continued searching for river sources, and so penetrated ever further into unexplored mountain zones. The geographical knowledge gained was thus anchored in precise territories, which enabled the establishment of maps. What did the mapped area of these imperial lands look like in 1914?

3.1 *The Pavie Mission, or the "Scientific Exploration of Indochina"*

The Europeans' annexation of the northern Indochinese Peninsula did not produce a major break in the process of constructing geographical knowledge of these spaces. Although the establishment of a colonial administration, and pacification, made for more frequent journeys and reconnaissance missions, and hence provided more opportunities for gathering information, there is no sign that the explorations conducted after the annexations differed qualitatively from those conducted before them. The methods used – observation and measurement – remained unchanged.

The Pavie Mission, however, which took place in many stages from 1879 to 1895, was an exception to this rule. Its importance has long been underestimated.[68] In France, the Mission was known first through the figure

68 There was initially little interest in Pavie, except in his home town of Dinan. See, for an interesting study which still focuses on his Breton roots, Hélène Simon, *Auguste Pavie, explorateur en Indochine*, Rennes, Éditions Ouest-France, 1997; and, particularly, a major recent re-appraisal of Pavie in the context of an exhibition on him, by Isabelle Dion, *Auguste Pavie, l'explorateur aux pieds nus. Cambodge-Laos*, Aix-en-Provence, Archives nationales d'outre-mer, 2010. See also the very instructive web site devoted to it, with interactive maps to reconstruct the Pavie Mission's itineraries: www.pavie.culture.fr.

of Pavie himself, a Breton who came to epitomise the colonial hero. He took part in the Cochinchina campaign between 1858 and 1862, and was then demobilised there, and entered the telegraphic service in 1872. While in this post he explored the south of the Peninsula, learnt Khmer, and developed links with the local population. Due to his in-depth knowledge of the area, he was appointed Vice-Consul in Luang Prabang in 1885, a position in which he advocated the annexation of Laos by France. He helped save Oun Kham, the King of the town, when the Black Flags ransacked it, in 1888–1889. In 1893, he led the blockade of Bangkok, which resulted in the French annexation of Laos, according to the terms of the Franco-Siamese Treaty of 3 October 1893. Pavie was therefore active both in the empire's confines and in its centres of power. His twin roles of diplomat and explorer enabled him to mount a large expedition, which bore his name, and was long-lived – until 1895. He then returned to France, got married, and devoted his time to writing up his notes and research.

The resulting oeuvre had two parts. The first, entitled *Géographie et Voyages* [Geography and Journeys], which was published in seven volumes, and an atlas, between 1900 and 1919, contained the accounts of the members of the expedition. The second part, entitled *Études diverses*, [Various Studies], which was published in three volumes between 1898 and 1904, presented the expedition's remaining scientific findings. In 1921, Pavie published a last work, *À la conquête des cœurs*, ["To Conquer Hearts"], with a Preface by Georges Clemenceau (the former prime minister of France), which ended on a sentence that added to his legendary status: "I experienced the joy of being loved by the people I encountered."[69] His humble origins, his brilliant career, and his talents as an explorer and a diplomat were much admired, as was his protection of the King of Luang Prabang in 1887–88, the territorial conquests against the Siamese and the English, and his rejection of violence. This perfect image of the "peaceful conqueror" contrasts with the oblivion into which he has fallen. People were struck by his great modesty, which may partially explain this neglect. His human qualities as an explorer and a diplomat were also noted: his first biographer referred to the "humaneness" of this "great human being of Indochina".[70] More recent research has brought to light Pavie's networks of sociability in the

69 *"Je connus la joie d'être aimé des peuples chez qui je passai"*, Auguste Pavie, *À la conquête des cœurs. Le pays des millions d'éléphants et du parasol blanc, les Pavillons noirs, Déo-van-tri*, Paris, Bossard, 1921.

70 Jean-Laurent Gheerbrandt, *Pavie, le grand humain de l'Indochine*, Paris, Éditions de l'Empire Français, 1949.

field, particularly in Tonkin.[71] His immense scientific œuvre is well known, yet it has not been studied in detail, despite the fact that the Pavie Mission gave decisive impetus to knowledge of the geography of the northern Indochinese Peninsula.

The Pavie Mission's map, first published by Augustin Challamel in 1893, was reprinted in 1895 and 1899, and revised in 1902 [Figure 7], with a reprint in 1909. The rivers were represented in blue, the roads in red, using solid lines or dotted lines according to the degree of certainty, and the relief was in bistre. The map featured many place names, from the capital city and large towns to villages. The key included abbreviations and translations of the main geographical terms.

The 1893 version of the map, drawn at the scale of 1:1,000,000, was commissioned by the Foreign Affairs Minister and the Ministry of the Colonies, and drawn by Capitains Cupet, Friquegnon and Malglaive. It had several sources: the Pavie Mission's surveys, the work of the *Bureau topographique des troupes de l'Indochine*, of the corps of engineer-hydrographers, the findings of the Mekong Exploration Commission, the itineraries of Jules Harmand, Paul Néis, Humann, Septans, and Yersin, and, lastly, certain English and Siamese documents. The map showed the region's frontiers, except for the border between Cambodia and Siam. Laos was mentioned as "Laotian principalities". The 1899 map, also at the scale of 1:1,000,000, did not differ greatly from the 1893 version, even though it was treated as the "definitive edition" of the previous editions. It added some names of Laotian ethnic groups, and removed the mention of "Laotian principalities", but it did not show any of their borders. As with the map of 1893, it represented the railway under construction between Mỹ Tho and Saigon. For the 1902 map, drawn at the scale of 1:2,000,000 this time, Captain Seauve supplemented the work of Cupet, Malglaive and Friquegnon. Additional information and documentation came from Friquegnon's maps, particularly of central Indochina, the explorations of Henri d'Orléans, Roux, and Aymonier, and some of Guillemoto and Vaulserre's geological expeditions. This map emphasised international and intra-Indochinese frontiers, which were represented by rather thick red lines. The border of Cochinchina was shown fully demarcated, as was the north of Tonkin and the frontier between Cambodia and Siam near Tonlé Sap. Territorial conflicts were also shown, particularly concerning the Siamese Cambodian provinces and the Laotian provinces. The 1909 map, also on a scale of 1:2,000,000, corresponded to the map revised by Capitain Tixier after the Boundary Commission's work on the

71 Philippe Le Failler, *La rivière Noire. L'intégration d'une marche frontière au Vietnam*, Paris, CNRS Éditions, 2014.

FIGURE 7 Map of French Indochina by Auguste Pavie, 1902
SOURCE: Auguste Pavie, *Atlas de la mission Pavie en Indochine*, Paris, A. Challamel, 1903 [http://gallica.bnf.fr/ark:/12148/bpt6k64603908/f25.image.r=pavie]

borders between Indochina and Siam, with additional input from the *Service géographique de l'Indochine* and the Siamese Chief of Staff. By this time, the map showed all the international borders, but some internal boundaries were still shown in dotted lines, and there were more railway routes than on previous versions.

So the Pavie Mission, due to its size, its duration, the extent of the territory covered, the quantity of information collected, and the quality of the maps produced constituted a qualitative leap forwards in the knowledge available concerning the Indochinese Peninsula. There was no equivalent on the British side.

3.2 *The Overlapping Paths of Other Expeditions*

The British concentrated their efforts on the recently annexed and little-known territories of Upper Burma, and neighbouring areas, although they did not officially organise any geographical expeditions. They explored the north-eastern spaces of their colonial possessions, much like the French, whose focus also moved north. On the British side, as with the French, the aim was to fill in the gaps remaining on their maps, according to their specifically European vision of the territory. Never satisfied with their colonial gains, both British and French pushed further north towards river sources, exploring the southern Chinese provinces and the Laotian principalities, which they eyed as their next potential conquests. The itineraries chosen by the Europeans tended to overlap more frequently than in the previous decades, a sign that the colonial frontiers were relatively contingent at the time. All sorts of hesitations and conjectures circulated before an imperial power would decide to annex or abandon new territories, especially ill-defined or much-coveted areas, crisscrossed by the explorers.

There was one particular area that both the French and the British had set their sights on since the 1870s, for economic reasons, and which was the object of intense rivalry between them by the end of the century. This was the province of Yunnan. Its link to China had been an issue from the start of European colonisation in the Peninsula, and became more acute as exploratory missions multiplied, and adjacent areas were better understood. The myth of Yunnan, following on from the myth of the Mekong and then of the Red River, started in France with Emile Rocher's book, *La province du Yunnan*, published in 1879.[72] Some explorers contributed to the mounting tensions by mentioning the progress made by the other colonial camp, and commenting on imperial policy. A good example here is the expedition of Colquhoun and Hallett, who

72 Émile Rocher, *La province chinoise du Yün-nan*, Paris, Ernest Leroux, 1879.

were commissioned in the 1880s to recommend a railway route from Burma to China, for which Colquhoun was awarded the Royal Geographical Society's gold medal in 1884. His work on the provinces of southern China, *Across Chryse*, was published in 1883, and immediately translated into French, under the title of *Autour du Tonkin* [Around Tonkin].[73] The title was re-used ten years later by Henri d'Orléans and Gabriel Bonvalot for the published account of their journey from Paris to Tonkin in 1889–1890, a book that was translated immediately into English – a sign of the ongoing topicality of the subject also for the British.[74] Henri d'Orléans, a descendant of the French Royal family, had to abandon his military career following the 1886 "Law of Exile", and so chose the life of an adventurer, in Asia; Gabriel Bonvalot, for his part, had been exploring Central Asia for some years, financed by the French Ministry of Public Instruction. He received an award for his work from the *Société de géographie de Paris*. Their journey from Paris to Tonkin, financed by Henri d'Orléans's father, eventually led them to Yunnan, and from there down the Red River to Hanoi. Five years later, in 1895, Henri d'Orléans returned to the northern Indochinese Peninsula, after a journey of exploration in Africa. Together with Émile Roux, an ensign in the Navy, and Briffaut, he left Hải Phòng for Calcutta, to explore the upper course of the Mekong (having already got close to its source in 1889–1890), then a section of the Salween River, finally reaching the sources of the Irrawaddy, which were the expedition's ultimate goal.[75] This discovery earned him the *Société de géographie de Paris*'s gold medal, and the maps he made prepared the ground for the future Yunnan Railway. From 1895 to 1897, the Lyon chamber of commerce launched an important expedition to China, called the *Mission lyonnaise* [the Lyon Expedition], which explored the three Chinese provinces bordering on Tonkin, and the territories beyond. The British also had their eye on Yunnan, and sent explorers there, who continued reconnoitring the Salween River and the Irrawaddy.

So in their drive northwards, the Europeans started with reconnaissance of the major rivers. The British explorers continued searching for the sources of

73 Archibald Ross Colquhoun, *Across Chrysê, being the Narrative of a Journey of Exploration through the South China Border Lands from Canton to Mandalay*, London, S. Low, 1883; Id., *La Chine méridionale*, translated by Charles Simond, Paris, H. Oudin, 1884.

74 Gabriel Bonvalot, *De Paris au Tonkin à travers le Tibet inconnu*, Paris, Hachette, 1892; Gabriel Bonvalot, *Across Thibet: being a translation of "De Paris au Tonkin à travers le Tibet inconnu"*, translated by C. B. Pitman, London, Cassell, 1891; 2nd ed. Henri d'Orléans, Autour du Tonkin, Paris, C. Lévy, 1894.

75 Émile Roux, *Aux sources de l'Irraouaddi, voyage de Hanoï à Calcutta par terre, illustré de cent dessins ou gravures directes d'après les photographies rapportées par l'auteur*, Paris, Hachette, 1897; Henri d'Orléans, *Du Tonkin aux Indes, janvier 1895–janvier 1896*, Paris, C. Lévy, 1898.

the Irrawaddy,[76] while the French, in the 1890s, brought back detailed maps from the Upper Mekong.[77] The mission report, and maps, from the expedition led by the junior naval officer Escande, in 1893–1894, refined the existing knowledge of the Red River, and opened it up as far as the China-Vietnam border.[78] Navigation on the Yangzi Jiang – the Blue River – became the new challenge of the day, and several explorers, such as Charles-Eudes Bonin, the Vice-Resident of France in Indochina, and Pierre Bons d'Anty, reconnoitred the river, starting out from Tonkin. Exploring rivers, and searching for their sources, thus continued to be the primary goal of expeditions, above all to find economic outlets, but also to push back the limits of known lands, and extend the field of geographical knowledge. There was some scepticism as to the treasures harboured by Yunnan, and the possibility of reaching them. The Lyon Expedition had indeed shown the limits of the navigability of the Red River, with the result that attention turned to the planned route of the railway, henceforth from Hải Phòng to Lào Cai, with an extension to Yunnan. However, there was also criticism – harsher on the British than on the French side – of the interest in Yunnan as illusory.[79]

Discussions were always based on current wisdom derived from the accumulated knowledge of the area and the progressively perfected maps. As in the previous period, corrections were made at every step, by comparing and contrasting exploration findings. Documentary sources increased and diversified. General maps now existed for all the imperial territories, compiled from the itineraries of explorers, or derived from their research.

3.3 *The Map of the Whole: Demarcating Political Dominion*

In the case of the French, explorers were the first to make general maps on the basis of direct observation in the field. Francis Garnier's map (1873), which he drew himself from his surveys, after several sessions at the *Dépôt des cartes et plans de la marine*, was the most sophisticated one until Jules-Léon Dutreuil de Rhins's map.[80] After sailing up the coast, de Rhins spent eight months in

76 RGS MGZ.56.12, "Diary of an expedition to the upper reaches of the Irrawaddy, and subsequently to the Indawgyi Lake", by Fenton; RGS MGZ.56.8, "Diary of a voyage from Bhamo to the upper reaches of the Irrawaddy, in the B. G. launch "Pathfinder", and from Sinbo to Mogaungand thence to Indawgyi Lake by way of the Indawchaung", by Barwick; RGS MGZ.56.9, "Diary of a journey to the Bor Khamti country, and sources of the Irrawaddy", by Gray.

77 ANV, GGI 3310; ANV, GGI 8868–8871; ANV, GGI 8872, year 1897–1898; BnF, SG F-7; BnF, GE D-21733; BnF, GE BB 58; BnF, GE DD 719.

78 *Notice sur le fleuve Rouge*, Paris, Imprimerie nationale, 1895.

79 This subject will be treated in greater depth in Chapter 6.

80 SHD, CCT, dossier Francis Garnier, pièces 7 et 16.

1876–1877 in Hué as captain of the Vietnamese fleet. He used the time available and his position to make several hydrographic maps. He then decided to produce a general map, for which he spent much time at the *Dépôt*, over a two-year period.[81] He published a "*Résumé des travaux géographiques sur l'Indo-Chine orientale*" [A Summary of geographical research on Eastern Indochina],[82] in which he made a list of the Europeans who had explored the Peninsula, and of all the general maps made to date. His map was published in 1879 at the *Dépôt de la marine*, with a reprint in 1881, and a revised edition in 1886. Unrivalled until the Pavie map (a first edition in 1893), its presence in several archives shows that it circulated widely within political, administrative and geographical institutions. It was in colour, with many blanks, and notably a large gap between the Mekong and the coasts, corresponding to the whole Annamite Range which seems to have been unexplored at this time.[83] De Rhins took a prudent stance on this, to show his scientific objectivity, stating that one "should not abhor a vacuum",[84] but should prefer blanks to mistakes. At the same time, the map contained a plethora of question marks, and irregularly marked routes, showing that de Rhins had difficulty leaving spaces unfilled, and was keen to make suggestions, for example, "probable direction of the watercourse", "hypothetical information", "unknown region", or even "position most uncertain". Any information which could be useful for subsequent work seemed preferable to blanks on the map. The latter more or less disappeared on the series of maps made by the Pavie Mission, by which time almost all the colonial territories had been covered, and demarcated, as well as the neighbouring territories. Apart from the map by Dutreuil de Rhins, the other general maps of this period were made subsequent to the two major exploratory missions in the Indochinese Peninsula, the Mekong Exploration Commission, and the Pavie Mission.

The British, by contrast, never made a general map, perhaps precisely because they did not undertake such large expeditions. One important exception was Henry Yule's map of 1857, which was compiled from many different itineraries, just like the French maps of the following decades [Figure 8]. The

81 J.-L. Dutreuil de Rhins, *Avertissement géographique et orthographique sur la carte de l'Indo-Chine orientale …, op. cit.*, p. 3.

82 *BSG*, Sixth Series, volume XIX, January-June 1880, p. 5–33.

83 On the subject of the spaces left blank on the map, and their implications for knowledge and conquest of the area, particularly in the context of colonial expansion, see Isabelle Laboulais (ed.), *Combler les blancs de la carte. Modalités et enjeux de la n des savoirs géographiques (XVIe–XXe siècle)*, Strasbourg, Presses universitaires de Strasbourg, 2004.

84 J.-L. Dutreuil de Rhins, *Avertissement géographique et orthographique sur la carte de l'Indo-Chine orientale …, op. cit.*, p. 7.

maps of British Burma, based on drawn itineraries, were mostly made by the Royal Geographical Society, or else by specialist London publishers. Examples include a *Burma with Parts of India, China and Siam* map published in 1886; and J. Arrowsmith's *Map of Burma and adjacent countries*, published in 1853, with a revised edition in 1875, for which various documents and manuscripts were consulted.[85] James Wyld, the Geographer to the Queen, published a *Map of Burmah and surrounding countries* following the Third Anglo-Burmese War [Figure 9], and a *Map of the Burmese Empire*, which used the findings of the Survey of India. General maps of Burma typically drew on material from the Survey of India, and used a relatively large, topographic, scale; indeed, most British general maps were produced by the Survey of India Office. In 1889, the Office borrowed Arrowsmith's title, *Map of Burma and adjacent countries*, for a 1:2,027,520 scale map, which was republished several times, also in a reduced version.[86] In December 1893, Henry Ravenshaw Thuillier, the Surveyor General of India, published a map of the whole Indochinese Peninsula entitled *Indo-China*, on the basis of the Survey of India for Burma, the Pavie Mission's maps of the French territories, and some Siamese maps. The fact that this map can be found in French archives suggests the importance of its representation of the whole region, embracing several sovereign territories.[87] Although these general maps focused on, and presented most information about, European imperial possessions (British Burma and French Indochina), they also showed the spaces beyond the marked borders, some of which were considered to be potential conquests.

Many representations of the Indochinese Peninsula, from a variety of sources, thus became available in the course of the nineteenth century. They fashioned the Europeans' mental image of the imperial territories, in the metropole and in the territories themselves. The maps made by missionaries, by engineer-hydrographers and by the first European explorers comprised the bulk of what was known when the conquest of the northern territories began in 1885. However, the information available was not necessarily consulted, and the information consulted was not always mentioned. Moreover, these sources were not all mobilised to the same extent, nor by the same people – a reminder that use of the "Colonial Library" took place in particular contexts that need to be defined if we are to avoid a false impression of homogeneity. This notwithstanding, it is clear that the French Mekong Exploration Commission led

85 BL, Maps 58765(10).

86 *Burma and adjacent countries*. Scale, 1 inch = 32 miles, or 1:2,027,520, Calcutta, Survey of India Offices, 1889. BL, Maps I.S.63.

87 BNF, GE C-3067 (1–2) and GE C-1880 (1–2).

FIGURE 8 *The New Map of Burma and the Regions Adjacent* by Henry Yule, 1857 (detail)
SOURCE: MEP, K MYA 3

FIGURE 9 James Wyld, *Map of Burmah and Surrounding Countries*, 1886 (detail)
SOURCE: MEP, K MYA 2

by Francis Garnier was a prominent book on these shelves. It was partially translated into English, and read in British schools, and it was also the work most quoted by explorers, British and French alike. They based their itineraries on Garnier's, correcting them where necessary. Garnier's work was thus arguably the most important example of the pooling of geographical documents between European powers, and of the convergence of French and British scientific knowledge as it was being produced at the time.

Common practice by explorers up to the beginning of the twentieth century was to correct available maps by eliminating errors and filling in the blanks. The political goal of understanding these territories was second to the scientific interest aroused, as suggested by the preeminent role played in explorations by the learned societies, particularly the Geographical Societies, as we have seen in the case of London and Paris. Driven by the immense scientific curiosity of their members, they encouraged and sometimes even financed the discovery of "unknown lands", and helped produce and circulate the scientific findings. They were a hub of relations between the colonies and the home country, bringing into contact politicians, enthusiasts, and explorers, whose research they collected and made available. Although one can sometimes note an "imperial turn" in the personality of their president or their Board members, it is not clear that they had any definite political project that they sought to apply. Geographical Societies had a real influence, and supported expeditions, but they did not initiate these, nor control the sometimes erratic order in which they occurred. In the construction of bodies of scientific knowledge about the Indochinese Peninsula, it was the explorers themselves who had the leading role, although they were guided in their work of observation by the learned societies.

All of these expeditions, whether they were individual or collective, and whether they were the fruit of circumstance or the outcome of an official decision, brought to light the fundamental role played by personal initiative in the cartography of the Peninsula. Although explorers often had similar approaches, and consulted the same "colonial library", much came down to their human qualities, the contingencies of their life-paths, and the chance events of their journeys. Some simply sketched their itineraries, while others created maps covering the whole territory; some explored only a few dozen kilometres, while others spent several years in the field, and covered vast distances. There was no fixed rule, and the freedom of the protagonists involved makes their results difficult to categorise. Gradually, however, standardised practices of cartography were introduced. The colonial knowledge developed in the framework of these late nineteenth-century expeditions enriched the findings from the southern parts of the Peninsula, enabling the whole area to be mapped, and the blank spaces on the maps to be filled in, instead of simply copying out linear itineraries. The maps reflected the volumes of knowledge acquired about Indochina, essentially thanks to the exploratory expeditions. However, these were small-scale representations, which were soon complemented by a more precise and systematic topographical coverage.

CHAPTER 2

Colonial Geographical Departments and Large-Scale Map-Making Programmes

> For the last fortnight I have been spending my time doing topographical sketches, which does nothing to sweeten the character. For is there anything more stultifying than drawing lines of different colours on pieces of paper, and spoiling one's eyesight trying to cram the largest possible number of hollows and protuberances, summits and valleys, into one square centimetre?[1]

The Geographical Departments' goal of producing large-scale maps of the colonial territories did not put an end to more personal initiatives, particularly those of colonial civil servants, who had specific objectives in mind. So it would be wrong to imagine an opposition between explorers and topographers, or a strictly chronological succession whereby the former simply gave way to the latter. Yet the production of topographical maps was also a turning-point in the large-scale mapping of the Indochinese Peninsula, accompanied as it was by standard-setting, and by the institutionalisation of mapmaking. Jules Hansen was thus the last official cartographer of the *Société de géographie* because the creation of a number of Geographical Departments at the end of the 19th century made this post obsolete.

Although the complete mapping of a colonial territory was a scientific operation, it was also the result of political will, as attested by the creation of appropriate institutions and by the scale of the enterprise, which some commentators have viewed as in itself an imperialist gesture, and an expression of domination.[2] The *Service géographique de l'Indochine*, unlike

1 Archives de l'Institut Pasteur (AIP), Fonds Fernand Bernard, BEF1, Dossier 2, Lettre de Fernand Bernard à son frère Armand Bernard, Hanoi, 21 February 1899.

2 This is Ian Barrow's position in presenting the cartography of India as a fully-fledged political programme, which succeeded in producing a representation of the sub-continent, inventing a territory, and imposing an image of it, with all that this implies for symbolic domination.

 | DOI:10.1163/9789004469853_004

its British counterpart in India, the Survey of India, has not been studied to date, although they shared the same aims of making a complete map of their colonial territories. Filling in this gap will enable us to compare British and French institutional practices through the prism of the personnel involved, the involvement of the indigenous population, and the type of maps produced. Some of the questions we ask are: does the creation of these Geographical Departments mark a break with what went before? Does it signal the end of the amateur enthusiast and the individual adventurer, replaced henceforth by a rationally organised administration, which alone could make colonial domination possible?

1 The Creation of Colonial Geographical Departments

1.1 *Towards the First Institutions*

Following the colonial conquests, the British and French urgently required large-scale maps, which were almost inexistent, while the general maps were too imprecise for military use. There was an evident, and declared, discrepancy between the need for maps, and the lack of resources. Consequently, one officer per column was made responsible for carrying out topographical surveys: "the officer shall accompany all expeditions and reconnaissance trips made by his batallion or his group, and shall make, on a scale of 1:20,000, the map of the terrain covered."[3] As the territorial conquest proceeded, these French Army officers drew the corresponding maps, since most of them had attended the elite training academies of the *École polytechnique* or *Saint-Cyr*, where there were classes in topography.[4] In Upper Burma, British topographer-officers were also attached to various expeditions, in order to cover the whole country.[5] Like their French counterparts, they had received some instruction in topography, which was part of British military training since 1804. Their work depended on the conquest and "pacification" of the territories, and thus on the army, which ensured security, appropriate logistics, and the overall conditions for obtaining

See Ian J. Barrow, *Making History, Drawing Territory. British Mapping in India, c. 1756–1905*, New Delhi, Oxford University Press, 2003.

3 SHD, 10H1, Dossier 2, "Conquête du Tonkin. Deuxième partie. Commandement du général Millot. 1884. Historique par les capitaines Sergent et Margueron, 1890", p. 72.

4 Bruno Belhoste, *La formation d'une technocratie. L'École polytechnique et ses élèves de la Révolution au Second Empire*, Paris, Belin, 2003. According to the author, topography was first taught in this School under Napoleon, during the first French Empire.

5 BL, IOR, V/24/3984, Report on the operations of the Survey of India, administered under the Government of India, during 1885–1886.

geographical information. The first British surveys, by Major Hobday in 1885–1886 in Upper Burma, were made in the course of three military expeditions, whose itineraries defined the corresponding topographical maps.[6] In Siam, likewise, while French and British conquests were taking place elsewhere, British and Siamese topographers accompanied a Siamese military expedition in 1884 against Chinese gangs, the Ho, in the region of Luang Prabang. The King of Siam, Chulalongkorn, requested that, for this first collaboration, all his commanders and officers help the topographers carry out their surveys and gather information on the areas under his jurisdiction. The maps, which were produced as the campaign progressed, were then sent to Bangkok. Colonial expansion thus provided a framework for topographical surveys, and the colonial military context abetted scientific progress; to this extent, knowledge and power were unquestionably linked, with geographers and military personnel being one and the same individuals, or belonging to the same organisation.

Once the itineraries and surveys had been completed in the field, the documents were centralised and worked on, either in the Peninsula, or in the home country. The first maps were thus compilations of reconnaissance sketches.

In early 1886, the British Army asked for all the sketch maps made during the conquest to be handed in.[7] A preliminary 1:125,000 scale map of the area around Mandalay was compiled from military reconnaissance expeditions, and submitted to Calcutta for publication. On the French side, sketches considered useful for military purposes were reproduced by the *Service topographique*, which became the *Bureau topographique* in 1886. The 1:200,000 scale map of the whole of Indochina was based on the surveys made by French officers; and the first sheets concerning Tonkin, on a scale of 1:100,000, were produced with instructions that they "should be constantly updated, by means of the itineraries of the troops and the reconnaissance expeditions."[8] The blueprints were initially sent to Paris, but after 1890, when a printing department was integrated into the *Bureau topographique*, the maps of Indochina were published in Hanoi itself.[9]

Like the Geographical Societies and the Hydrographic Departments, the geographical institutions in the colonies had an important role in collecting and centralising geographical documents, and so became sorts of "centres of

6 *Ibid.*, p. 65.

7 *Ibid.*, p. 53.

8 "Travaux du bureau topographique des troupes de l'Indo-Chine", *Annales de géographie*, 7/36, 1898, p. 457–462, here p. 457.

9 At the time, this Office had an annual budget of 35,000 francs, of which 18,000 came from France, and the rest from the local budget.

calculation" [*centres de calcul*].[10] In India, this role was by no means self-evident, because until the end of the eighteenth century, no one was responsible for the goegraphical archives, which thus ended up in the head geographer's office, but without any attempt to collect the maps from other departments. One of the objectives in creating the post of Head of the Geographical Department of the Survey of India in 1814 was precisely to assemble the information gathered over the previous decades, and send it to London. Although it was clear that the project could not be exhaustive, it could at least make accessible and thus useful the largest possible body of geographical resources.

This was also the primary objective of the *Service géographique des colonies*, a much overlooked institution which actively contributed to the overall goal of compiling geographical and cartographic documents.[11] The *Service hydrographique de la marine*, which conserved and produced hydrographic maps, was created in 1887, as was the *Service géographique de l'armée*, which replaced the *Dépôt de la guerre*, founded in the late 17th century to centralise and publish the military maps needed by the French army.[12] Two years later, in 1889, it was the turn of the *Service géographique des colonies*, which seems to have been created in order to constitute a purely colonial "centre of calculation", designed to collect and centralise all the geographical and cartographic documentation on the French colonies.[13] It had become clear that the records from the colonies, which were classified separately, were of no use in the home country, whereas, if collected and classified methodically, they could constitute a "very rich collection of priceless geographical documents."[14] The new institution's role as a documentation centre was thus highlighted, its other role being to publish new maps on the basis of the information gathered. As with the British colonies, full-coverage maps were to be built up from compiling the sketch maps sent in. However, these tended to be unintelligible to officials in the home country without some prior treatment. This is why explorers were recommended to contact the *Service géographique des colonies* before

10 B. Latour, *La science en action, op. cit.*, 1989.

11 The fact that French Indochina had a Geographical Department of its own reduced the importance of the *Service géographique des colonies* in this region. However, it deserves at least a mention, since it has been little studied to date.

12 Patrice Bret, "Le Dépôt général de la guerre et la formation scientifique des ingénieurs-géographes militaires en France (1789–1830)", *Annals of Science*, 48/2, 1991, p. 113–157; Martina Schiavon, *Itinéraires de la précision. Géodésiens, artilleurs, savants et fabricants d'instruments de précision en France, 1870–1930*, Nancy, Presses universitaires de Nancy/Éditions universitaires de Lorraine, 2014, chap. 7.

13 Georges Antony, *Le Service géographique et des missions. Historique, organisation et publications*, Levallois-Perret, Impr. Crété de l'Arbre, 1901, p. 2.

14 *Ibid.*

setting out, in order to receive instructions which would ensure best practice by all geographers in the field, just like the advice given to explorers by the Geographical Societies.

Despite these precautions, the first maps were not considered to be up to scratch. The ones drawn by French officers, who were involved in a variety of different types of expedition, were of variable quality:

> [...] a bundle of geographical tatters which we attempted to harmonise and cobble together in a framework which was itself uncertain, since it was the result of the chance assembly of geodetic surveys made many years apart, without any overall plan, and according to methods of most unequal value.[15]

Although this assessment by Lubanski, the Head of the *Service géographique de l'Indochine*, can be understood as a way of underlining the inadequacy of the work performed prior to his arrival, it is not hard to imagine a mass of drafts dashed off in a hurry, which were rarely considered worth saving: "These rapidly executed sketches rarely match each other. Many of the original drawings have been lost; this is hardly to be regretted."[16]

At the opposite extreme from this project of collecting and compiling documents was the desire to carry out systematic, programmed work, in order to guarantee the scientific quality of the outcome. This was precisely the role of specialised institutions.

1.2 *Models: The Maps of France, Algeria and Ireland*

These mapmaking institutions did not emerge *ex nihilo*, but grew out of a process which went back to the 18th century, and the world of Enlightenment scholarship.[17] The geographical practices of the time privileged observation, measurement and experiment, which were deemed the only ways to ensure high-quality results. Field work was championed: the geographer could no longer remain cloistered in his cabinet, as in Vermeer's painting, rather, his book knowledge had to be supplemented by research in the field. Unlike the situation in England, where the need for maps was not so keenly felt, and where no nation-wide mapmaking programme existed until the 19th century, the

15 Jules-Clément-Ladislas Lubanski, "Note sur le Service géographique de l'Indo-Chine", in Paul Doumer (ed.), *Situation de l'Indo-Chine (1897–1901)*. Rapport, Hanoi, FH Schneider, 1902, p. 475.

16 "Travaux du bureau topographique des troupes de l'Indo-Chine", *art. cit.*

17 J.-M. Besse, H. Blais, I. Surun (eds.), *Naissances de la géographie moderne, 1760–1860, op. cit.*

French adopted these principles, which inspired many map-making projects in France, and laid the foundations for a long tradition of map-making. Maps, initially for military purposes (they were useful for waging war), also had a political dimension, as instruments for the orderly administration of a country.

This was the background to the idea of mapping the whole national territory, which resulted in the Cassini map.[18] The project was launched in 1747, under Louis XV, and was the first systematic cartographic programme on the scale of the whole nation. It was brought to a halt in 1793, with the French Revolution, and was replaced, despite some plans to finish and adapt it, by the *Carte d'état- major*, at the beginning of the nineteenth century. This was also a lengthy undertaking.[19] The surveys were carried out by an elite corps of geographer-engineers, and later by military staff officers. Reductions to a scale of 1:80,000 were made in the *Dépôt de la guerre*, a major cartographic institution until the 1870 defeat at the hands of Prussia. Topographical activities then slumped, until the foundation in 1887 of the *Service géographique de l'armée*, which took over the prerogatives of the *Dépôt de la guerre*, namely geodetic surveys, triangulation, and the topographic work necessary for the map of France.[20]

This map inspired the topographer-officers working in Indochina, who wanted to achieve the same degree of scientificity. In 1901, Lubanski proposed a method for mapping mountainous regions, which derived from that used for the map of France. He wanted to apply "the general methods invented in France and applied throughout the world, to extend the geodetic network in Indochina."[21] He thus situated his Indochinese project in the long tradition of French cartography, and used French officers trained in these techniques. The model invented in mainland France had spread to other countries, as illustrated by the mapping of Algeria and Tunisia, which was also entrusted to the *Dépôt de la guerre*.[22] It was the first application of topographical techniques developed in the home country to colonial territories.

18 Monique Pelletier, *Les cartes des Cassini. La science au service de l'État et des provinces*, Paris, Éditions du CTHS, 2013.

19 On the *Carte de la France* and the *Dépôt de la guerre*, see H. Blais, *Mirages de la carte. L'invention de l'Algérie coloniale, op. cit.*

20 Triangulation is the method used by geodetic scientists to fix the geographical coordinates of points in terms of latitude and longitude. It provides a reliable grounding for further large-scale geographical work, for example topographical surveys, which are positioned within the geodetic control network of fixed coordinates. These subjects will be treated in greater depth in chapter 3.

21 Rouget, "Étude sur la cartographie de l'Indochine française", *Annales de géographie*, 15/79, 1906, p. 26–42, here p. 37.

22 H. Blais, *Mirages de la carte. L'invention de l'Algérie coloniale, op. cit.*

The map of France was not a model for the French Empire alone: the British, who had no tradition of topographical coverage of their national territory comparable to the Cassini map, felt the need to emulate their neighbours, and have maps of their own. Adopting the French model, they started to map not England, but Ireland, often called "the Crown's first colony". This project, which started in 1824 and ended twenty-two years later, was based on detailed triangulation, and was Britain's first attempt at systematic cartography of a territory.[23] The British then exported this formula to India, where the terrain was certainly very different, but where the methods used were similar, even if they were not applied at a uniform rhythm.

The British example in India was important for French Indochina, because the terrain was similar, unlike the map of France or Algeria: "the successful use of these methods in British India was a sure sign that they would be suitable in our colony in the Far East, which is comparable in terms of climate and vegetation."[24] Borrowings thus went in both directions, between the British and the French.

1.3 *The Uneven Development of Geographical Departments: The Raj, Siam and Indochina*

In order to base the production of maps on an identical framework, that is, on the triangulation of the territory, governmental Geographical Departments were set up in each country, at different times, as the need arose. Although these institutions developed along separate lines, it is significant that they existed or were created in all the countries discussed in this book. Due to the large amounts of time and money required for triangulation, such Departments could not be established without political and institutional support, and their existence thus implies a strong political will, and ageed objectives.

When Upper Burma came into British hands, it simply extended the Raj, where the Survey of India already existed, so there were not too many false starts when Upper Burma began to be mapped. Although the Survey of India only received its name in 1878, it was the oldest Geographical Department in the colonies. Matthew Edney has written a history of it for the period 1765–1843: Rennell's exploration of the Bengal region between 1765 and 1771 gave

23 John Harwood Andrews, *A Paper landscape: the Ordnance Survey in Nineteenth-century Ireland*, Dublin, Four Courts Press, 2002; Gillian M. Doherty, *The Irish Ordnance Survey: History, Culture, and Memory*, Dublin, Four Courts Press, 2006; Stiofan Ó Cadhla, *Civilizing Ireland: Ordnance survey 1824–1842*, Dublin, Portland, Irish Academic Press, 2007.

24 ANOM, SGEOCOL 18, Annexe au Rapport sur le fonctionnement du Service géographique de l'Indochine du lieutenant-colonel Claude de Mussy, chef du Service géographique: "Travaux géodésiques et astronomique", Hanoi, 1st August 1903, p. 22–23.

rise to the first rules, which were drawn up in Calcutta, and valid until 1815, at which point an Office of the Surveyor General of India was created in Madras. This did not, however, imply unified mapping methods, because there were still three centres – Calcutta, Madras and Bombay – each of which had its own mapmaking activities. 1802 saw the start of the Great Trigonometrical Survey, which aimed to triangulate the whole of the territory in order to give a solid basis to topographical surveys, and to make possible an *Atlas of India*. The idea behind it was to create a "scientific" map of the whole territory, by converting the available itineraries and topographical surveys into a coherent cartographic system. Over this period, which is well researched, cartographic activities were headed by the British East India Company, which wished to keep control of geographical data, and of the hierarchy of personnel involved.

The situation changed, as it did for other Departments, in 1858, when the East Indies Company was replaced by the India Office. The Indian Geographical Department would henceforth be under the direct authority of the Government of India, whose headquarters were in Calcutta. The goal of systematic cartographic coverage remained unchanged, confirming the "cartographic turn" whereby the collection, compilation and archival organisation of observed facts were replaced by structured studies, according to an overall plan. The Great Trigonometrical Survey was to provide the general framework for cartography in India, into which all information could be entered; it was simply to be extended to Upper Burma, in order not to neglect any part of the British Empire.

At the start of the campaign to annex Upper Burma, there was no topographical brigade to carry out cartographic work, and so it was the Andaman Party, later supplemented by the forces of the Burma Topographical Party, which carried out the first triangulations, around Mandalay.[25] The resources already present in the south of Burma were easily mobilised, and two detachments of topographers were created for the territories annexed in 1886–1887, such that the cartographic work could begin almost immediately. Thus the Raj's cartographic infrastructures proved to be a well-oiled machine.

The situation was different for the French, because they had no supporting cartographic institutions already in the area. Cochinchina was to Tonkin nothing like what India was to Burma, and there were no preexisting institutions, and no human or material resources that could simply be moved into place. The *Service géographique de l'Indochine* was set up by a ruling of 5 July 1899, thus completing the process of institutionalisation which had begun in 1886.

---

25 BL, IOR, V/24/3984, Report on the operations of the Survey of India, administered under the Government of India, during 1885–1886, Calcutta, 1887, p. 53.

This was considered to be a fundamental step forward for mapping the territory, a long time after the British. The *Service* was integrated into the military staff, it was directly accountable to the commander-in-chief of the Troops of Indochina, and, as such, its expenses were part of the general budget for Indochina, except for the salaries of the officers of the permanent and the auxiliary cadres. It had the following assignments:

1. Astronomical surveys and geodetic triangulation to provide the basis for geographical maps.
2. Surveys for large-scale maps.
3. Extensions and improvements to provisional maps.
4. Special topographical studies for the colonial administration.
5. Drafting and publication of maps and studies for the colonial administration.[26]

As with its Indian equivalent, the creation of the *Service géographique* was an attempt to standardise cartographic activities. The map of Tonkin, made by the *Bureau topographique*, and called "provisional", was replaced by regular surveys. Indochina was a special case in the French imperial territories because its *Service géographique* gradually separated from the metropole, and became relatively autonomous, as also occurred in Madagascar. These two French colonies, along with French West Africa from 1903,[27] were thus exceptions to the rule. Two particularly distinguished directors of the Indochinese Department were Lubanski, who headed the institution from its establishment in 1899 to 1902, and Friquegnon, as from 1903.

The Geographical Department in Siam was also a special case, due to the exceptional status of this independent monarchy in the Indochinese Peninsula. King Chulalongkorn, enthusiastic about Western progress, welcomed

26 ANOM, GGI 17538. This ruling was signed in Hanoi by Paul Doumer, the Governor General of Indochina, Desbordes, the General Commander in Chief of the Troops of Indochina, and Langlois, the Head of the Military Office.

27 Madagascar's Geographical Department, which was created on 11 April 1896, is less well known than its Indochinese counterpart. Its budget was 135,000 francs (90,000 francs from the colony and 45,000 francs from the metropole). It published a complete map of the island on a scale of 1:500,000. When Gallieni left the Department in 1905, this first subsidy came to an end, and the institution's budget thereafter was less than a fifth of the *Service géographique de l'Indochine*. Yet it continued publishing four sheets per year of the systematic 1:100,000 scale map, and began the on at 1:200,000. Like the other colonial Geographical Departments, its staff were trained in drawing, engraving and printing, and comprised both Europeans and indigenous personnel. It was the only such Department to be discontinued provisionally due to lack of funds (on 31 December 1911). See "Disparition du service géographique de Madagascar," *Annales de géographie*, 21/116, 1912, p. 180.

European experts, and adopted Western techniques of mapmaking.[28] A corps of topographers was created in 1875, called the "Military Engineers of the Royal Guard", headed by the King's private secretary Henry Alabaster, and two Siamese officials. Like its French counterpart, and to a lesser extent the British Geographical Department, this was a military body.[29] Of the 50-strong unit, however, only a few followed Alabaster's training through to the end.

James McCarthy, a British surveyor, was a central figure in the institutionalisation of Siamese cartography. He was appointed as Government Surveyor by the Siamese Minister of War in 1880, after heading a topographical brigade for the Survey of India, whose work of triangulation ended in Bangkok. In late 1882, Prince Damrong offered him the post of Head of the Geographical Department that he wanted to set up. He founded a mapmaking training school, which taught the European discipline and techniques, including mathematics, astronomy, the use of scientific instruments, the calculation of coordinates and other measuring techniques in topography. It trained an elite corps of future assistants to European topographers, drawing its students mainly from the families of the upper ranks of the Siamese civil service. In its early stages, prior to 1885, as the Department found its feet, roles and institutional affiliations were fluid, and expeditions combined all sorts of interests. Thus in 1883–1884, McCarthy concentrated on topographical surveys to resolve border problems in the Malaysian Peninsula, the North-West of Siam, and the North-East. In October 1884, he entered the service of the Siamese government as a Captain of the Royal Guard, a military more than a topographical post, although he also led a joint scientific expedition to Luang Prabang with a member of the Survey of India.

By a royal decree of 3 September 1885, a sharp distinction was introduced between topographers and royal guards. The Royal Siamese Survey Department which emerged from this was to carry out all the government's surveys and mapmaking, in order to establish a Siamese cartography that would no longer be solely a European science. Despite this, the limits of the different bodies were never clearly defined: the geographical division was under the authority of the Ministry of Agriculture, except between 1897 and 1899, when it came under the Finance Ministry. The Directors were all European experts:

28 See T. Winichakul, *Siam Mapped: A History of the Geo-body of a Nation, op. cit.*

29 On the early history of the Royal Survey Department of Siam, see R. W. Giblin, "Royal Survey Work", in *Twentieth Century Impressions of Siam*, London, Lloyds Greater Britain Publishing Company in London between 1901 and 1914, 1908, p. 121–127; "The Royal Survey Department Siam: A Retrospect", whose author is probably Ronald Giblin, the Department's head from 1901 to 1909, quoted in T. Winichakul, *Siam Mapped: A History of the Geo-body of a Nation*, n. 33.

James McCarthy until 1901, then Ronald Giblin until 1909. Giblin was born on 3 January 1863 in Hobart, Tasmania, where he grew up. He studied topography in New South Wales, Australia – another British dominion – and his career gravitated between different centres and peripheries of the British Empire. After being singled out by a University Professor in Sydney, who was in contact with McCarthy, he set out for Siam in December 1894, where he spent the rest of his career. He replaced McCarthy as Director during the latter's periods of leave in 1896 and 1898, and he was appointed Director by the Siamese government when McCarthy retired in 1901.

These Geographical Departments were thus created with clear political ends in view, and, except for the Siamese Department, they prolonged the European tradition of State-sponsored cartography and of collecting geographical data. However, their early production was barely distinguishable from the older explorers' sketches, of very variable quality, and the itineraries drawn by actively serving officers were not much better than those of hurried travellers. We shall examine whether any real change took place in the scientific production of these departments when their goal became systematic territorial coverage. Can one conclude that they successfully implemented the objective for which they had been created, namely mapping all the imperial territories in order to make them easier to administer?

2 Systematic Mapping of the Whole Territory: A Qualitative Advance?

All the different elements in the mapmaking process were centred on producing a standardised map of the whole territory. Some operations took place at the Head Office – principally drawing, and map production – but the backbone for the whole process was the fieldwork. The Survey of India thus had three branches, each of which was linked to an activity in the field: geodetic surveying, topography, and cadastral surveys. The Royal Siamese Survey Department imitated this division, whereas the French incorporated only the first two (the land survey office was separated off).

2.1 *The Stages in the Project of Full Cartographic Coverage up to 1914*

Mapping a whole territory is a long process, which can take several decades. The first stage involves making a geodetic reference frame, that is, a set of geographical coordinates into which all the detailed topographical information can then be integrated.

In French Indochina, officers began triangulating Tonkin in the 1870s, resumed their work in the 1880s, and continued in the following decade. Although some of the groundwork was completed, the triangulation was not yet systematic. Captain de Gaudel, on an assignment to check the triangulation work from Cao Bằng to Hà Giang, in 1898, observed that "only the triangulation of Lang Son and the chain Lang Son-Cao Bang give adequate results, but these are separate from the triangulation of the delta, because the two triangulations have not yet been linked via the region of Cai Kinh."[30] The discrepancies, where different coordinates existed for the same point, made it impossible to compile the itineraries available into a coherent map of Tonkin.

The establishment of the *Service géographique de l'Indochine* in 1899 produced a qualitative improvement. The systematic cartography of Tonkin could begin, according to a precise work plan, which had not existed previously. The first measurements concentrated on Tonkin and Annam, particularly the delta of the Red River, from Hanoi, then moving south-eastwards.[31] The first campaigns, which triangulated the delta areas, ended in 1904, and led to the idea of starting on the mountain ranges.

Three years earlier, in 1901, Lubanski had given an important impetus to mapmaking when he introduced a new method for mapping the neglected mountain regions, for which the Government General was now requesting maps. He presented it as a new *credo*, although it actually drew on the innovations of the map of France, and had been used in British India already. The first brigade to be trained in these techniques operated during the campaign of 1901–1902. Called the *Brigade topographique de la Haute Région* or the *Brigade de cent millième de la Haute Région*,[32] it worked in the area between Cao Bằng, Bảo Lạc and Hà Giang, headed by Lieutenant Lachèvre, who had previously been employed in the *Service géographique de l'armée*, and who was accompanied by six topographer-officers, three of whom had recently arrived in Tonkin (Lieutenants Rouget, Bellot and Andréa), and two other officers, who made up an "astronomical pair". This clear division of labour was one of the reasons for the brigade's efficiency. First, the "astronomical pair" determined the latitude of a number of points very far apart, in order to form an astronomical grid.

30 ANOM, SGEOCOL 18, Annexe au Rapport sur le fonctionnement du Service géographique de l'Indochine du lieutenant-colonel Claude de Mussy, chef du Service géographique: "Travaux géodésiques et astronomiques", Hanoi, 1st August 1903, p. 15.

31 Despite the many maps of Cochinchina, no triangulation had been carried out, contrary to the situation in India. So mapping Tonkin did not involve extending a network already in existence, but founding a new one.

32 J.-C.-L. Lubanski, "Note sur le Service géographique de l'Indo-Chine", *art. cit.*, p. 479.

Then the head of the brigade made a rapid triangulation using this set of geographical positions. Lastly, the six topographers carried out the surveys, which were fitted into the triangulated territory. The brigade thus carried out the whole range of astronomical, geodetic and topographical operations, thanks to the smooth functioning of the astronomical pair and the skill of the operators making surveys, who demonstrated their fine technical training and thorough knowledge of the relief of the region. The work epitomised the successful adaptation of rules to the realities of the terrain, which was necessary due to the lack of geodetic surveyors. This *modus operandi* was later generalised, in order "to undertake and successfully conclude, extensively and rapidly, the publication of fundamental documents essential to giving a solid basis to our system of military protection along the Chinese borders of the colony."[33] The idea was thus to create four brigades (called *cent millième*) to survey the border provinces and thus finish the military map of Tonkin. [Figure 10]

At the time of its annexation, Upper Burma, unlike its equivalent, Tonkin, was really an extension of the existing British Empire, since most of India had already been triangulated. The geodetic work continued, with the establishment of primary and secondary triangulation networks in Upper Burma [Figurre 11]. It was initially difficult to distinguish clearly the different tasks of each brigade within the Survey of India, because pragmatism over-ruled prior planning. For example, a triangulation team might actually carry out topographical work, as was the case for the Andaman Party in 1886, and geodetic, geographical and topographical work were often combined in the results.

Quite quickly, however, two brigades – nos. 11 and 21 – became specifically responsible for topographical surveys in Upper Burma. In 1887, this area had almost no history of topographical reconnaissance. Ten years later, most of the work was taking place in the region of Mandalay, and the area covered by topographical reconnaissance had expanded, because by 1894 the Head of the Survey of India considered that the general topography of Upper Burma was known. By 1910 the whole Burmese territory, except the border zones, had been systematically mapped, from Moulmein in the South to Bhamo in the North and the Mekong in the East. [Figure 12 & Figure 13]

In Siam, by contrast, there was no systematic territorial coverage, and the establishment of the Royal Siamese Survey Department in 1885 did not seem to bring about the qualitative change which occurred in the French and British cases. The first geodetic measurements there were made in 1880, when James McCarthy calculated the position of the most important towns in the country astronomically. The Western frontier was the first to be triangulated, at the

33 J.-C.-L. Lubanski, "Note sur le Service géographique de l'Indo-Chine", *art. cit.*, p. 480–481.

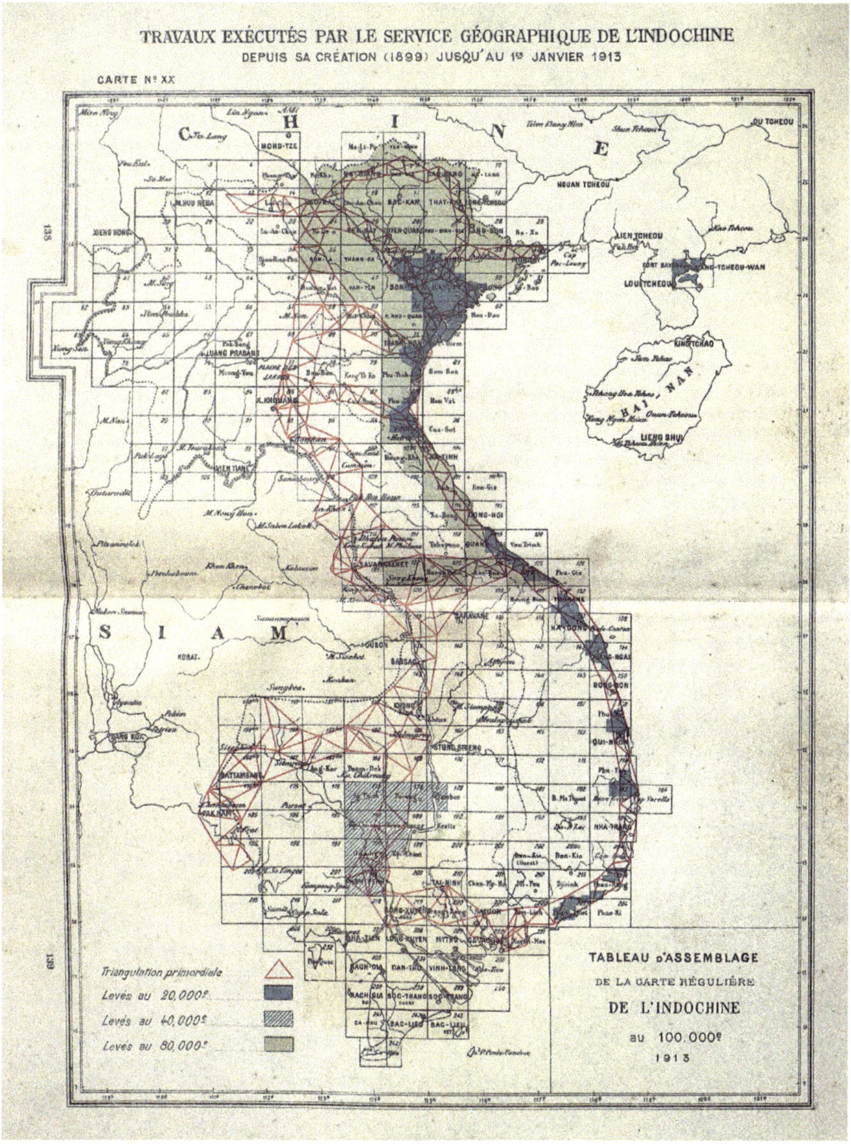

FIGURE 10 The triangulation of Indochina, 1913
SOURCE: Henri Brenier, *Essai d'atlas statistique de l'Indochine française*, Hanoi, Imprimerie française d'Extrême-Orient, 1914, p. 139. [https://gallica.bnf.fr/ark:/12148/bpt6k1100178q/f138.item.zoom]

FIGURE 11 The triangulation of British India (detail)
SOURCE: *Index Chart to the Great Trigonometrical Survey of India*, 1912

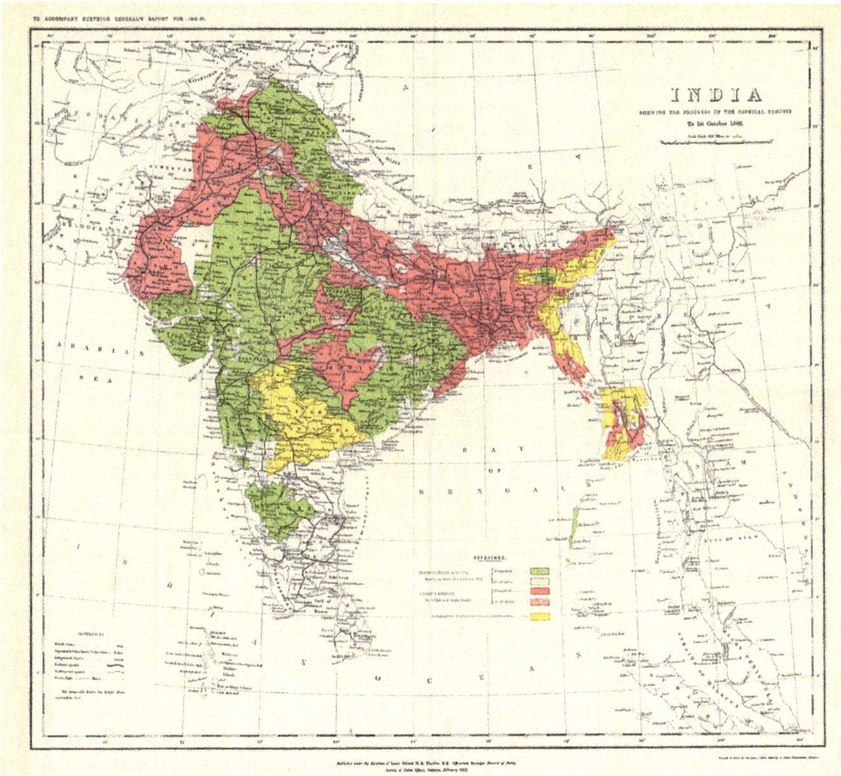

FIGURE 12 Land surveys carried out in India in 1886
SOURCE: BL, IOR, V/24/3984, Report on the operations of the Survey of India during 1885–1886, by Lieutenant-Colonel H. R. Thuillier, Calcutta, 1887

request of the Government of India, which wanted to complete the map of the Raj. These final triangulations from the Indian side thus laid the basis for the map of Siam, which was entrusted to McCarthy. Once the border regions had been measured, the Siamese topographers worked on one-off projects, depending on the demands of the moment, such as the construction of the telegraph line between Siam and Burma in 1881, or the building of the railroad to which McCarthy contributed in 1888–1889, when he helped plan the route for the Bangkok – Chiang Mai train line. Apart from these projects, most of the topographical work took place in Bangkok and its surroundings. The goals were very similar to those of the European powers, namely determining the best routes for communication infrastructures, and demarcating borders. Similar methods were also used, insofar as the surveys were carried out in the

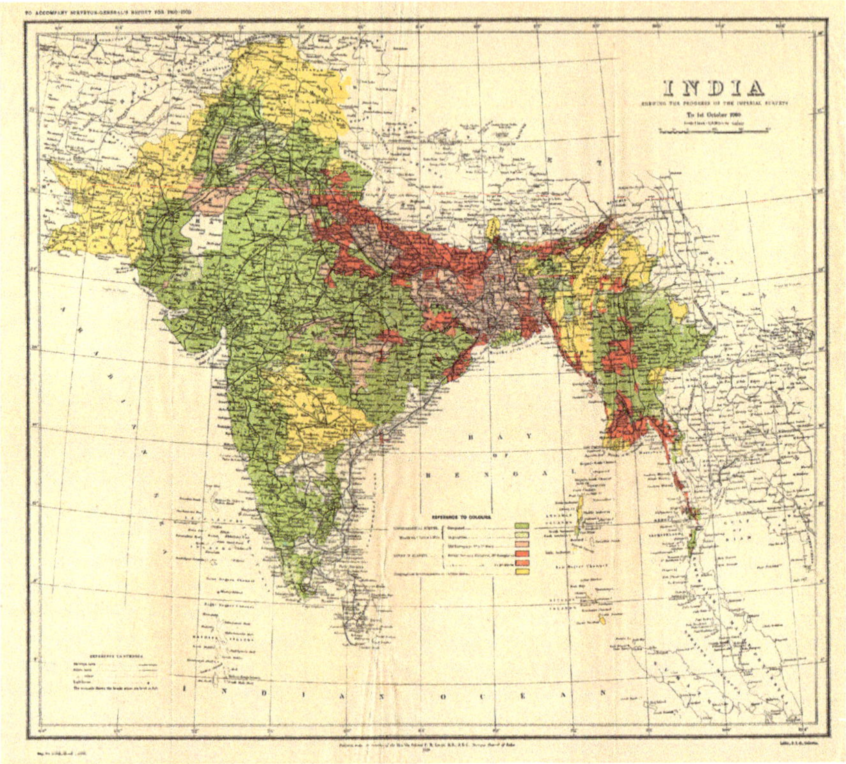

FIGURE 13 Land surveys carried out in India in 1909
SOURCE: BL, IOR, V/24/4007, General report on the operations of the Survey of India during 1908–1909, by Colonel F. B. Longe, Calcutta, 1910

context of military expeditions, which alone could ensure the scientific standards of the work. But unlike in French Indochina and British India, no coherent plan was drawn up to survey the whole Siamese territory in a consistent fashion. Common to all these Geographical Departments, however, was the choice of the scale to be used, depending on the objectives of the topographical survey, and the territories concerned.

2.2 *The Choice of Scale*

The choice of scale was closely linked to how the colonial territory was to be represented. A tension emerged between the wish to show a unified whole, on a small scale, and the need for documents that would be useful in the field, on a large scale. The first maps were compilations of itineraries, while later work was based on regular topographical surveys. The issues were slightly different in Siam, because apart from McCarthy's work of triangulation to demarcate the border, the Royal Siamese Survey Department concentrated on surveys for the Land Registry, which produced very precise maps in order to help streamline land administration.[34]

In the French case, before the *Service géographique* was created, the maps produced were all general maps, usually on a scale of 1:200,000, which had the advantage of giving an adequate degree of detail while not having to rely on a very precise geodetic network. 12 sheets were needed to cover Tonkin, 13 for Annam and 15 for Cochinchina. Given the rather unwieldy nature of these maps, reductions were often made (to scales of 1:500,000 and 1:1,000,000). When the *Service géographique* was founded in 1899, systematic large-scale mapping began, with topographical surveys using a scale of 1:20,000 for the deltas of Tonkin and Thanh Hóa, with a view to publishing a reduced map at 1:25,000; and the mountain regions surveyed at a scale of 1:80,000, for a resulting map at 1:100,000. This difference in scale and hence precision can be explained by the ease of access to the delta areas, and their greater economic value.

Throughout this period, the conflict between economic imperatives and the requirements of national defence never found a satisfactory solution. In 1920, for example, the Director of the *Service géographique* criticised the 5-year plan of his predecessor, both in principle – the impossibility of foreseeing the work to be done over that time-span – and, particularly, concerning the distribution of tasks.[35] In his view, there was too much emphasis on "the uninhabited regions of Upper Laos and Upper Tonkin", whereas they had waited "30 years to get a large-scale map of Cocinchina, the richest area of the Union."[36] One of the reasons for these choices was the large number of Land Registry surveys already carried out in Cochinchina, so there was less need for further surveys there than in the provinces where no regular topographical work had been carried out. Be that as it may, the balance between economic and military priorities generally gave rise to tensions and clashes.

---

34 McCarthy's 1:2,000,000 scale map of 1887, considered by the Thai today to be the first modern Western map of their country, was based on triangulation, but not on systematic topographical surveys. T. Winichakul, *Siam Mapped: A History of the Geo-body of a Nation*, *op. cit.*

35 It is interesting to note that the British Government of India also adopted 5-year plans, under pressure from the local levels of the colonial administration. These included, for 1903, the topographical surveys to be carried out by the Third and Tenth Brigades in Burma, and the surveys of the Shan States, on a scale of 1:63,360, which were to continue up to the frontier. See BL, IOR, V/24/4001, General Report on the operations of the Survey of India, administered under the Government of India, during 1902–1903, Calcutta, 1904.

36 ANOM, GGI 17565, Programme de travaux à poursuivre par le Service géographique, 1920.

The British seem to have been less affected by this problem. They focused on areas reputedly rich in resources – the Province of ruby mines, for example – and on frontier zones. The first provisional map of Upper Burma to be published, on a scale of 1:1,000,000, was produced from the compilation of sheets prepared by Major Hobday on a scale of 1:250,000.[37] Topographical surveys were then carried out on a scale of 1:63,360 [1 inch = 1 mile], although larger-scale versions were also produced, 1:45,000 and 1:30,000, for example. In 1910–1911, 115 sheets of the Map of India drawn on a scale of 1:63,360 were reproduced and printed. For Upper Burma, the focus was clearly on mapping the borders – the first sheets to be prepared – at the cost of large-scale mapping of the interior, and the in-depth knowledge to be gained from this. Maps of Tonkin featured a similar, but less extreme, distinction. The traditional division between the delta and the higher mountainous regions was not operative, because although the delta area was surveyed and mapped on a larger scale than the rest of Tonkin, the border regions, despite being mountainous, were also measured very precisely. Overall, however, the systematic mapping of Upper Burma and Tonkin resulted in large-scale maps for the urban centres and border areas, while other spaces were not treated in such detail.

Apart from these regular mapping activities, the Geographical Departments kept on producing other essential documents and re-drawing the provisional maps. Hence the replication of European-style overall cartographic coverage on the basis of triangulation was not a miracle solution, and older methods, such as sketch maps and rough surveys, were still used. Systematic mapping was a lengthy and arduous task. In the words of General Geil, writing in 1910, "completing the definitive map of Indochina, which has been delayed through lack of funds and staff, will need about 50 years."[38] Making a map of a whole territory was not a smooth process, and in French Indochina the same disagreements surfaced as had beset India in the first half of the nineteenth century: there were those who favoured compiling available geographical data, and those who advocated a systematic approach based on triangulation. The former wished to preserve the bureaucratic system as it was, and to ground cartography in the archives of geographical documents, placed within a general

---

37 BL, IOR, V/24/3985, Report on the operations of the Survey of India, administered under the Government of India, during 1886–1887, Calcutta, 1888.

38 ANOM, GGI 17555, Lettre de Geil, général de division commandant en chef le groupe des troupes de l'Indochine, au gouverneur général de l'Indochine, 1st April 1910.

framework; the latter, taking a longer view, wanted to produce a systematic, uniform and definitive coverage, which might well cost more, but which alone could guarantee a high-quality result.[39] Thus no consensus emerged on the standardisation of cartographic activities.

However, as General Geil complained, there were fewer and fewer cartographic documents sent to the *Bureau topographique*, and, after the creation of the *Service géographique* in 1899, none at all.[40] Alongside its work on the final version of the map of Indochina, the *Service géographique* updated the provisional maps, which were still needed, and which depended on "the precious efforts of the other officials, who work all over Indochina at the expense of the General Budget". He gave an exhaustive list of the different departments involved, a sign of the collaboration required: departments of public works, administrators, heads of provinces, Land Registry departments, the Navy's department of hydrography, and so forth. The *Ecole française d'Extrême-Orient* played a particularly important role in providing information "for the marginal territories of Indochina", that is, for areas beyond French control, to which in principle the *Service géographique* had no access. The appeal to willing hands was understandable in an invariably understaffed colonial situation, and in addition, there were the budgetary restrictions, which plagued French and British initiatives alike, both in the home country and in the colonies. So how did Geographical Departments manage to function correctly and achieve their objectives?

3 Meeting the Challenge of Limited Staff, Budgets and Material

If one looks through the tasks assigned to the different Geographical Departments, one might be led to think that the map-making programme was a vast enterprise undertaken methodically and efficiently by appropriately trained staff. The reality was actually quite different, and behind the high-flown discourses on the advances made in cartography and the need for a map covering all the colonial territories, lay a reality of limited staff on a tight budget.

39 M. H. Edney, *Mapping an Empire: The Geographical Construction of British India, 1765–1843*, *op. cit.*, p. 123.

40 ANOM, GGI 17555, Lettre de Geil, général de division commandant en chef le groupe des troupes de l'Indochine, au gouverneur général de l'Indochine, 1st April 1910.

3.1 *The Bare Essentials: Topographers, Geodetic Surveyors and Draughtsmen*

In the British case, the core number of topographers was no more than 15 for the whole of India, in 1800, rising to 30 by 1840. Few officials spent their whole career in this role.[41] When the India Office replaced the British East India Company in 1858, the recruitment process for colonial civil servants changed also. At the time, the India Office consisted of the Imperial Service, which incorporated people from the the Indian Civil Service who had passed a competitive exam in Britain, and the Provincial Service, which drew its civil servants from India alone. This difference in status was supplemented by a civil-military hierarchy: civilians were generally confined to the Provincial Service, whereas the military personnel, especially the most talented officers, joined the Imperial Service. The French *Service géographique de l'Indochine* likewise had an internal hierarchy: a permanent body of geodetic surveyors, astronomers and topographers, all of them officers; a civilian head of the reproduction workshops, European non-commissioned officers, corporals and soldiers; and local draughtsmen, secretaries and interpreters. Additionally, there was a support staff of officers and non-commissioned officers from the Troops of Indochina, on special assignment to the *Service géographique*.[42]

For British and French alike, the programme management, and the manufacture of the maps, were concentrated at the Head Office [Figures 14a–c]. The Head of the Department had a few officials and an office at this disposal. In the British case, the officials were the three heads of the map-making branches. The Head Office, in Calcutta and Hanoi respectively, housed the cartographic and engraving sections; apart from the heads, all the staff were recruited locally.[43] In the French case, a military officer and a civilian were selected from the *Service géographique de l'armée* for their experience at the drawing-board and in the reproduction workshops. The British Department had more officials than its French equivalent, but this was also due to the larger amount of work to be done in India.[44]

41 M. H. Edney, *Mapping an Empire: The Geographical Construction of British India, 1765–1843*, *op. cit.*, p. 133.

42 Article 4 of the ruling of 5 July 1899, which founded the Service géographique de l'Indochine.

43 J.-C.-L. Lubanski, "Note sur le Service géographique de l'Indo-Chine", *art. cit.* p. 477.

44 The *Service géographique de l'Indochine* had an area of 737,000 km2 to cover, whereas Burma alone (Lower and Upper) had around 678 500 km2, to which should be added the more than 3 million km2 of the surface area of India. The task of the British Survey was thus gigantic compared to that of the French Department.

a Photography and lithography office, built in 1889, Calcutta
SOURCE: BL, IOR, V/24/3987, Report on the operations of the Survey of India, administered under the Government of India, during 1888–1889, Calcutta, 1890. Cover page

b Survey of India offices, Calcutta, January 1895
SOURCE: BL, IOR, V/24/3992, Report on the operations of the Survey of India, administered under the Government of India, during 1893–1894, Calcutta, 1895

c Royal Thai Survey office, Bangkok
SOURCE: R. W. Giblin, "Royal Survey Work", dans *Twentieth Century Impressions of Siam*, London, Lloyds Greater Britain Publishing Company, 1908, p. 122 [https://archive.org/stream/twentiethcentury00wrigrich#page/n125/mode/2up]

FIGURE 14 Photographs of the headquarters of the Geographical Departments

Just as explorers would work on their maps at the *Dépôt de la Marine*, so topographers returning from the field occupied the drawing workshops there, where they could go over the surveys they had made and retrieve the useful information. When Hobday was surveying in Upper Burma, he spent time in Mandalay compiling the sketch maps and itineraries already collected. The same year, A. J. Wilson and M. J. Ogle, who had joined a detachment in the Hukong Valley a few months previously, perfected their work in the drawing workshops.[45] In 1904, the Head of the *Service géographique*, Mussy expressed his satisfaction with the fact that the reproduction process was closely linked to the fieldwork: "topographer-officers, who operated during the October 1902 to May 1903 campaign, finished writing up their work and went back to join their respective divisions."[46] After drawing came printing, which was done in Calcutta for the Burmese maps, and in Hanoi after 1890, for the maps from the *Service géographique de l'Indochine*.

The Siamese Department likewise included the draughtsmen's workshop, where 30-odd Siamese officials were employed in 1908, ranking from the first to the fourth grade. The unit was headed by two Europeans, the deputy head being J. R. Bell (the post of head of the workshop was vacant that year).[47] The photozincographic workshop for printing the maps was headed in 1908 by P. Mackenzie, and employed 25 Siamese staff, including an engineer and seven interns.

In the French territories, four geodetic surveyors and six topographers, who had graduated as officers from the *Service géographique de l'armée*, where they had acquired their technical skills and experience, carried out the fieldwork. They formed the core of the triangulation and topographical brigades. The other members were recruited from the Occupation Corps: topographers were exclusively European, whereas the geodetic surveys involved Europeans and also the indigenous population (as aides, secretaries, signal positioners, and so forth). In the British territories, the brigade head was a member of the Imperial Service, accompanied by a few Provincial officers and some 30 topographers. Most of the roughly 20 Europeans working at the Siamese Survey Office in the early 20th century carried out surveys as well as managing the brigades. The

---

45 BL, IOR, V/24/3986, Report on the operations of the Survey of India, administered under the Government of India, during 1887–1888, Calcutta, 1889.

46 ANOM, SGEOCOL 18, Compte rendu des travaux intérieurs exécutés par le Service géographique de l'Indochine pendant l'année 1903, Hanoi, 1er février 1904, par le lieutenant-colonel, chef du Service géographique, Claude de Mussy, p. 14.

47 R. W. Giblin, "Royal Survey Work", *art. cit.* p. 125.

director and his deputy were supplemented by seven superintendants[48] and eight deputies.[49] The brigades themselves numbered 240 people, of which roughly 180 dealt with Land Registry surveying, which was a major feature of the Siamese Department's fieldwork. A dozen Indians were among the staff, employed mostly on the topographical surveys.

3.2 *Staff Hierarchies: The Role of the Indigenous Personnel*

As we have seen, the Geographical Departments employed both European and indigenous staff. The presence of the latter is documented, but the number of sources available vary between the British, the French, and the Siamese.

The Director of the Siamese Department, Giblin, stated that "the Siamese have proved they are reliable topographers in the field, precise draughtsmen and excellent printers,"[50] and he also mentions their competent handling of all the other steps in the mapmaking process. They were trained at the Royal Survey School, where two Siamese professors, Praya Kam-nuan and Pra Sakon, trained about 35 students. After some teething problems, the Department thus functioned efficiently, and indigenous staff played an important role in it.

Indian officials likewise played a substantial role in the Survey of India. In 1863, the British created a training school for Indians at Dehra Dun in order to familiarise them with European techniques and instruments. In 1892 the geographer Louis Raveneau, who at the time was Head of the *Annales de géographie*, mentioned "a sort of preparatory school for indigenous explorers", set up by Colonel Montgomerie.[51] Pavie likewise expressed his enthusiasm for the pandits, that "admirable body of 300 Hindus who are remarkably well-trained in mapping routes, and who considerably simplify the task of the officers, such that they can devote themselves entirely to triangulation."[52] The Survey of India's annual reports mention their work in the field, their name, and sometimes comments from their superiors. They were praised for their technical skills, their human qualities, especially their enthusiasm, their motivation, and their hard work, beyond the call of duty. Those who had proved their capacities were given a very free hand, sometimes carrying out assignments alone, as was

---

48 Verdon, Lowe, Dumbelton, Masterman (civil engineer), Kemp, Michell and Collingwood (civil engineer).

49 Warner Shand, Byrne, Thompson, McCormick, Edwardes, Gairdner, Sawkins, Rust.

50 R. W. Giblin, "Royal Survey Work", *art. cit.*

51 Louis Raveneau, "Notre connaissance de l'Asie", *Annales de géographie*, 1/2, 1892, p. 160–176.

52 Auguste Pavie, *Mission Pavie. Indo-Chine. 1879–1895. Géographie et voyages. II, Exposé des travaux de la mission (troisième et quatrième périodes, 1889 à 1895)*, Paris, E. Leroux, 1906, p. 364.

the case with the Maharashtrian topographer Bapu Jadu,[53] and of Lachman Jadu, who was assigned between 1897 and 1899 first to the British officer Longe, then to Ryder, to help demarcate the Burma-China border. During the first campaign of surveys, he was sent to the border alone, where the defection of the Chinese officer who was supposed to assist him hindered his work,[54] but he received warm praise for his accomplishments in the second season: Ryder wrote that he "worked unflaggingly, and energetically, in a country he did not know, where the inhabitants were not always welcoming, and his traverse work proved, after checking, to be very good, across the board."[55] He did not let the difficulties of the terrain nor the sometimes unhelpful local populations get in his way.

These pandits, or deputy topographers, thus played an important supporting role for the British officers, and sometimes even replaced them, when they were sent alone on an assignment. In disguise, and using instruments considered less than standard by the Europeans, they could visit places where a British presence would have been immediately noticeable, and this was another reason why they were extremely useful.[56] Their valuable contribution sometimes earned them a mention on the frontispiece of British maps, and hence a form of recognition they never received on the French side, where little information exists concerning the role of local populations in the surveying work on the ground. At the Head Office, however, for the British and French alike, the sources show us that the draughtsmen were drawn almost exclusively from the indigenous populations.

In the British Department, the most thankless tasks like copying the maps or printing them were assigned, after 1850, to Eurasians – mostly Anglo-Indian orphans – and to "poor whites", a practice common to several other public agencies in India.[57] This use of orphans and *métis* (often one and the same person) was most probably also the case in the French *Service géographique*, since

53 BL, IOR, V/24/3986, Report on the operations of the Survey of India, administered under the Government of India, during 1889–1890, Calcutta, 1889, p. 201.

54 BL, IOR, V/24/3996, General report on the operations of the Survey of India, administered under the Government of India, during 1897–1898, Calcutta, 1899, p. v.

55 *Ibid.*

56 Peter Hopkirk, *Trespassers on the Roof of the World: The Race for Lhasa*, London, J. Murray, 1982; Derek J. Waller, *The Pundits. British exploration of Tibet and Central Asia*, Lexington, University Press of Kentucky, 1990.

57 M. H. Edney, *Mapping an Empire: The Geographical Construction of British India, 1765–1843, op. cit.*, p. 133.

this category, a thorn in the side for the colonial powers, was often employed in the lowest ranks of the administration.[58]

Several French sources mention the Vietnamese draughtsmen employed in the *Service géographique*, their pay and promotion. As with all the other colonial departments, a rigid hierarchy was in place there. The drawing and engraving room was staffed by 15 European draughtsmen, who were military personnel from the Occupation Corps, and they managed 54 indigenous draughtsmen. Among the latter, there were 12 scholar-bureaucrats, who climbed through the ranks on recommendation from the General Commander in Chief to the Resident Superior of Tonkin, and who received a civil servant's salary and a retirement pension. In 1906, the salaries ranged from 840 piastres for the top-grade chief draughtsmen to 240 piastres for the fourth-grade auxiliary draughtsmen.[59] By1911, the number of such draughtsmen had risen from 12 to 21.[60]

The skills sought in these staff were above all linguistic and technical. For example, Đỗ Quang Tú was described as a "well above average scholar due to his in-depth knowledge of [Chinese] characters and of Quốc Ngự [*sic*]", as well as his "skill at drawing on zinc, and particularly his drawing work using a lithographic pencil, which he executes with rare perfection."[61] Apart from these talents, the person's overall attitude was also mentioned. Hard work and above all enthusiasm, as much as his skills, led to Đỗ Quang Tú's promotion to tenured draughtsman of the sixth grade at the age of 35, after more than 13 years in the *Service géographique de l'Indochine*.

58 Emmanuelle Saada, *Les enfants de la colonie. Les Métis de l'Empire français entre sujétion et citoyenneté*, Paris, La Découverte, 2007.

59 ANOM, GGI 17546, Augmentation de la solde des dessinateurs indigènes, arrêté du 9 juillet 1906. By way of comparison, the monthly salary of a first class indigenous governor of a province and head of a provincial administration, was 3,000 piastres; an intern in the indigenous administration, 380 piastres. The lower echelons of staff working for the indigenous upper civil service earned between 72 and 360 piastres a month. A second class indigenous guard earned 0,2 piastre per day, 6 piastres per month. Personal tax was around 3 piastres per year. One night in a hotel cost 2 piastres, as did dinner. See Henri Brenier, *Essai d'atlas statistique de l'Indochine française*, Hanoi, Imprimerie française d'Extrême-Orient, 1914; Claudius Madrolle, *De Marseille à Canton, guide du voyageur: Indo-Chine, canal de Suez, Djibouti et Harar, Indes, Ceylan, Siam, Chine méridionale*, Paris, Comité de l'Asie française, 1902.

60 ANV1, GGI 1804, Propositions pour l'avancement en faveur des dessinateurs indigènes du Service géographique, 1911.

61 ANOM, GGI 17545, Promotion du 14 juillet 1906 dans le personnel indigène des dessinateurs; ANV1, GGI 1804, Propositions pour l'avancement en faveur des dessinateurs indigènes du Service géographique, 1911.

Outside of this hierarchical structure, employment was less stable. The roughly 42 interns received a merit-based salary, as decided by the Head of the Department, and they could join the corps of scholar-bureaucrats if their name was put forward to the Resident. Thus the apprentice draughtsman Hứa Mạnh Lân was paid 4,8 piastres monthly, and the Head of the Department regretted that the over-stringent regulations prevented him from rewarding this young man as he deserved.[62] A temporary solution was found, when he was hired for a one-off assignment, on 17 February 1913, to draw the *Atlas graphique et statistique de vulgarisation de l'Indochine*, which had been commissioned by the Secretary General. His initial monthly wage for this work was 6 piastres, rising to 8 piastres as from 1 April, since his performance was deemed wholly satisfactory.[63] So a handful of opportunities did exist for the indigenous population working in the *Service géographique* to escape the administration's rigid regulations.

Regarding employment and promotion of the indigenous population, we should therefore avoid over-simplifyng the picture. Indian topographers and the work they did in the colonial situation are amply documented in British sources. The French also employed indigenous workers, at least as draughtsmen, and the French sources too provide a wealth of details on the subject, belying the received idea that the archives are silent in this sphere.

3.3 *Costly Procedures, Insufficient Budgets*

The use of a local workforce was not sufficient, however, to solve the budgetary problems encountered by the colonial Geographical Departments. The obsession with a "low-cost" colonisation, and the ever-present budgetary constraints, put stringent limits on large-scale colonial enterprises. As such, the declared objective of making a complete map of the colonial territories remained a pipe dream, since sufficient funds were never forthcoming to carry it out. Cartography was a costly operation in human and material terms, and, what is more, Geographical Departments were not a priority for the colonial administration.

The budget of the *Service géographique de l'Indochine* increased year after year, as the unit became established, but the funds were always considered by the Department to be well below its needs. In 1890, the new *Bureau*

62 ANV1, GGI 7459, Recrutement des dessinateurs pour les travaux de *l'Atlas graphique et statistique de vulgarisation de l'Indochine*, 1912–1913.

63 ANV1, GGI 7459, Rapport du chef du service des Affaires économiques au gouverneur général de l'Indochine, Hanoi, 28 April 1913.

topographique received some 18,000 francs, and double that amount in 1901.[64] Staff costs were a large proportion of the budget, even though the topographer-officers were usually paid by another Ministry, the Ministry of War. However, the Head of the Department, and the European and indigenous draughtsmen, were on the *Service géographique*'s payroll, which put a strain on the budget, as was the case in the Survey of India also.

The materials needed for map-making also weighed on colonial budgets. In 1902, the *Service géographique de l'Indochine* received two chronometers with electrical switches from the Army's Department, for 8,000 francs, that is, roughly a quarter of the Department's whole budget.[65] Having instruments and materials supplied by the home country guaranteed their quality, but the drawback was unreliable delivery times: in 1896, the *Bureau topographique* in Hanoi was late in printing its maps, because it "had not yet received all the necessary material from France", and "the Office's press [was] obliged to lie idle for over a month."[66] Not only was the technical material expensive, and sometimes late, but it was also fragile, and required frequent repairs, in which different colonial Departments collaborated. The repair and maintenance Department for the Survey of India's instruments also worked for the Army's Geographical Department.[67]

In addition to the costs of the activities at the Head Office, there were the costs of work in the field. Using topographical brigades was very expensive, but it:

> [...] has so many advantages that it has to be used in the major colonies where communication networks have to be built, and where markets are well developed, with considerable trade. In these sorts of circumstance, the benefits to be gained from a precise knowledge of the country fully make up for the colony's budgetary sacrifices in establishing a topographical brigade.[68]

---

64 G. Antony, *Le Service géographique et des missions. Historique, organisation et publications*, *op. cit.*, p. 79.

65 ANOM, SGEOCOL 16, Lettre du ministre de la Guerre au ministre des Colonies, 5 September 1902.

66 ANOM, GGI 23827, Lettre du général Dodds, commandant en chef les troupes de l'Indo-Chine, au gouverneur général, Hanoi, 13 May 1896.

67 BL, IOR, V/24/4009, Report on the operations of the Survey of India, during the survey year 1910–1911, Calcutta, 1912.

68 G. Antony, *Le Service géographique et des missions. Historique, organisation et publications*, *op. cit.*, p. 27.

This argument seemed convincing enough to be adopted in French Indochina, and before that in the British Raj. The high cost of the topographical brigades was partly due to the extra pay received by their members, and those accompanying them, and partly to the system of allowances and compensation: the peasant who provided accommodation or a sampan, the porters recruited or their families, had to be compensated, if a material or human loss arose. The material losses were often due to the hazards of the journey. Thus on 7 August 1911, Lieutenant Jacquemet received 30 piastres for the owners of the canoes he had used on a solo mission the previous month in the region of Đồng Hới (Quảng Bình): returning from the delta with coolies and rice, a sudden rise in water level sunk two canoes, and the village which owned them demanded compensation.[69] Human losses also occurred, particularly due to accidents during the clearing work necessary for geodetic measurements. There are some rare traces of these in the French archives: on 28 June 1910, 18 piastres were allocated to families after an accident involving two coolies, and on 24 July 1911, the family of a coolie who died, and one who was injured, received 12 piastres.

The high costs of topographical brigades was also due to the large distances which had to be covered to reach the area to be surveyed. It was not until 1906, for instance, that the members of Brigade no.11 established summer quarters in Taunggyi, the capital of the Southern Shan States, where they had several years' work ahead of them.[70] Prior to this date, they had regularly gone back to Bangalore in India, a journey of three to four weeks, and spent the same length of time returning to Upper Burma the following season, with all the costs this implied.

Lastly, apart from the staffing and operational costs of these brigades, expenses were incurred due to the project's scientific objectives, namely to survey the territory in its entirety. Thus in March 1911, in the province of Quảng Bình, the 1st topographical brigade was faced with "a rocky range almost entirely devoid of economic interest, but which we still have to measure because we are being exhaustive."[71] This notwithstanding, scientific goals did not always win out over economic considerations, and the brigades often found a pragmatic solution in the light of the lack of staff, and the need to focus efforts as

69 ANOM, GGI 17562, Secours ou indemnités accordées aux victimes d'accidents ou à leurs familles ou aux victimes de pertes d'objets, 1911.

70 BL, IOR, V/24/4004, General Report on the operations of the Survey of India, administered under the Government of India, during 1905–1906, Calcutta, 1907.

71 ANOM, GGI 17557, Lettre du résident supérieur en Annam au gouverneur général de l'Indochine à Saigon, Hué, 25 March 1911.

efficiently as possible. But generally, over the whole period, the restricted staff numbers and funds available to both the British and the French Geographical Departments set significant limits on their work, in common with the constraints on the colonial administrations as a whole.

•••

So colonial Geographical Departments were established to introduce orderly and systematic procedures of scientific map production to replace the compilations, sketch maps and provisional maps which had gone before. The resulting maps were based on prior triangulation, and were made by experienced topographers who built on a long tradition of State-sponsored map-making, orientated this time towards the needs of the colonial administration. The Geographical Departments were conceived first and foremost as "centres of calculation", that is, as places to which all geographical documents concerning the relevant territories were sent. Their goal was to produce a map of the whole territory, as had already been done for France, Algeria and Ireland.

Significant differences existed between these institutions. The Royal Siamese Survey Department, closely modeled on the British Survey of India, and headed by Europeans, was an exception to the rule in the Indochinese Peninsula. It did not aim to map the entire country with scientific accuracy. Instead, it focused on surveys for the Land Registry, on border demarcation, and on communication routes. Both the British and the French wanted to possess scientifically produced maps, but the British in Burma already had a century of experience in this, in the form of the Survey of India, whereas the French had to create their *Service géographique de l'Indochine* from scratch. Yet the development of both Departments, as institutions, was uneven and subject to chance events, just as it was for their counterparts in the home country.

Although there was a general consensus in Europe that it was important to have maps and a large-scale representation of the territories conquered, the Geographical Departments were rarely a priority for central government, nor for the colonial administrations. The cartographic priorities were not always unanimously accepted either, and some people criticised making maps of spaces considered marginal. So there was no enduring consensus on how to implement the idea of detailed and systematic mapping of the colonial territories, quite apart from the fact that the real conditions were often unaccommodating. Budgets and staff were in short supply – an endemic feature of colonial administrations – and the operations met with countless difficulties. Structural budgetary restrictions inevitably slowed down the process of map production.

Scientific, large-scale map-making thus proceeded at a slow and irregular pace in the colonies, hampered by the difficulties encountered by the brigades, and by the problems mentioned earlier. As such, we would do well to reconsider the idea of domination through cartography, especially since the confines of the colonial territories remained uncharted. Can one really speak of symbolic domination when substantial areas are never mapped scientifically, but can only be imagined through compiling different routes taken? In the Indochinese Peninsula, as in other imperial territories, the Europeans did not have the means to dominate, symbolically or otherwise, through the maps they made. Although one can note a qualitative improvement between the first compilations of sketch maps, and the topographical sheets produced later, even then, the latter were still based on individual itineraries and often left much blank. So maybe we should look elsewhere for the political power of geographical institutions, and not put undue emphasis on the plan to map colonial territories exhaustively, nor on its execution. Maybe it is elsewhere that the power of maps – if power they had – has left its mark.

CHAPTER 3

Geographical Institutions: Collaboration, Competition, and Confrontation

As we have seen, the geographical institutions in the Indochinese Peninsula did not embody the dominance of cartography, nor the exercise of power through maps. On the contrary, they encountered substantial difficulties in the systematic topographical coverage of the territory (the main goal of two of them). What strategies were adopted, and what solutions were found? In other words, how did geography become a site of collaboration and not simply of confrontation? When considered as the imperial science *par excellence*, geography has been treated in terms only of the rivalries it underpins. In this interpretation, "Geography is useful primarily to wage war",[1] to conquer territories and to ensure they can be controlled, through knowledge of the terrain. This knowledge gives the coloniser an advantage over the colonised country, and generates cartographic rivalries between Empires in their rush to grab land. Where this analysis falls down is its emphasis on geography as solely an agent of discord. For it can also lead to collaboration, on various levels.[2] Geographical data has an important role to play in the imperial administration, in relations between the powers, and even more so through the circulation of published maps and their reception.

Turning first to the role of geography within colonial administrations, the main mapmaking institutions, always under-funded and short-staffed for the scale of their task – producing a map of the whole territory – had no choice but to collaborate with other Geographical Departments. This was initially a structural necessity, which fell away later on. We shall examine how these exchanges functioned and developed. One of the questions arising is whether colonial Geographical Departments simply imitated their counterparts in the home country, or whether they attempted to adapt to their particular location, leading to the idea that the colonies were "laboratories".[3] This notion can be applied to all areas of experimentation in the colonies – political, legal,

1 Yves Lacoste, *La géographie, ça sert, d'abord, à faire la guerre*, Paris, Maspero, 1976.

2 Volker Barth, Roland Cvetkovski, *Imperial Co-operation and Transfer, 1870–1930: Empires and Encounters*, London, Bloomsbury Academic, 2015.

3 Helen Tilley, *Africa as a Living Laboratory: Empire, Development, and the Problem of Scientific Knowledge, 1870–1950*, Chicago, The University of Chicago Press, 2011.

 | DOI:10.1163/9789004469853_005

scientific and environmental. New ideas and practices were transferred to the centre, and not solely from the centre to the periphery. The circulation took place in both directions, and while we shall show, in the case of mapmaking, a certain reliance on central administrative models regarding the staff, standards and models involved, we shall also address how the colonies themselves contributed to the gradual institutionalisation of the geographical sciences at the time. Additionally, the Geographical Departments overseas had to contend not only with other geographical institutions, but also with other colonial departments which produced maps, particularly the Land Registry. We know that in Algeria there was no love lost between these two bodies, but can the same be said for French Indochina? And for the British colonies in the Indochinese Peninsula?

Secondly, the role played by geography in international relations has been insufficiently studied. Much has been written on Franco-British antagonism, and some examples of this rivalry concerned maps, namely the dispute at Thaeng in 1888. This was where the French confronted the Siamese, who were supported by the British, in present-day Điện Biên Phủ. It was the area of the Indochinese Peninsula most intensely coveted by both European powers, somewhat like, though with much less serious consequences, the later incident of Fachoda between the British and the French in Sudan in 1898.[4] What role did geography really play in this Franco-British quarrel, and was it simply a battle of maps? If we accept that cartography is an instrument of diplomacy, how was it used to this end? Only by examining how models, people and maps circulated between different institutions, within an Empire and internationally, can one understand the role played by cartography at the time in international relations, and can one assess whether it is possible to speak of "the power of maps".

1 Rivalry and Cooperation between Map-Makers

1.1 *The Distribution of Roles between Different Geographical Departments*

The connections between the Geographical Departments in the home country and the colonies varied in nature and intensity, depending on the circumstances. We have already seen that the *Service géographique de l'armée* and the *Service géographique de l'Indochine* had close links in matters of staffing and materials. The relations between the *Service géographique des colonies* in

4 T. Winichakul, *Siam Mapped: A History of the Geo-Body of a Nation*, *op. cit.*

Paris and the *Bureau topographique* in Hanoi, however, were variable. In the 1880s and 1890s, collaboration with the *Bureau topographique des troupes de l'Indochine* seemed to be effective, and there were good relations between the *Service géographique des colonies* and the Governor General, with a view to:

> [...] ensuring the common purpose necessary for making maps of Indochina, in France or Hanoi, which it was urgent to publish. It was agreed, at the same time, that the *Service géographique des Colonies* would be kept abreast of all the publications of the *Bureau topographique d'Hanoi.*[5]

Despite this unusual situation, the relations seemed relatively affable. However, they did not remain so, and the rivalry between the Ministry of the Colonies and the Ministry of War erupted on several occasions. In a dispatch of 24 February 1899 to the Governor General of Indochina, the Minister for the Colonies, Guillain, expressed his dissatisfaction at not having been notified personally of the upcoming reorganisation of the *Service topographique de l'Indochine*. He had learnt from the annual report of the *Société de géographie de Paris* that geodetic surveyors from the *Service géographique de l'armée* would soon be taking over triangulation in Indochina. His reaction shows above all his wish to assert his authority: "No large-scale work can be undertaken in the Colonies without prior ratification by the Minister, and countersigned by the *Service géographique et des Missions*."[6] At issue, in his view, were the unity and good coordination of the different topographical departments, including Madagascar's, and there was no reason for Indochina to be an exception either. These rivalries, which surfaced within the national administration itself, call into question the idea that cartography incarnates one-sided domination, since several agencies, with sometimes conflicting positions, are involved.

The Geographical Departments were important in formulating cartographic standards, which were being developed nationally, thus contributing to the geographical sciences' institutionalisation.[7] The *Service géographique des colonies* in Paris gave detailed advice, reflecting this wish to rationalise

5 G. Antony, *Le Service géographique et des missions. Historique, organisation et publications*, *op. cit.*, p. 8.

6 ANOM, GGI 17539, Dépêche, no. 238, 24 février 1899, ministre des Colonies au gouverneur général de l'Indochine.

7 On the institutionalisation of geography and the introduction of standards, see J.-M. Besse, H. Blais, I. Surun (eds), *Naissances de la géographie moderne, 1760–1860: lieux, pratiques et formation des savoirs de l'espace, op. cit.*

and monitor geographical productions.[8] Much ink flowed on the subject of representing relief. The 1:100,000 scale map of Tonkin espoused the model of the map of Algeria and Tunisia, and used the technique of shading.[9] Methods of representation changed again with the innovations of Lieutenant Colonel Aubé in 1910, who said he had abandoned "the colour bistre, which lacks the subtlety to depict contours and rocks."[10] The British likewise hesitated over the standards to be used to represent relief. After examining the Swiss, French and American map symbols, they decided to represent hills through contouring, and used a wash to give the impression of oblique, even horizontal lighting.[11] As surveys became more precise, and lithographic cartography developed, after 1895, so relief was represented by contour lines at intervals of roughly 50 feet. The same process was adopted by the French, but certain differences remained, like the case of bistre. Thus new cartographic standards came into force, based on comparing other methods, some of which became prototypes. There was a concerted reflection on maps, on their legibility – their ease of use, and their visual qualities – and at this stage, when cartographic standards were still evolving, substantial changes could occur.

Geographical departments also corresponded abundantly about engraving and printing maps. When the colonial departments were first set up, they had no choice but to collaborate with their counterparts in the home country. Thus, for the British, before engraving was transferred to Calcutta for the *Atlas of India*, in 1867, it was carried out in London.[12] The maps from the *Service géographique des colonies* and from the *Bureau topographique des troupes de l'Indochine* were initially printed in Paris by the *Service géographique de l'armée*, since it took four years for printing services to become operative in Hanoi, in 1890. Thus despite setting up geographical departments in the colonies, specifically to serve the empire, other geographical departments – of the Army, the Navy and the Ministry of Foreign Affairs – continued to be used, for essentially financial reasons. The maps prepared *in situ* thus had to be sent to the national capital for printing, generating shortages in the colonies, when

---

8 G. Antony, *Le Service géographique et des missions. Historique, organisation et publications*, *op. cit.*, p. 29.

9 Édouard de Martonne, Jean Martin (eds), *La carte de l'empire colonial français*, Paris, G. Lang, 1931, p. 119.

10 ANOM, GGI 17556, Lettre du lieutenant-colonel Aubé, chef du Service géographique, au gouverneur général de l'Indochine, Hanoi, 22 June 1910.

11 BL, IOR, W/6503, Notes on Survey of India maps and the modern development of Indian cartography, lieutenant-colonel W. M. Coldstream, RE, Superintendent, map publication, published under the direction of colonel CHD Ryder, officiating Surveyor general of India, 1919, p. 5.

12 *Ibid.*

the maps were not sent back, and complicating the circulation of documents. The general map of Indochina, on a scale of 1:1,000,000, and the first sheets of the map of Tonkin, on a scale of 1:200,000, were sent to Paris in 1867 to be printed by the *Service géographique de l'armée*, and only returned to Hanoi two years later.[13] This contradictory situation, where the maps useful to the colonial territory remained blocked in the home country, due to the constraints of centralisation, was resolved only when the geographical departments began using the local printing companies at their headquarters in Calcutta or Hanoi.

Turning now to the use or not of European methods for cartographic reproduction, the problem was not essentially one of available skills, since the European staff in the imperial territories had all received a Western training. However, the specificities of the location complicated cartographic production. Tight budgets led to lack of staff and materials. In Burma as in French Indochina, most of the maps were reproduced using a lithographic technique, that is, engraving on stone, but this required large, cumbersome stones. However, it was suited to large prints, which increased with photolithography, as used in Calcutta. The French considered this technique to be inadequate for large-format maps, and it was replaced by zincography, which used a zinc, instead of a stone, plate. The workshop in Hanoi required a Head Printer, five Assistant Printers and lithographers, all of them Europeans from the occupation forces, and 12 indigenous coolies to polish the plates, move and maintain the presses, and so forth.

This is where another location-specific problem emerged, whose impact should not be under-estimated: the problem of the climate. The Head of the *Service géographique de l'Indochine* explained in his 1904 Report that "maintaining the zinc plates of the different maps is more difficult in Tonkin because of the constant humidity there; the plates oxidise and spot".[14] Not only did this generate extra costs, but some procedures could not be used because of the climatic conditions. This problem was cleverly circumvented in India by the use of a heliozincographic method imported from Britain and specially adapted.[15] Given this need to adjust methods and find tailored solutions, the idea of a "colonial laboratory" makes perfect sense. The geographers were highly

13 É. de Martonne, J. Martin (eds), *La carte de l'empire colonial français, op. cit.*, p. 97.

14 ANOM, SGEOCOL 18, Compte rendu des travaux intérieurs exécutés par le Service géographique de l'Indochine pendant l'année 1903, Hanoi, 1st February 1904, par le lieutenant-colonel, chef du Service géographique, Claude de Mussy, p. 12–13.

15 BL, IOR, V/24/4041, Extracts from the Report of Mr. Pope, the Assistant to the Surveyor General, in charge of the Office of Photography and Lithography, Calcutta, 1901–1902: "Though similar to the Southampton method in its results it is quite different in principle, and is, indeed, entirely novel", p. 2.

creative, motivated by the challenges of their environment, and the need to find solutions. The process did not work the other way round, however, for the simple reason that such innovations were of no use in the home country.

1.2 *Rivalry with the Land Survey Office*

The links between Geographical Departments and Land Survey Offices were often stormy, and the distribution of tasks between the two establishments was not the same from country to country. For the British, the Land Survey Office was a branch of the Survey of India just like the geodetic surveying and topographical units. However, there seems not to have been much communication between them, and the local authorities responsible for raising taxes, rather than the Survey of India, tended to supervise the cadastral mapping.[16] This changed the relations between topographical and land data. Town plans at a very large scale could serve as a basis for topographical maps, which had always been the case in certain locations, but this practice became problematic when local authorities concentrated on the plan's finality, from their point of view, namely tax collection, and paid little attention to the specificities of the terrain and the topographical details.

The two institutions were more closely linked in Siam where, when the need for land surveys arose, for administrative as much as for tax reasons, the geodetical control work was momentarily suspended so that the personnel could devote themselves to large-scale land measurement and demarcation of landed property.[17] In the case of France, where the norm was the separation, and rivalry, between the cadastral and geographical offices, it was precisely the issue of scale that crystallised disagreements and was the main argument used by the opponents to any unification.

In Tonkin, the Land Survey office was a section within the Agriculture Department until 1906, when it came directly under the authority of the Resident Superior. This Land Survey and Topographical Department was reorganised several times, and fulfilled several roles: it organised, carried out and monitored triangulation, topographical surveys, land measurements and the production of cadastral maps. Europeans worked at all levels of the Land Survey offices until 1914, from chief surveyor to assistant, and from office manager to clerk. Half of the trainee surveyors were military sub-officers, and the

16 BL, IOR, W/6503, Notes on Survey of India maps and the modern development of Indian cartography, lieutenant-colonel WM Coldstream, RE, Superintendent, map publication, published under the direction of colonel CHD Ryder, officiating Surveyor general of India, 1919.

17 R. W. Giblin, "Royal Survey Work", *art. cit.*, p. 121–127.

other half were civilians who had passed a recruitment exam in the colony. The merging of the two Departments, envisaged for Cochinchina in 1905, was strongly recommended by the Minister for the Colonies, for whom the Geographical Department should "*make general control surveys, within whose network detailed topographical maps can be inserted*."[18] There were several dissenting voices, however, to this fusion. The Secretary General Broni and the Resident Superior in Cambodia, Jules Morel, echoed by the Governor General of Indochina, sought to show how the two Departments were in their nature and objectives essentially different: the Geographical Department, they maintained, was primarily a scientific institution, whereas the Land Survey office was an administrative and legal entity, in the case of court action. Then there was the question of scale, in the reply to the proposal from the Ministry of the Colonies:

> It seems no more beneficial to require land surveyors to have a knowledge of astronomy, for example, than it does for the officers seconded from the army's geographical department to have precise ideas about land values in Indochina from the point of view of their place on the property tax rolls and the mode of land tenure.[19]

The real sticking point was, finally, that the staff of the Geographical Department were military, whereas the Land Survey Office was 50% civilian. The Governor General, before coming round to the dominant viewpoint, had initially annotated the report of his Secretary General to stress that the cadastral staff could use the geodetic control surveys without becoming part of the Geographical Department staff. But his position was short-lived, suggesting that the tensions on this issue were too great. The model which won out was the one used in mainland France, justified by the fact that the *Service géographique d'Indochine* is:

> a purely military body, similar to the *Service géographique de France*, under the aegis of the army's chief of staff, which is responsible for keeping our1:80,000 scale map up to date. In France, alongside it, there are other, similar organs of the civil administration which also carry out special topographical assignments, but these are absolutely separate from

18 ANOM, GGI 17541, Minute du secrétaire général au gouverneur général de l'Indochine, Hanoi, May 1905. Underlining in the original.

19 ANOM, GGI 17541, Lettre du gouverneur général de l'Indochine au ministre des Colonies, 20 June 1905.

> the military. The results obtained by this division into two distinct elements have proved sufficiently conclusive for a similar organisation to be facilitated, and for us to wish to maintain it, in Indochina.[20]

The organisation of the colonial departments should thus exactly replicate those of the home country. The had shared goals – map-making – but at different scales, requiring different skills and actors. The project of merging the *Service du cadastre et de la topographie* with the *Service géographique* thus failed, even if, after 1905, it was possible to place military personnel in civilian posts in the colony,[21] but not the other way round, were civilians to be employed in a purely department. Thus, when Commander Scherdin, the head of the *Service géographique*, applied for the post of Head of the *Service du cadastre du Tonkin et de l'Annam*, as part of a restructuring, only the the fact that the post was already filled made this impossible.[22]

Staff seemed to move more easily between Land Survey Departments in the empire than between the two Departments – cadastral and geographical – of the same colony. The case of Monsieur Raynaud is interesting in this respect. He was born in the Hérault, France, on 25 March 1879, and left when still young for Indochina. He was a Draughtsman for the *Service du cadastre et de la topographie* in Saigon from 1897 to March 1901, but was then obliged to return to France to perform his military service, until September 1903. The following year, he was employed by the *Service topographique d'Algérie*, where he stayed for five years, first as an auxiliary draughtsman in Algiers for four months (June–October 1904), then as a student topographer in Oran for a year (October 1904–October 1905), and finally as Ordinary Topographer of the Fourth Class in the *département* of Oran (October 1905–January 1907), then in the *département* of Constantine (Januaty 1907–April 1909). He left the *Service du cadastre* to become a land surveyor in Sétif, and then settled in Paris in May 1911. In November of the same year he was recommended for a post at the *Service du cadastre et de la topographie du Tonkin* by his employer, who highlighted "his years of practice with the tacheometer, his speed and precision, which one may legitimately expect from a good operator."[23] To return

20 ANOM, GGI 17541, Lettre de Jules Morel, résident supérieur au Cambodge, au secrétaire général de l'Indochine à Hanoi, 13 June 1905, Phnom Penh.

21 ANV1, GGI 6716, Au sujet de la nomination des militaires classés pour le Service du cadastre et de la topographie de l'Indochine, 1906–1920, 93 p.

22 ANV1, GGI 6703, Lettre du secrétaire général Van Vollenhoven à A. Pichon [président du Conseil], Hanoi, 9 February 1913.

23 ANV1, GGI 6703, Lettre de Philippe Jarre, ingénieur topographe, ancien élève de l'École polytechnique, Paris, 19 November 1911.

to Indochina, Raynaud thus exploited the possibilities made available by an imperial administration, and comes to represent a sort of "imperial surveyor". However, the mobility of land surveyors had certain limits, partly because of the small number of staff involved, s (15 in Tonkin in 1914, a few more in Upper Burma),[24] and partly because the authorities increasingly tried to hire local people for their Land Survey brigades.

Both the British and, later, the French turned to local staff for land measurement operations. This was sound common sense, given the huge amount of work to be done, and the fact that it could be delegated, and thus spare the costs of a scarce European labour force. Giblin also stressed that a local labour force, with its knowledge of the language and of local customs, was irreplaceable because it simplified practical aspects of the work such as procuring transport.[25] The idea was to have just a few European officers in charge, and to use the local population for the measuring work. Its implementation proved more difficult, however. The British, for example, were obliged to import labour from India in the early years, due to the time needed to recruit and train local workers. The Indian workers were employed in both Burma and Siam, and moved between them, but everything was done to replace them rapidly with a local workforce, such that in the later years Siamese topographers could almost entirely replace their Indian and Burmese counterparts.

When work began on the ground in Burma, in 1889, the British attempted to ensure the presence of Burmese staff at every stage.[26] However, their work, although adequate, was not as good as that of the Indian surveyors from the North Western Provinces, and they were considered less strong physically. They were also less productive due to their local ties: "Nothing could induce them to refrain from leaving their work *en masse* whenever a market, a game, a funeral or some other amusement was taking place nearby."[27] They managed to measure half as much land as the Indians from the North Western Provinces, who covered one square mile per month; and the Burmese cadastral surveys cost 40 rupees more per square mile. For the ensuing season, some Burmese workers were attached to the Indian teams, to copy their example. In general, the Burmese worked well, as long as they could stay near their village, but as soon as they had to go further afield, and pay for their food, they deserted. This placed another limit on local recruitment. The following year, the British hoped

24 ANVI, GGI 6587, Travaux d'avancement du personnel du service du cadastre et de la topographie du Tonkin, 1910–1914, 45 p.

25 R. W. Giblin, "Royal Survey Work", *art. cit.*

26 BL, IOR, V/24/3988, Report on the operations of the Survey of India, administered under the Government of India, during 1889–1890, Calcutta, 1891.

27 *Ibid.*, p. 35.

to recruit enough Burmese in the villages to economise the costs of bringing 200 Indians from Iindia. The plan worked, with 900 *khalassis* brought in 1889, as against 500 in 1894,[28] representing an economy, for the transport alone, of 2000 rupees. This was an enduring solution, however, because Burmese labour costs were increasing, making the use of Indian workers more attractive, and because few Burmese from the district to be measured agreed to work there. There was no advantage to be had employing Burmese from Lower Burma in Upper Burma, instead of Indians, because they cost more. Yet attempts to train Burmese staff continued, by having them accompany the Indian brigades, and employment contracts often awaited them at the end of their training, but this did not solve all the problems.

Efforts were also made in French Indochina to have the Vietnamese from the region work on the cadastre, rather than sending Cochinchinese from the South. Lubanski reported an attempt in January 1901 to "use the main guards of the local militia for elementary measuring tasks, with a view to determining with precision the cultivated surfaces in the indigenous districts, which are liable for land tax."[29] The individuals selected, who were trained by an officer of the *Service géographique*, worked in the province of Sơn Tây – the first to have been mapped on a scale of 1:25,000. The results were deemed satisfactory because a rough land registry map was produced, which could be improved, and serve as a basis for establishing the rolls. This was an important step, and pioneered the use of indigenous military personnel to establish the cadastral map, and hence the amounts of tax due, starting from the 1:25,000 scale map. The idea behind the indigenous brigades was that this would free up the European labour force for posts of management and verification. The excellent results in Sơn Tây in 1901 showed that the method could be generalised, with four indigenous workers being supervised by one European. In 1914, M. Roussotte, a land surveyor in Cochinchina, expressed a similar view in a draft ordinance: the young talents trained by the French should not only aim to enter the civil service, he wrote, and so he welcomed the alternative of entering the *Service du cadastre et de la topographie*.[30] A school was set up within the Public Works Department of the colony by an ordinance of 10 February 1915, in order to train these indigenous surveyors. Its programme was ambitious: in addition to French, Chinese (especially the characters) and vernacular languages, there

28 BL, IOR, V/24/3991, Report on the operations of the Survey of India, administered under the Governement of India, during 1892–1893, Calcutta, 1894, p. 8.

29 J.-C.-L. Lubanski, "Note sur le Service géographique de l'Indo-Chine", *art. cit.*, p. 479.

30 ANV1, GGI 5403, Lettre du résident supérieur du Tonkin au gouverneur général de l'Indochine, Hanoi, le 21 April 1914.

were the technical subjects of geodetic surveying, geometry, mechanics, drawing, surveying, and basic notions about roads, canals, irrigation works, bridges, railways and also arithmetic.[31] A school of measurement and surveying was also created in Burma, attached to the Deputy Commissioner's office. In 1889, the Deputy Superintendent of the Survey of India asked to visit it, and himself gave instructions for the use of the plane table, and was present at the exam supervised by the local government.[32] Thus, in French Indochina as in Burma, no effort was spared to ensure that marking the boundaries of properties was carried out by the populations directly concerned by the tax burden that ensued. To keep the costs of colonisation down, and with a view to the colony's financial autonomy, the indigenous population had to be involved, and this affected the domain of cartography also. Increasingly, indigenous labour was used, and the work done by Europeans was delegated to the local population, although this was more widespread in the cadastral than in the geographical departments.

1.3 *The Relations with Other Colonies in the Empire*

The technical training required in geographical departments gave rise to exchanges between different institutions in the Peninsula. More generally, models, personnel, and modes of functioning circulated within colonial empires. What worked in one colony tended to be copied in another, and skills already proven in one context were sought after elsewhere, producing the phenomenon of "imperial careers". Some research has been done on these in Britain, but not yet in France.[33] Imperial careers could simply demonstrate that colonial administrations tended to apply similar methods in different colonies. Some protagonists made comparisons between their different posts, and we can also examine the transfer of practices produced by this movement of staff within the empire.

Geographers, just like other colonial agents and civil servants, as we have seen with the staff of the cadastral offices also moved between the colonies. We shall ask whether any particular career path can be identified, or whether

---

31 We are unable to say at this point how many local people were employed in cadastral surveying, and this is not immediately the focus of our research, although we could take it further in that direction.

32 BL, IOR, V/24/3988, Report on the operations of the Survey of India, administered under the Governement of India, during 1889–1890, Calcutta, 1891, p. 34.

33 Anthony Kirk-Greene, *Britain's Imperial Administrators, 1858–1966*, Basingstoke, Macmillan, 2000; David Sunderland, *Managing the British Empire: The Crown Agents, 1833–1914*, London, Royal Historical Society, 2004; Roger Owen, *Lord Cromer. Victorian Imperialist, Edwardian Proconsul*, Oxford, Oxford University Press, 2004.

imperial careers are the fruit of chance circumstances that cannot be foreseen in advance.

Some of those concerned had prior experience in a colony that was more advanced in cartographic techniques. Arthur Irwin, for instance, the Deputy Director of the Siamese Geographical Department in the early years of the twentieth century, was born in Ireland, and trained at the University of Dublin, graduating in civil engineering in 1889. Between 1891 and 1897, before his nomination in Siam, he carried out engineering work and surveys in Ireland and abroad. He thus moved from "the Crown's first colony" to the informal empire, to apply his skills of land surveyor in this different context. Two Directors of the *Service géographique français* likewise had prior cartographic experience elsewhere. Jules-Clément-Ladislas Lubanski, a graduate of the *École polytechnique*, left for Algeria at the age of 27, where he was employed as a topographer from 1881 to 1887, then in the geodetic survey section until 1893. After these 12 years, he returned before leaving then left for Indochina on 15 April 1899, for the post of Head of the *Bureau topographique du corps d'occupation*. He created the *Service géographique de l'Indochine*, and was its Head until May 1902, when he returned to France. He was replaced by Lieutenant-colonel Gueneau de Mussy, who had been responsible for surveys for the map of Algeria, which was overseen by the *Service géographique de l'armée*.[34] The first two heads of the *Service géographique* in Hanoi had thus learnt the ropes in Algeria.

These movements did not go only in one direction. Ireland and Algeria were mapped before Indochina, and were thus taken as models, but the autonomy of Indochina's geographical department meant that it could also have a certain influence in Algeria. This is illustrated by one man, Colonel Servière, who took part in the conquest of Tonkin, and became the Head of the second military Territory in 1891 (Lạng Sơn). He directed the boundary commission for the China-Vietnam border in 1891–1892, from which a 1:50,000 scale map was published in 1894. He left his post in November 1893 to "demarcate the border between Yunnan, Tonkin, Siam, the Shan States and the famous Buffer State."[35] He turned up in Algeria in 1909, as General and Commander of the 19th Army Corps, where he asked for a Topographical Department to be set up that was as independent as the Indochinese Department, which he held up as a model. In a situation where geographical documents were often lacking, because they were all sent to France, he recommended collecting them and publishing a monthly report on the progress made in mapping Algeria.[36] So geographers

---

34 Rouget, "Étude sur la cartographie de l'Indochine française", *art. cit.*, p. 36.

35 Lucien de Reinach, *Lettres d'Indochine, 1893–1899*, Paris, Dubuisson, 2010, p. 56.

36 H. Blais, *Mirages de la carte. L'invention de l'Algérie coloniale*, *op. cit.*

did not always start their imperial careers in colonies where cartography was well established. Often the transfers were simply due to the way the imperial system functioned – in this respect, it is not so surprising that a colonial officer or governor should move from one colony to another. At all events, these transfers elucidate how models circulated and practices were shared, such as to create a certain unity in the Empire, and to create links between different colonies through applying similar or comparable solutions.

Transfers of geographers were thus accompanied by transfers of proven methods. In 1891, Lieutenant-Colonel Holdich, from the Survey of India, gave advice on applying the Indian model to the British colonies in Africa.[37] He recommended establishing a geodetic network and systematic topographical coverage, even though he recognised, rightly, that the African colonies would never have the same value for the British Empire as did India. Yet he thought there were sufficient similarities between the Indian and African territories of the Empire to warrant applying in Africa the scientific methods tested in India. The recruitment of local land surveyors had also been a success in India, in his view, and he proposed the same system for Nigeria.[38] The French made similar comparisons, between the topographical brigades in Europe and in the French colonies, in order to justify moving officers from the army corps of the Troops of Indochina to these brigades, for geodetic and topographical work: "The maps of France, Algeria, Tunisia, and all the major countries in Europe could not have been made otherwise".[39]

For French Indochina, the *Service géographique de l'Indochine* looked to the models developed in North Africa, and transfers occurred between the two areas, among others. One of the Department's original aims was to produce and publish triangulation data annually for use by cadastral surveyors. This aim was justified as follows: "This procedure is followed in the *Régence de Tunis*, where it gives satisfactory results".[40] The new institution thus drew partly on successful practices from elsewhere in the Empire. In fact, this idea probably originated in a request made by General Borgnis-Desbordes, in May 1900, in his correspondence with the Resident Superior in Tonkin concerning the preservation of geodetic markers, which the local population were

37 T. H. Holdich, "African Boundaries, and the Application of Indian Systems of Geographical Survey to Africa", *Proceedings of the Royal Geographical Society and Monthly Record of Geography*, 13/10, October 1891, p. 596.

38 Peter Collier, "The Colonial Survey Committee and the Mapping of Africa", *International Symposium on the History of colonial cartography*, 1750–1950, Utrecht, 21–23 August 2006.

39 ANOM, SGEOCOL 18, Rapport sur le fonctionnement du Service géographique de l'Indochine, p. 15.

40 J.-C.-L. Lubanski, "Note sur le Service géographique de l'Indo-Chine", *art. cit.*, p. 476.

neglecting. He broached the issue by remarking that in Tunisia, greater efforts than in Indochina were being deployed to strengthen these markers, and make them permanent features. He then mentioned the *Description géométrique de la Tunisie*, a booklet published annually by the *Service géographique de l'armée* for use by Tunisia's cadastral department. It contained the exact coordinates of the trig points, the measurements of the angles and the lengths of the triangles. In conclusion: "Might there not be a case for envisaging an operational method similar to the one used in Tunisia?"[41] A few months later, this idea was adopted.

Relations between departments in a given colonial administration changed over time, and they also differed between empires. In the French case, the relations between different geographical departments were relatively good, despite some tensions between the *Service géographique de l'Indochine* and the *Service géographique des Colonies*, and positively stormy relations with the cadastral office. In the British case, the Survey of India included the cadastral office, which effectively solved the problem, and was more of a colonial experimental laboratory, as evinced by the new printing techniques developed there. Officers and colonial administrators moved around within the Empires, making for a certain standardisation of cartographic practice through the transfer of know-how. How did the rivalries and collaborations we have identified within the British and French Empires express themselves internationally?

1.4 *Institutional Collaboration between Geographical Departments in the Indochinese Peninsula*

Taking as our vantage-point the boundary commissions, in which cartographers with different training and techniques worked side by side, we can see that the activity of map-making in the Indochinese Peninsula gave rise to not always complimentary comparisons between different practices. For instance, the Europeans were almost unanimous in considering Chinese land surveyors to be unworthy of the name. In 1885, during the Franco-Chinese boundary demarcation, Paul Néis mentioned "a dozen young Chinese taken from different arsenals in the empire, whom they called officer topographers".[42] He clearly thought the title to be ill-suited because, as he continues, apart from two of them who understood a bit of English because they had spent some time in the United States, none of them seemed to know how to read a map at all. The British perspective was no less damning: Litton, the consul at Tengyue

41 ANVI, RST 73614, Lettre du général Borgnis-Desbordes au résident supérieur au Tonkin, Hanoi, 21 May 1900.

42 Paul Néis, "Sur les frontières du Tonkin", *Le Tour du monde*, 55, 1888, p. 321–416.

in Yunnan, China, mentioned the presence of "three 'topographers' from Yunnanfu, thus named by virtue of the rule of *lucus a non lucendo*, for, excepting one perhaps, none of them had the faintest notion of topography."[43] This categorical judgement on the Chinese cartographers with whom they had to work was echoed by all the Europeans involved.

The work of Siamese surveyors was considered better, by their French and British counterparts, even though they were not deemed up to the standard of the cartographers in the British Geographical Department. During the Anglo-Siamese border demarcation of 1889–1890, Captain Jackson expressed his astonishment that the triangulation of Siam had not progressed beyond Bangkok, and that the work of the Siamese Geographical Department was thus based on astronomical observation for the latitudes, on the chronometer for the longitudes, and on compass-calculated traverse lines for the distances.[44] The position of Chiang Mai was thus defined not by the geodetic network and triangulation, but by astronomical measurements, and thanks to the telegraph link to Bangkok. In 1892–1893, during another border Commission for the frontier between Siam and Burma, Walker made the same point, but was more enthusiastic. After accompanying the Siamese geographer, Luang Sakol, for a while, he observed:

> He carried out the traverse measurements using a compass and a chain, which was the only type of surveying they practised (apart from taking the latitude from time to time). They had a very good map, mostly made in this way, which showed a wholly satisfactory precision for distances, etc., even if one cannot rely upon the topography[45]

So the techniques used by the indigenous population were applauded when they approximated to those, considered to be scientific, used by the British themselves. The Siamese were thus rated, unsurprisingly, higher than the Chinese, since the staff of the Siamese Geographical Department had Western trainers. The French were less impressed by the skills of the Siamese surveyors.

---

43 BL, IOR, L/PS/20/D239, Report on a joint survey of the Burma-Yunnan boundary, north of latitude 25° 35' north, by British and Chinese officials, March to May 1905, by Litton, p. 1.

44 BL, IOR, V/24/3988, "Report on the Survey Operations with the Anglo-Siamese Boundary Commission, by Captain H. M. Jackson, season 1889–90", in Report on the operations of the Survey of India, administered under the Government of India, during 1889–1890, prepared under the direction of Colonel H. R. Thuillier, Surveyor General of India, Calcutta, 1891, p. li.

45 BL, IOR, L/PS/20/92, Anglo-Siamese Boundary Commission, Southern Party, 1892–1893, by Captain Walker, p. 10.

In 1905, when a Border Commission was working on the Laos-Siam frontier, the French Minister for Foreign Affairs received the following letter from the French representative in Bangkok:

> The Siamese government has just informed me that it will assign the future work of the border commissions exclusively to French surveyors, who will be the only ones entrusted with operations in the field, without any monitoring by the corresponding specialists in the Siamese commission. Without wishing to minimise the dimension of friendship in this token of trust, which was issued in a most courteous manner, it is principally due to the difficulty encountered by the Siamese in finding competent assistants for the first period of the work. Those they had were notoriously inferior, and in reality only the work of our officers was considered trustworthy.[46]

Thus collaboration was not always the rule, and the production of maps, for example, was dealt with solely by the French, after various attempts made by French diplomats, and by Edward Strobel, the American Advisor to the Siamese Government from 1903 to 1908,[47] in view of the need to replace Siamese personnel with Westerners for the topographic work:

> Given the lack of resources on the ground, should we turn to Dutch officials borrowed from Java, or experts in this field? It was feared they would be too costly. Or ask the help of English civil servants in India? Without directly contesting this idea, I objected that before my arrival reservations had been expressed concerning similar choices, when the commissions were being constituted, at the beginning of the year.[48]

This quotation highlights how experts circulated between different territories, and the use of foreign personnel, who were sometimes considered to be too costly or ill-suited to the tasks. It is interesting that using English specialists is

---

46 ANOM, GGI 37779, Lettre de M. Riffault, ministre de France au Siam, au ministre des Affaires étrangères, Bangkok, 31 May 1905.

47 This post, which was created in 1892, was "supraministerial" insofar as the incumbent had to advise the King of Siam on his home affairs policies, his foreign affairs and also on legal issues. The first post-holder, the Belgian jurist Rolin-Jacquemyns, was replaced by the American, Strobel. See Thamsook Numnonda, "The First American Advisers in Thai History", *Journal of the Siam Society*, 62/2, 1974, p. 121–148.

48 ANOM, GGI 37779, Lettre de M. Riffault, ministre de France au Siam, au ministre des Affaires étrangères, Bangkok, 31 May 1905.

envisaged, even if strategic considerations probably prevented this ever really taking place. We should not forget, in this respect, that McCarthy had been the Director of the Siamese Geographical Department from 1885 to 1901, and that British influence was dominant throughout those years, even if one cannot really speak of collaboration between the Indian and Siamese Geographical Departments.[49] Yet the French also tried to get a foothold in Siam. Although no Frenchman ever headed the Siamese Geographical Department, the Hanoi Department trained several Siamese officials in 1906, as a way of gaining influence. This formula was deemed a success, because the same training was then proposed to Chinese officials, although the outcome remains unclear.[50] The appointment of Albert Gallois, a member of the *Service géographique de l'Indochine*, to the post of Professor of geodetic science and cartography at the Peking Military Academy's Topographic Survey Unit, in 1916, did not work out, because he had health problems, such that this post remained vacant within the Chinese military hierarchy.[51]

Thus we have seen that Europeans were present in the Siamese and Chinese Geographical Departments, and that, reciprocally, officers from these institutions were trained in French institutions. The French clearly sought to counterbalance the British influence in Siam in this way. For this policy was adopted later, however, and to a lesser degree than the British, after the settlement of territorial disputes, that is, after 1907, once the borders had been demarcated and it became easier to plan scientific and cultural exchanges, and Franco-Siamese institutional collaboration, across them. Meanwhile, McCarthy was an active figure in the border delineation discussions as from the 1880s, and although the French offered training in cartography to Siamese officers, no Frenchman was appointed to head their Geographical Department, unlike the British case. Moreover, the few examples of French influence should be viewed with caution because it was so limited, although real enough. Perhaps we shall find that the circulation of the maps themselves, rather than of the map-makers, is the

49 On this subject, and for further details on what follows, see our article "La géographie, un outil diplomatique? Rivalités impériales et collaborations internationales dans le nord de la péninsule Indochinoise (fin XIXe–début XXe siècle)", *Relations internationales*, 167, 2016/3, p. 37–52.

50 ANOM, GGI 18274, Rapport sur l'envoi d'officiers chinois à l'Institut géographique de l'Indo-Chine à Hanoi, du lieutenant-colonel Brissaud-Desmaillet, conseiller militaire du gouvernement chinois, qu'il souhaiterait soumettre à Yuan Shikai, transmis par la lettre de M. Conty, ministre de la République française en Chine, à Albert Sarraut, gouverneur général de l'Indochine, Beijing, 10 September 1913.

51 ANOM, GGI 18286, Lettre du capitaine Salel, chef du Service géographique, au gouverneur général de l'Indochine, Hanoi, 16 November 1916.

real vehicle of scientific and political influence, demonstrating the power of cartography and the particular territorial representation it imposes.

2 The Circulation and Uses of Maps

Not only people, but also cartographic documents, circulated between Geographical Departments, and constituted another form of collaboration. The question of how these maps were used is closely related to that of their real capacity to impose a specific representation of a particular territory: who used them, how many were available, and within what networks? By examining the sales figures, one can ascertain the actual dissemination of maps, and the links between different institutions, not only Geographical Departments; and looking at the addressee tells us something about how maps were used. Their publication in textbooks or atlases is a whole topic in itself:[52] who devised the textbook, what representations it contained, and which maps it drew on, calls upon the whole history of geography and, on a larger scale, of education. This subject is important for the home country, because it can reveal how the Empire conceived the schooling of British and French youth.[53] It is also relevant to schooling in the colonies, showing how the Europeans inculcated a sense belonging to the Empire by introducing a colonial programme of geography and map-rcading from the earliest years.[54] By examining the circulation of maps within colonial administrations, between them, and between learned societies, and also their reception by individuals, as a private market for maps developed for the public at large, we can achieve a more accurate idea of how these documents were really used, and hence of their possible influence.

52 Isabelle Avila, "Les cartes de l'Empire dans les atlas britanniques, (1884–1914)", in H. Blais, F. Deprest, P. Singaravélou (eds), *Territoires impériaux. Une histoire spatiale du fait colonial, op. cit.*, p. 169–188.

53 The question of the place of the Empire in the hearts and minds of the British in the British Isles has given rise to an important historiographical debate between those who maintain that there was a genuine awareness, fostered by propaganda and culture, of Britain as an Empire, and those who claim that the Empire mattered little to the average Briton. See John M. Mackenzie, *Imperialism and Popular Culture*, Manchester, Manchester University Press, 1986; Bernard Porter, *The Absent-minded Imperialists: Empire, Society, and Culture in Britain*, Oxford, Oxford University Press, 2006.

54 Alexander Campbell, "Education in Burma", *Journal of the Royal Society of Arts*, 49, 1946, p. 438–448; Juliane Schober, "Colonial Knowledge and Buddhist Education in Burma", in Ian Harris (ed.), *Buddhism, Power and Political Order*, London, Routledge, 2007, p. 52–70.

2.1 *The Circuits between the Home Country and the Colonies*

Among the obstacles to the circulation of these materials was the distance between the place of production of the maps, and the place where they would be used. Colonial civil servants waited impatiently for maps made in the home country, while Ministries and learned societies experienced the same frustrations for maps printed in the colonies.

The maps produced in the colonies were used there, and also sent to the home country, for example, the map of Indochina whose sheets were drawn up, revised and printed by the *Service géographique de l'Indochine* in Hanoi, or in Calcutta for the Survey of India. This was also the case for the maps from the boundary commissions. All the French Geographical Department's maps took the same course: the office of the Governor-general of Indochina received them at no cost, and sent them to France, to the different secretariat departments, and to the military headquarters for defence purposes.[55] They could also be sold to the public services and to individuals (with the exception, for the latter, of maps marked "confidential", which included the maps of the borders). Individuals paid the price printed on the sheet – about a piastre – whereas the public services and officials received a reduction of 25%. The colony's bookshops could also be granted this reduction, but they had to sell the map at the full price. Outside the colony, in France or abroad, the maps were sold by bookshops at the same price as in the colony, plus 10%. The Geographical Societies were faithful clients, even if they were interested primarily in the maps of the whole area. In 1898, before the creation of the *Service géographique*, Doumer first sent "a collection of maps of Indochina to the Geographical Society which had requested it via the Topographical Office".[56] Generally, receipt of the documents triggered a flood of letters of thanks, in which Deloncle's map figured prominently. The Survey of India also worked for societies such as the Astronomical Society of India and the Asiatic Society of Bengal.[57] The *Service géographique des colonies* in Paris also sent its publications for free to various Geographical Societies, which sent their Bulletins in return. This exchange of courtesies created flows of geographical information in the home country itself.

We have already mentioned that the maps made by the *Bureau topographique de l'Indochine* were printed and distributed in the home country by the *Service géographique de l'armée*, until 1890, before being sent to the colonies. The many

55 ANV1, GGI 1807, Arrêté du gouverneur général Paul Beau, 25 November 1904.

56 ANOM, SGEOCOL 16, Lettre de Paul Doumer au ministre des Colonies, Hanoi, 11 August 1898.

57 BL, IOR, V/24/4009, Report on the operations of the Survey of India, during the survey year 1910–1911, p. 29.

requests for maps by colonial departments often met with the Topographical Office's response that since the maps were printed in France, there would be a delay.[58] Sometimes Ministries and Geographical Departments worked with publishing companies, for both the final versions and the intermediary stages, to reduce the very high costs of manufacturing the maps: "The publisher bore all the costs, on the condition that the Administration agreed to a buy a certain quantity."[59] Thus a larger number of maps could be printed without the government having to spend more, a procedure already employed at the time of the British East India Company, when the famous London publisher Arrowsmith printed a large number of maps compiled from the Company's documents.[60] In 1893, the publisher Challamel printed the first edition of the French Pavie map, the *Carte générale de l'Indo-Chine*, on four sheets, made by Capitains Cupet, Friquegnon, and Malglaive; it paid all the expenses in return for the guarantee of a substantial number of purchases.[61] It also published a reduced version for a single sheet, of which 200 copies were printed for a cost of 700 francs.[62] However, the limited number of maps seemed to be a chronic problem for those who used them, regardless of who published them, and where.

2.2 *The Recurrent Shortages of Topographical Maps*

The constant requests for maps received by Geographical Departments and publishers indicate both the lack of provision and the level of need. In the early stages particularly, there were no documents available, for example in Upper Burma, where "the want of maps giving the relative positions and names of villages caused some inconvenience and delay, but to have waited for them would have caused needless waste of time."[63] The Residents of the provinces of Tonkin repeatedly asked for maps, a dozen copies at a time, above all of their

58 ANV1, RST 73618–5, Demandes d'établissement des cartes marines et provinciaux formulées par des résidents des provinces du Tonkin et du résident supérieur du Tonkin (1889–1900).

59 G. Antony, *Le Service géographique et des missions. Historique, organisation et publications, op. cit.*, p. 9.

60 M. H. Edney, *Mapping an Empire: The Geographical Construction of British India, 1765–1843, op. cit.*, p. 228.

61 G. Antony, *Le Service géographique et des missions. Historique, organisation et publications, op. cit.*, p. 73.

62 ANOM, SGEOCOL 16, Rapport au ministre, de la part du Service géographique des colonies, Paris, 12 October 1892.

63 BL, IOR, V/3988, Report on the operations of the Survey of India, administered under the Governement of India, during 1889–1890, Calcutta, 1891, p. 33.

province and of the whole of French Indochina,[64] as did people involved in trade and industry. In February 1888, the Topographical Office was inundated with requests addressed to it directly, after 609 copies of various maps had been distributed to the administrative departments since June 1887.[65] General Bégin complained that he had neither the financial resources nor the printing capacity to continue meeting requests, and he proposed that these be addressed to the Resident, who would prioritise them and respond to the most urgent. A printer, Schneider, was given exclusivity for the Topographical Office's work before it could use its own presses, and it published the Office's output for 1890, thus obviating the need to send them to the *Service géographique de l'armée* in mainland France.

Two years later, in 1892, the consumption of maps seemed to fall off, and some maps were in good supply, such as the provisional maps used by the military. For the 1:500,000 scale map, the "annual consumption of each of these sheets in the preceding years hardly reached an average of 50. The consumption for 1892 will drop by 20 copies."[66] There were stocks of 70 to 100 items, figures far exceeding the demand. For the 1:200,000 scale map, there were half as many requests, and for the 1:100,000 scale map, the average annual consumption was 30 sheets in 1890 and 1891, and, since the provinces which had been supplied with maps rarely made further requests, the consumption for 1892 was barely ten copies. These figures show a situation very different from the shortages of the earlier years, as though receipt of the first maps had stemmed further requests. However, this pause, to the extent that it existed, was short-lived. In 1905, the Head of the Political Affairs Office of the Government-General suggested that the maps in its possession be bound "in order to be able to consult them more easily and especially to prevent the sheets getting dispersed, since they are not always returned by the people who consult them or attach them as supporting documents."[67] The penury of maps was a general phenomenon, common to the French and the British, but also to the Siamese.

---

64 ANVI, RST 73618–4, Demandes de cartes du Tonkin et de l'Indochine formulées par des résidents des provinces du Tonkin (1886–1888).

65 ANVI, RST 73618–4, Lettre du général Bégin au résident général au Tonkin, Hanoi, 29 February 1888.

66 ANV1, RST 73618–5, Résidence supérieure du Tonkin, Service des cartes et plans. État des cartes de l'État-Major et administratives en magasin (au 28 juillet 1892) portant indication des quantités approximatives nécessaires pour un an, Hanoi, le 28 juillet 1892, par le conducteur principal chargé du bureau des cartes.

67 ANOM, GGI 25663, Note du chef du bureau politique au gouverneur général de l'Indochine, Hanoi, 19 September 1905.

2.3 *For Officials in the Field or Diplomatic Networks, a Similar Use of Maps*

Colonial civil servants used maps, of different scales, to administer the conquered territories. The Government General needed a constant supply of general maps of Indochina, as well as thematic maps, for example, maps of the troop positions in Tonkin and in Cochinchina.[68] The other departments often needed very large scale maps, which were provided by the *Service géographique*. The Survey of India likewise reproduced maps, plans and illustrations for other departments – forestry, agriculture, the telegraph services, the official architect, the Indian Geological Department, the Engineering Department, and the Indian Museum – and sometimes produced tailor-made maps.[69]

Maps were also used within the diplomatic networks. The French and British consuls in Asia requested maps, and supplied them, just like other information. In 1888, the French consul in Hong Kong requested, and received, two copies of a map of Tonkin for other French nationals.[70] This shows that cartographic information was exchanged within diplomatic networks in the region. These circuits sometimes persisted despite tensions elsewhere. Thus while France was in lengthy border discussions with Siam, the French Governor General in Siam requested the 1:760,320 scale, 12-sheet map, which Prince Damrong had commissioned, which was not for sale, and for which the French Administration lacked sheets 1, 3, and 5.[71] Prince Devawongse, the Minister for Foreign Affairs, sent him a copy via the French diplomat in Bangkok.[72]

The circulation of cartographic documents internationally intensified at the end of this period. In the late summer of 1908, the Office of Foreign Affairs of Yunnanfu [Kunming] requested that the *Service géographique de* l'*Indochine* provide it with recent and complete maps of Yunnan, showing the Tonkin and Burmese borders, adding that the English had already sent their own maps of Burma.[73] The Head of the *Service*, Friquegnon, sent back five copies of the sheets of Mongtze and Cao Bằng, drawn on a scale of 1:500,000, and advised the Governor General not to show anything about the Burmese frontier to the

68 ANOM, GGI 1805, Demandes et envois des cartes géographiques, 1911.

69 BL, IOR, V/24/4009, Report on the operations of the Survey of India, during the survey year 1910–1911, p. 29.

70 ANV1, RST 73618–4, Lettre du consulat de France à Hong Kong, à M. Parreau, résident général de France à Hanoi, Victoria, 14 September 1888.

71 ANOM, GGI 25663, Lettre du gouverneur général de l'Indochine Broni au ministre de France au Siam, Doson, 21 November 1905.

72 ANOM, GGI 25663, Lettre de M. Roy, Vice consul faisant fonction de consul de France à Bangkok, à M. Beau, gouverneur général de l'Indochine, Bangkok, 30 November 1905.

73 ANOM, GGI 19714, August-September 1908.

Chinese authorities. This could have been simply prudence, and a wish not to meddle in bilateral relations, or else a concern that the map might be of low quality, or had mistakes in it; but it could also be due to the fear that the maps could have real consequences in the frontier zones. Be that as it may, standard practice was to recognise the cartographic sovereignty of each power over its own territories.

In this overview of cartographic circuits and use, large scale maps of frontiers were a rule of their own. They could be used to prove borderlines in disputed cases, and they showed the extent of the territory appropriated by the Europeans from the Chinese, at the local level of the frontier, as well as on the scale of international relations. However, they generated less interest than other general maps, and circulated among a more restricted group. They were extremely precise, and rather rare, due to the sensitivity of what they represented. Sometimes the scarcity of maps was a problem even before the stage of manufacture, as was the case with the sketches of the border between Indochina and the kingdom of Luang Prabang, to which a French civil servant added a postscript: "Since there is absolutely no drawing or tracing paper, I have been obliged to draw the attached sketches on Shan paper, and in the course of a single expedition", a fact confirmed by a marginal note on the attached sketch.[74] The rarity of the document limited its circulation, even if copies could be made later. Sometimes, on the other hand, the circuits were available, but there was little interest in the product. When the head of the Franco-Siamese Border Commission, Lieutenant-colonel Bernard, submitted the first sheet of his 1:200,000 scale map in 1907, he recommended that it be printed in large quantities, for diplomatic purposes, and also because he deemed that it would useful for a long time to come.[75] The head of the *Service géographique*, Friquegnon, did not agree. In his view, 100 copies for sale would be more than enough:

> Once the interested Departments have acquired these very specific documents, the latter are never requested; they generally concern only a

74 ANOM, GGI 24917, Lettre de Paul Macey, commissaire du gouvernement chargé des agences de Pak Lay et de Xieng Khan, au commandant supérieur du Haut-Laos, Pak Lay, 2 août 1898 ; ANOM, GGI 14491, croquis au 1:100 000 de la "Frontière historique du royaume de Luang Prabang établie il y a 150 ans environ lors de l'invasion birmane", annexé à la lettre no 10 du 2 août 1898.

75 ANOM, GGI 14490, Lettre du lieutenant-colonel Bernard, président de la Commission de délimitation franco-siamoise, au ministre des Colonies, Paris, 16 July 1907.

> narrow strip of territory, and so are not useful to individuals. [...] In short, these maps are unsalable.[76]

The situation changed over the period, and the maps became out of date, as priorities changed. The organisation of infrastructures took over from settling border disputes; gradually the course of the railroads attracted more attention than the path of the borders.

Thus topographic maps circulated essentially between the various colonial Departments, and constituted a fundamental tool for the Europeans in their attempts to get their bearings in the colonies, and administer the territories. However, they arrived late in the day, and in small numbers, compared to the needs expressed. The less precise maps, but which gave an overall representation of the territories, also circulated through diplomatic channels. Cartographic documents also entered into a whole series of networks in the home country, for learned societies and individuals wishing to keep abreast of the the latest publications. Not all maps were considered in the same light. Geographical Societies were as eager for large-scale maps of itineraries as for small-scale maps covering a whole territory, whereas individuals tended to be interested only in the latter, and very precise maps such as those of border delimitations seemed to have no use other than for settling border disputes. Hence a map's use and reception should always be examined in its context, and in the period from 1885 and 1914, for example, there were moments of international tension as well as of relative calm. Analyses of the "power of maps" tend to distort the reality if these sorts of changes, which are not taken into account. A key diplomatic instrument at one moment may be of little use the next. And not all maps met with the same interest from the Europeans, nor were they always used in the same way. Moreover, few records exist of how maps were received by the indigenous populations, except in diplomatic circles, and so it is impossible to assess to what extent cartographic representations of their territory were influential. The most likely hypothesis is that these populations did not have access to such documents and, if they did, that they attached little weight to them.

•••

Map-making thus gave rise to networks of exchange between institutions within imperial administrations, then internationally, in response to particular

76 ANOM, GGI 14490, Lettre de Friquegnon au gouverneur général de l'Indochine, Hanoi, 11 March 1909.

difficulties, or in order to make the most of a particular opportunity, rather than according to a clearly defined programme.

Cartography required close links between the home country and the colonies, and it could lead to the circulation of people, standards and models between the colonies themselves. The links with other map-making bodies reveal differences between the French and the British approaches, particularly for cadastral documents, which were separate from the *Service géographique*, and in rivalry with it, in the French case, but integrated into the British Survey of India. The inter-departmental rivalries show that the Empire was not a homogeneous whole, and that the multiplicity of protagonists was a recipe for a multiplicity of divergent views. Despite institutional differences, the practices were similar, for instance the use of local labour to keep costs low. This relative unity was reinforced by imperial careers, with proven models and methods being applied in different locations, regardless of whether the person's mobility was due to circumstances, to administrative requirements, or to a personal wish.

The Indochinese Peninsula was to some extent a testing-ground for new techniques and inventive solutions in cartography, especially printing, due to the specificities of the climate, and the distance from the home country. Although the inventions could not be put to use in Europe, they demonstrate the creativity, and relative autonomy of scientific experimentation in the colonial context.

The British and the French also attempted to have their methods adopted in Siam and in China, as a form of "soft" diplomacy, driven by circumstances rather than by a structured strategy. The rivalry between the powers was thus compatible with international collaboration, and the latter even seems to have depended on the former: thus the strong British influence in the Siamese Geographical Department from the 1880s motivated the French to match it with their own influence at the beginning of the following century. Scientific endeavours were prompted in Siam and China by this Franco-British rivalry, with results visible in maps used in diplomatic circles, although this phenomenon should not be over-estimated.

Franco-British rivalry was itself not wholly consistent, insofar as scientific imperatives sometimes overrode political considerations. The scientific information shared, inter-connection of the geodetic networks over the whole Indochinese Peninsula, and the circulation of maps through diplomatic channels alone prove that the purpose of geographical information was not always to wage war. The chronic shortages of maps and their long publication times affected their impact, which also depended on the context of reception. Other factors also limited any real domination through cartography: the production difficulties, the fact that mapmaking was not a priority in the colonial budgets, and the sometimes unclear functions which the maps ultimately had. To these institutional and diplomatic aspects should be added the problems of the work on the ground: the Europeans were working in territories mostly unknown prior to the colonial period, and alongside local populations whom they only gradually came to know.

CHAPTER 4

Triangulation "from the Mountains to the Sea"

> We had only managed to climb three quarters of the total height of the mountain; we would have reached the summit with ten times less difficulty and exhaustion if, instead of starting with the most inaccessible side, we had taken on the East slope, which has a gentler and less uneven gradient. These sorts of errors, which are inevitable when one is without a guide in an unknown region covered in dense vegetation, should not be cause for regret: through them we acquire, more accurately than all the information and descriptions, an idea of the topography of the area, and they are absolutely necessary for understanding the map, or, in other words, for finding its geographical formula.[1]

Francis Garnier's reference to the high mountains, the difficult ascent and the abundant vegetation highlights certain enduring difficulties for the cartographers of the northern Indochinese Peninsula. The mountains, and the vast, and sometimes impenetrable forests hindered the Europeans' steady progress. These aspects of the terrain were mentioned time and again in reports by travellers, explorers and topographers. Each brought to bear a different viewpoint: travellers simply described the landscape, explorers' itineraries sought to link up different points, and topographers measured the land in order to find its "geographical formula" and represent it cartographically. For these surveyors, observation was fundamental, and more valuable than "all the information and all the descriptions" – a significant comment, given the number of documents they had consulted before going into the field. Observation and measurement are indeed at the heart of modern European science, and are considered the best guarantees of scientificity. This priority notwithstanding, the Europeans had no prior experience of the spaces they were exploring, with the result that

1 F. Garnier, *Voyage d'exploration de l'Indo-Chine, op. cit.*, p. 117.

 | DOI:10.1163/9789004469853_006

they were often at a loss, and left to their own devices, when it came to mapping the terrain.

In this endeavour, the Survey of India Department's motto, "from the mountains to the sea" (*A montibus ad mare*), was partly a gauntlet thrown down to other comparable institutions, but mostly an encouragement to explore the furthest reaches of the country. The Survey aimed not only at marking routes, but also at surveying the whole country, regardless of the vicissitudes of the terrain. In the light of these ambitions, how was this alien terrain actually apprehended, observed and measured? The topographers had been trained in their home country in the use of European instruments and techniques, but despite the superiority which they had in the colonial situation, they had to adapt to unfamiliar situations. So to what extent were they really dominant, and what degree of success did they have in the hoped-for full cartographic coverage? The questions we shall address in this chapter are therefore how the topographers worked in practice, what they were able to perceive of the colonial territory, what their blind spots were, and how they dealt with the specificities of the northern Indochinese Peninsula. How did the topographers work, and what were their blind spots, given their interpretive framework? They which gave them a certain superiority, but they were faced with an unfamiliar territory, to which they had to adapt. This raises the question of their concrete dominance over the territory, and whether could actually be carried out. We also shall examine whether the British and the French had similar practices, and whether these were identical to those used in the metropole, in order to assess whether specifically colonial scientific practices can be identified.

1 Tough Working Conditions and a Dangerous Terrain

The surveyors' natural and human environment affected their operations and, even though security gradually improved, and some areas were safer than others, their real conditions should make us reconsider the thesis that cartography is an activity of domination. A few isolated examples, of both human vulnerability and material obstacles, can give us a glimpse of the reality of the terrain: a report states that in 1903, in certain parts of Tonkin, "we have had to take precautions because the area is infested with tigers", and the horse of one of the geodetic surveyors was snatched from the shed of an outpost South of Lang Som.[2]

2 ANOM, SGEOCOL 18, Compte rendu des travaux sur le terrain exécutés par le Service géographique de l'Indo-Chine pendant la campagne 1902–1903, p. 29.

1.1 *Relief and Climate*

Reports from French and British cartographers contained a wealth of detail on the hindrances to movement: the slopes were too steep, and were made slippery by the rain, the forests were dense, and work was sometimes halted due to the suffocating heat or the dense fog. The form of the sources – the daily log kept by the cartographers on the terrain covered – encouraged reports on the relief and the climate, since these were their constant preoccupation, and everything deserved to be noted, to swell the geographical information available (as well as sometimes to justify poor results).

Topographical activity was hampered not only by the characteristics of the terrain, but also by the climate, particularly fog and rain. These affected the rhythm of the work to the extent that, in order to avoid the worst seasons, campaigns took place roughly between October and May, while the summer months (rain and searing heat) were reserved for working on the information gathered. Notwithstanding this, some seasons could be rainy, which slowed the work down. The humidity, the vegetation and the mountainous terrain, which were the cartographers' daily fare, created constant difficulties for the scientific work, affecting the route taken, the measurements, and the observation. Sometimes the work completed got destroyed, as reported by Captain Walker, a British member of the Anglo-French Buffer State Commission of 1894–1895, when the mule carrying the bamboo crates containing all his surveys and sketch maps disappeared in the confusion following a sudden storm.[3] Illness also struck in this tropical climate. Between 1902 and 1914, 16 Europeans, "not counting the indigenous population, paid with their lives the arduous and unsung labour they had undertaken."[4] The climatic conditions thus had a significant impact on the work, and largely determined the mission's success or failure.

Topographers, whether regular cartographers or explorers, arrived in the Indochinese Peninsula with their own ideas and interpretive framework, which was not always appropriate to the area. They conveyed a particular and partial view of the territory, depending on what they were capable of observing, and they invariably compared what they saw with European landscapes. This can be understood as a *topos* for travellers and explorers, rather than as a reaction specific to the colonial situation: the foreigner is delighted to find, in the land stretching out before him, familiar realities that he can thus also

---

3 BL, L/PS/20/83, Report on the Anglo-French Buffer State Commission, 1894–1895, by Captain Walker, p. 31.

4 H. Brenier, *Essai d'atlas statistique de l'Indochine française*, *op. cit.*, p. 136.

communicate to his readers.[5] A "system of comparisons, analogies and differences" enabled the Westerner's vision to adapt to the landscapes of, here, the Indochinese Peninsula, and make the two commensurable.[6] The most common comparison in this case was that between the mountains in the north of the Peninsula and the Alps, which also mark national borders (France, Switzerland and Italy).

The recurrent comparisons thus express the perceived alien nature of these territories, which the topographers are discovering, and trying to make sense of for the first time. The reality cannot always be adapted to the cartographic standards in force. The 1:25,000 scale maps mostly used conventional European signs. The drafts and sketches had to be on a scale of 1:20,000 – and one can see from the request for a very precise scale that there was a drive to establish standards. The sketch maps made by the surveyors – of varying sizes and quality, some drawn in graphite pencil, others in colour – all used the same symbols: the roads were in red, the rivers in blue, relief was bistre, vegetation was green, and dwellings were in black. The sketch maps also showed important features in the landscape such as pagodas, forts, or telegraph transmitters. Anything that caught the eye was mentioned, such as a "pagoda, noticeable due to a coconut tree",[7] to distinguish it from others (its name was probably not known). This method shows how the French topographers attempted to understand the territory, which was largely alien to them, and one can note Western realities, such as the telegraph station, alongside elements new to the Europeans, such as pagodas. Markets also figured since, as in Europe, they were places of economic activity, of sociability and of exchange. Their almost systematic inclusion shows their particular importance in the Indochinese Peninsula, and their significance for military forces in war-time, since markets were a place where rebel gangs came to get supplies. The maps' keys contained both particular symbols for local features such as rice paddies, dykes, pagodas, and isolated traditional dwellings, and standard European signs [Figure 15]. The symbols were often accompanied by written comments, to explain unfamiliar elements. The resulting maps were very detailed in part, but the rest was undeveloped, for instance the landscapes.

5 The same phenomenon can be found in Chinese travel writings describing the colony of Formosa [Taiwan]: Emma Teng, *Taiwan's Imagined Geography. Chinese Colonial Travel Writing and Pictures, 1683–1895*, Harvard, Harvard University Asia Center, 2004.

6 Marie-Noëlle Bourguet notes a similar phenomenon in the 18th–19th-century expeditions in the Mediterranean: M.-N. Bourguet, B. Lepetit, D. Nordman, M. Sinarellis (eds), *L'invention scientifique de la Méditerranée. Égypte, Morée, Algérie*, *op. cit.*, p. 21.

7 SHD, 10H6, Dossier 14, Rapport de la reconnaissance faite par le commandant du "Léopard", Bord, 3 June 1884.

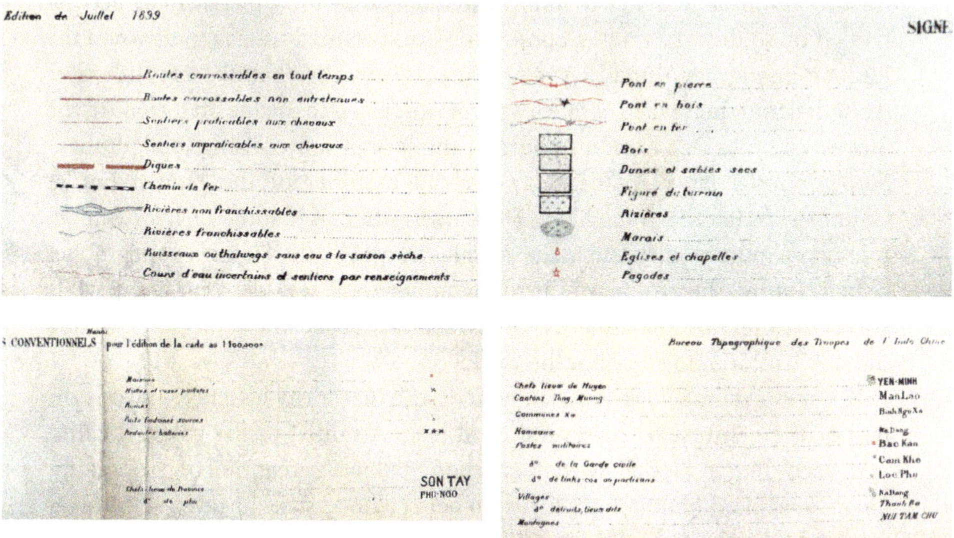

FIGURE 15 Conventional signs used for the topographical map of French Indochina, edition of 1899
SOURCE: ANOM, Asie 165, Indo-Chine [Carte topographique régulière] au 1:100 000, Hanoi, Service géographique de l'Indochine, 1899–1924

1.2 *Constant Threats: "Pirates", "Dacoits", and Local Resistance*

To the unfamiliarity of the territory and the difficulties posed by the natural environment should be added the turbulent political context of the topographers' work, which started during the conquest, and continued during pacification. In Tonkin, which was not fully pacified until the end of the 19th century, the threats came from resistance to the French presence, from local leaders who had not yet been subdued, and from breakaway rebel gangs.[8] French descriptions often mention "pirates" for the Black Flags, just as British accounts use the term "dacoits", or brigands. Although this vocabulary blurs the distinctions between the groups, it shows that insecurity affects all the topographers in the northern Indochinese Peninsula. In Burma, where armed brigands were rife, the British also had to contend with ethnic groups which did not accept the Westerners' presence. And in Siam, the first geodetic and topographical surveys took place while accompanying the Siamese army in their fight against the Ho, the Chinese gangs which made incursions from China into north-west Siam.

The problem of insecurity is mentioned in all the reports, since it had a direct impact on the surveying. It obliged the topographers to accompany

8 B. C. Davis, *Imperial Bandits. Outlaws and Rebels in the China-Vietnam Borderlands*, *op. cit.*

troop columns, and hence to follow the routes they took, or else to be accompanied by soldiers or militia. Some expeditions ended tragically, as when the junior topographer Ahmad Sayad was surprised by Shan bandits on 17 July 1886, wounded and then killed, while his escort managed to escape. In its report, the Survey of India Department mourned the loss of a promising topographer.[9] Some months later, Haïtce, a 27-year-old member of the French Border Commission for the Guangdong (China) frontier, and some of his companions, were murdered.[10] And in Siam, in January 1891, the Siamese topographer officer Noi Tat, the brother of the topographer Seng, was also murdered, while surveying the Sibsongbanna [Xishuangbanna], because the local population did not want him to carry out the work.[11]

Things developed over time, of course, and insecurity lessened towards the end of the century, when several local leaders rallied to the colonial camp, and, according to McCarthy, in 1895, when the local perception of cartography changed in Siam from being a threat to being a necessity. However, some areas remained dangerous, as illustrated in May 1911, when the soldiers Perrin and Médard from the 9th regiment of the colonial infantry, the rifleman Bach Lai from the 3rd regiment of the Tonkinese Rifles, and three indigenous guards were assassinated.[12] They were all on secondment to the *Service géographique*, and were all members of the first geodetic survey group, when they were surprised by rebels while working on the Annam-Laos border.[13] Perrin and a militiaman were hacked to death with an axe, on 30 April, and Médard and two others were killed the following day, while they were looking for their companions. The last body, of a militiaman, was found at the foot of a 30-metre high tree, where he had been attaching a geodetic signal, when his assailants made him fall. Personal effects, money, and supplies were stolen. The first explanation

9 BL, IOR, V/24/3984, Report on the operations of the Survey of India, administered under the Government of India, during 1885–1886, Calcutta, 1887, p. 54.

10 Charles Fourniau, "La frontière sino-vietnamienne et le face à face franco-chinois à l'époque de la conquête du Tonkin", in P.-B. Lafont (ed.), *Les frontières du Vietnam, op. cit.*, p. 85–103.

11 J. McCarthy, *Surveying and exploring in Siam*, London, John Murray, 1900, p. 1–2; James George Scott, *Gazetteer of Upper Burma and the Shan States*, Rangoon, Government printing, 1901, p. 308.

12 They are mentioned, along with others, in the long list of staff who died during the work. See É. de Martonne, J. Martin (eds), *La carte de l'empire colonial français, op. cit.*, p. 103–104.

13 ANOM, GGI 42511, Assassinat des soldats de la mission géodésique par des Moï du bassin de la Se Ban Hien (a frontier region between Annam and Laos).

advanced was that the soldiers had violated the sacred nature of the mountain in question.[14] The head of the brigade refused this simplistic version:

> For the Moi of this region, every mountain is inhabited by the spirits of their ancestors, and as such is considered to be taboo. If we accepted this argument, we would be unable to construct a single signal. Besides, when we leave, the Moi ward off the evil that could befall them due to the violation of the mountains by sacrifices of cattle, which are very effective, according to them. We should not consider this case to be an outburst of fanaticism.[15]

In his view, the real reason was that this population hated Europeans and foreigners, and were afraid of losing the independence they had enjoyed until them; the reasons were therefore political, and not religious. The sacred character of certain places, which was certainly true, was a pretext for violence against foreigners, Europeans and any indigenous individuals working for them. A foreigner is someone who is different, and whose strange behaviour can introduce unpredictable events;[16] he or she is strange, abnormal and even barbarian. So this is how the Europeans probably first appeared to the inhabitants of the Indochinese Peninsula. The foreigner could, however, also be familiar, and even close: the indigenous assistants to the Europeans were treated hardly any better than the Europeans themselves, as though they too embodied the figure of the foreigner. Consequently, relations with the local populations were always a challenge, as were relations within the topographical brigades, in which men from different horizons were brought together and had to work harmoniously together.

1.3 *Topographical Brigades: The Colonial Encounter in Miniature*

Topographers could not work alone. They had to be accompanied by a large escort – several dozen soldiers – to carry out their work. The brigade members, from many different backgrounds, had to live side by side throughout the campaign, which lasted several weeks, and even months. As such, the brigades are arguably paradigmatic of the colonial encounter, and its difficulties. These

---

14 ANOM, GGI 42511, Télégramme du résident supérieur de l'Annam au gouverneur général à Saigon, Hué, 1 June 1911.

15 ANOM, GGI 42511, Rapport du lieutenant Dubuisson, chef du 1er groupe géodésique, à M. le lieutenant-colonel, chef du Service géographique, Hué, 10 June 1911.

16 Jean-Baptiste Duroselle, *Tout empire périra. Une vision théorique des relations internationales*, Paris, Publications de la Sorbonne, 1981, p. 38.

included, as the archives show, the problems linked to the fact that the brigades were formed exclusively of men, who sometimes had recourse to prostitutes:

> Officer heads of brigades allow sub-officer operators to have *congai* with them, taken from the brothels of the neighbouring province, women who mostly serve as intermediaries and interpreters between their lovers and the local authorities.[17]

The few Europeans present were accompanied by men who had a similar technical training, but who knew the territory as little as they did. The Indians who accompanied the British in Burma, for example, fell into this category, and were as foreign to the area as their employers. Porters were also essential brigade members, to carry the equipment and instruments for the expedition, as were men to mark out the paths and clear the summits to build the geodetic station. Coolies were often recruited locally, but could also come from further afield, as could the military escort required to protect the team. Guards were recruited locally. Thus there were always decisions to be made, between hiring staff before the expedition, and doing so locally. In the former case, the Europeans could be accompanied by individuals they considered competent and trustworthy, with whom they may have worked before. But the costs were high, and sometimes the men did not want to go too far, and abandoned the work halfway through. Recruiting locally, on the other hand, ran into the problem of the availability of men to carry materials, or to guide the brigade, and of finding men with the right skills. There was also a lesser issue, that of distance, since some of the hired staff did not want to be away for too long. Recruited staff thus came together from almost everywhere: some from Europe, others from Indian villages, still others from Burmese towns a hundred miles away, and some from neighbouring villages.

This diversity, sometimes described in a racial discourse, was a source of disagreement and resistance.[18] Thus Colonel Woodthorpe, on assignment in the Kubo Valley, on the Chindwin, mentioned a problem of interpreters because those who had accompanied him from Manipur had wanted to go home. He had had to appeal to the few local speakers of Burmese, and this had led to many misunderstandings. One of the *chaprassi*,[19] Jhanu, had been led to the

17 ANOM, GGI 17548, Rapport du résident supérieur en Annam au gouverneur général, Hué, 20 October 1906.

18 Both the British and the French believed that certain "races" were better suited than others to warfare, or to serving the Empire in some other way, due to their innate qualities. Some of the European military topographers adopted these stereotypes.

19 In the British Empire, the *chaprassis* were messengers or employees of the India Office.

wrong place, to position a signal, and he then refused to go to the summit. Here the problem was not recruitment, and it concerned an indigenous Indian, not a European. Hence the terms of the colonial encounter here are not simply the coloniser and the colonised, but at issue is an Indian "sub-imperialism".[20] The Indians of the Survey of India Department had no real strategy or power of their own, but they had authority over the Burmese to the extent that they were employed by the colonisers. This position was, however, a failure because of the unbridgeable gap between the two groups. The relations were a daily challenge for the topographical brigades. Woodthorpe expressed his thanks to the Khasia coolies[21] for their greater goodwill than the *sepoys*,[22] who used the excuse that they were not built in the right way to scale mountains, and then the Burmese, who carried lighter loads. These distinctions were operative in Woodthorpe's brigade, even if they were stereotypes elsewhere.

As a result of these many obstacles to the project of steadily mapping the new territory – the climate, terrain, illness, insecurity, and conflicts internal to the brigades slowed down or halted the work – the objectives set could not always be reached, despite the forward planning (several years), and the instructions given. The cartographers had to modify their practice to accommodate the prevailing conditions.

2 Topographers and Geodetic Surveyors in Action: Adapting Standard Practice

2.1 *Topographical Instruments and Techniques*

As we have seen, the topographical brigades, whether British or French, had to adapt to the environment of their work, namely the relief of the terrain to be mapped, and the climate. Additionally, they were constrained by the pace of the party, making regular surveys difficult to achieve. When topographers accompanied a military campaign, they had to keep to the speed of the column, in order not to compromise the troops' security or the success of the conquest. In the annual reports from the Survey of India Department, the topographers often thanked the military commander for supporting their

20 The term was first coined to refer to Indian trading communities, who had played their cards so well in the imperial economic system that they had acquired a dominant role in the economies of other countries, for example in South-East Asia or East Africa: Thomas R. Metcalf, *Imperial Connections: India in The Indian Ocean Arena, 1860–1920*, Berkeley, University of California Press, 2007.

21 The Khasi Hills are in north-east India along the border with Assam.

22 A *sepoy* is an infantry soldier in the Indian Army.

work, particularly when this slowed down the advance of the troops, or when a soldier was sent to accompany a topographer. Sometimes the speed with which the troops advanced was used to justify gaps in the topographical surveys. The general rhythm was set by the pack animals, if there were any – generally mules or elephants. The topographers carried out their work independently, at their own pace, but they were obliged to keep the military column constantly in view, and so were tied to the rhythm of the animals. Time pressures were always present: due to the constant insecurity, the hostility of the local population, and the fear that the rainy season would arrive before the surveying was completed, the expedition marched rapidly, and could not always keep to the planned path. As a result, the topographers simplified their techniques as much as possible.

From time to time, the topographers would split up for greater efficiency. They would leave for several hours or days, with a second-in-command and a small escort, to map a particular road or reconnoitre a mountain summit, when exploring, working on Border Commissions, or even for regular surveys. Their fieldwork required a good understanding of their own bodies, for which they trained, in order to know the length of their stride, and to educate their vision to notice salient features and imagine those out of sight. The British had the idea of teaching these measuring techniques to Indian auxiliaries, who were then sent to far-flung places where the presence of a European would have aroused suspicion.[23] The character of Kim in Rudyard Kipling's novel of the same name was inspired by this practice, which was known and admired by the French. Kim is a young boy who secretly carries out surveys for the British Survey of India, while officially following an old lama in his spiritual quest.[24] However, since the resources of the human body as a work tool are limited, technical instruments were essential for full surveys, due to their greater precision than the naked eye.

Hallett recalls, in all its complex detail, the topographer's daily work in the late 1870s, when he was carrying out surveys for the path of a planned railway line to link Burma and China, via Siam:

> Surveying by time-distances and a prismatic compass, when on the march, requires a steady hand, a quick judgment for selecting an object

23 P. Hopkirk, *Trespassers on the Roof of the World, op. cit.*; D. J. Waller, *The Pundits. British Exploration of Tibet and Central Asia, op. cit.* 24. BL, IOR, L/PS/20/162, The Trans-Salwin Shan State of Kiang Tung, by Lieutenant George John Younghusband, Calcutta, 1888, Preface, p. iv.

24 Rudyard Kipling, *Kim*, London, Macmillan and Co, 1901.

> for your angle, and a good memory. If the hand is unsteady, the ring of the compass, which is balanced on a needle, will not come to rest. In a jungle-clad country you must watch the foremost elephant as it winds to the trees, and rapidly select the point for your next angle as the animal is just passing from view. A good memory is required, otherwise in noticing the trees, rocks, by-paths, width and depth of streams, breadth of fields, size of villages, and taking sketches of, and angle to, neighbouring hills, you will forget the object, twig, branch or trunk of tree, that you have aimed at. Having taken your angle, you must catch up the last elephant – for you are taking your distances by the time it takes in passing over the ground – and observe the time and your next angle on arriving at the object you had formerly chosen. This constant observation, continuing from dawn to dark with one interval for refreshment, is a great strain upon one's attention, and when joined with the necessity of taking heights from the aneroid barometer, and temperatures from the thermometer at every change of level, makes one thankful for a halt at the foot or summit of a mountain pass, where one has to check the height by the boiling-point thermometer.[25]

The topographical brigades did not use elephants as much as the explorers did, but their methods and instruments were comparable. Hallett describes here his techniques for recording an itinerary, from which could be produced a sketch map, and then a topographical map. Distances were measured in terms of the time spent covering them, for which a regular walking pace was essential. Whenever the road changed direction, the new angle was measured. For each measurement, the time and altitude were jotted down, and everything visible on either side of the road was noted and drawn.

Among the instruments used, Hallett mentions the prismatic compass, the aneroid barometer and, crucially, the thermometer. There seem to have been few technical advances over the period, for thirty years later MaCarthy could still describe the aneroid barometer in the following terms:

> In our difficulties, the little aneroid showed what a useful companion it could be, for, if one had any notion of the height of the mountain one was in search of, this valuable little instrument saved a good many moments of doubt and anxiety.[26]

25 Holt S. Hallett, *A thousand Miles on an elephant in the Shan States*, Edinburgh, W. Blackwood and Sons, 1890, p. 9.

26 J. McCarthy, *Surveying and exploring in Siam …*, *op. cit.*, p. 131.

When the surveys had to be secret, the scientific procedures were often complex, although the same methods were used. The discretion of the pandits who operated during the Great Game was the key to their success, as in the case of René Caillé (1799–1838), who managed to enter Timbuktu disguised as a muslim. However, there are few such examples from the Indochinese Peninsula, where topographers rarely needed to be concealed. An exception was Lieutenant George Younghusband, on an assignment in the Shan State of Kengtung in 1887, when he joined a caravan of Yunnan traders to complete the last part of the route, because his interpreter had run out on him.[27] He managed to carry out surveys unobserved, by dismounting every mile, as soon as the caravan was out of sight, and measuring the angle. The Indian topographers from the Survey of India Department learnt to handle the European instruments, and they worked on Burmese soil even more discreetly than the British, with a high degree of autonomy.

In 1886–87, the Marathi topographer Bapu Jadu accompanied General Gordon's column along the Kubo Valley and the River Chindwin, as part of a detachment of topographers headed by Woodthorpe, who were surveying in the north of Upper Burma. In January 1887, he joined groups of refugees to carry out a survey of the river he had travelled along, and so managed to complete the geographical information already gathered in the territory. For the routes and for traversing, he used an angle-measuring instrument:

> This consists of a small telescope mounted above a prismatic compass, the whole fitting on a light stand. […] A table in the observer's note-book gives the distances corresponding to the angles subtended by the rod, and the compass gives the direction.[28]

This instrument, which Woodthorpe and Bapu Jadu had already used on the Gilgit Mission (1885–1886), won Woodthorpe's admiration, because with it one could survey 18 to 20 miles per day, and together with Bapu Jadu – who was thus himself no stranger to this instrument – they had managed to carry out some 800 miles of traversing.

Yet not everyone was so enthusiastic about this instrument when compared with the one used by the French. A note of 21 February 1895 by Captain Walker

27 BL, IOR, L/PS/20/162, The Trans-Salwin Shan State of Kiang Tung, by Lieutenant George John Younghusband, Calcutta, 1888, Preface, p. iv.

28 Robert Gossett Woodthorpe, "Explorations on the Chindwin River, Upper Burma", *Proceedings of the RGS*, XI, 1889, p. 201.

from the Franco-British Border Commission, on Monsieur Thomassin's equipment, observed that:

> His traversing was done with "Colonel Peigné alidade compass", a clever invention superseding the ordinary prismatic compass in all the French *polytechniques* and used by French explorers for all reconnaissance work and rapid surveys.[29]

Comparing French and British performances was thus wholly acceptable, for although the two camps used the same techniques, their instruments differed. The superiority of "Colonel Peigné's alidade", invented in 1873, is highlighted here, and the first thing that strikes Captain Walker in the French camp, on the same expedition, is the contrast between the almost total absence of tents for the men and "A huge red and white umbrella (which was capable of affording shelter to at least one person), for Captain Rivière's theodolite" (4 January 1895).[30] In this particular case, but also more generally in Indochina where conditions were extreme, the French seemed to lavish more attention on their instruments than on their men. Yet although the former are essential for measuring, the latter have to observe and select what they want to show of the territory they are charting. At the end of the day, it is the topographer's powers of observation that determine the representation produced, rather than how the scientific instruments are transported, or the other constraints on the expedition as a whole.

Thus the surveys were dependent on the itineraries taken, and many areas were consequently neglected, or measured with much less precision. Spaces which could not be measured or observed scientifically were left as blanks on the map, contributing to the overall sense of vagueness. The conclusion drawn by Malte-Brun for the whole of the Indochinese Peninsula in 1810, when only the coasts were known, still applied to certain circumscribed areas: at the time, the interior was the subject of "vast conjectures."[31] The spaces with the densest vegetation and the highest mountains remained inaccessible to the European map-makers, alongside areas being gradually charted by the topographers. The topographical coverage was simply the sum of the individual routes taken, and only made sense in relation to all the others. Although the topographers scoured the territory, taking notes and measurements, to be compiled with

29 BL, IOR, L/PS/20/83, Report on the Anglo-French Buffer State Commission, 1894–1895, Captain Walker, Rangoon, November 1895, p. 22.

30 *Ibid.*, p. 11.

31 C. Malte-Brun, *Précis de la géographie universelle, op. cit.*, p. 145 (in the 5th edition of 1845).

others and result in a map, the latter's quality was dependent, in turn, on the insertion of all these partial geographical representations into a geodetic control network, which itself had to be measured.

2.2 *Measuring the Earth*

The aim of geodesy, etymologically "the division of the Earth", is to provide reference points on a territory in relation to which the topographical work can be situated. The method used for the *Carte de la France* took on a different meaning when redeployed in a colonial context. For as soon as the different triangulation networks had been linked up, the whole could be correlated with, and compared to, the territory of the home country. We shall examine how this work was performed in the colonies, and to what extent the practices were different from those of the metropole.[32]

Triangulation involves creating a network of triangles whose vertices are interdependent. First, a line is measured with a ruler or a length of steel wire, and the positions of its two end-points are calculated astronomically. This forms the baseline of the triangle. The angles of the triangle are then calculated by taking up a position at each end of the baseline, and establishing a third point, defined by a feature in the landscape – the top of a hill, a tree, a pagoda, etc. This third point can also be built for the purpose, and the sources mention geodetic "beacons", "marks" and "signals". To measure the triangle's angles, and the difference in height of the terrain between the points of the triangle, a theodolite or a repeating circle is used. Once the length of the baseline is known, and the value of the two angles, one can calculate the lengths of the two other sides through trigonometry. The operation can then begin again, by taking one of the sides of the triangle as a base for another triangle, and so on. The measurements result in a primary network of triangles. A second, third, etc., network can be built on the first one, each level being increasingly precise, with an increasing number of points, to ensure the accuracy of topographical calculations.

Following Delambre and Méchain's first triangulation, in 1792, to measure the Dunkerque-Barcelona meridian arc, geodetic surveys were carried out on several continents, including the Indochinese Peninsula, with a view to measuring the Earth.[33] In 1889, on a mission to triangulate Upper Burma, Eccles

32 M. Schiavon, *Itinéraires de la précision. Géodésiens, artilleurs, savants et fabricants d'instruments de précision en France, 1870–1930*, *op. cit.*, 2014.

33 See Id., "Les officiers géodésiens du Service géographique de l'armée et la mesure de l'arc de méridien de Quito (1901–1906)", *Histoire & mesure*, XXI/2, 2006, p. 55–94, or Chapter 2 of M. Schiavon, *Itinéraires de la précision ...*, *op. cit.*

explained how the instrument he was using, the new 12" Troughton and Simms theodolite, worked.

> It has two microscopes 180° apart, and is furnished with micrometer eye-pieces on the horizontal circle as well as a third index microscope, a similar system being also attached to the vertical circle. [...] The performance of the instrument on the whole was good: at first it gave me some trouble, but thus was the result, I think, of my own inexperience and of the unsteadiness of the signals due to atmospheric causes, rather than of the instrumental defects. The only real defect I found in the instrument was, that the horizontal clamp sometimes got loosened by the motion of the alidade, and this has since been remedied.[34]

He recommended using the instrument in a particular way, namely multiplying the number of measurements taken with the vertical circle in order to reduce the margin of error. This shows that fieldwork could invent more precise methods of observation, and that the geodetic surveyors adapted to the conditions of their work. His approach foreshadowed the repeating circle, which precisely aimed to iron out error by taking the same measurements several times. When Captain Turner began a new first-order network, the *Great Salween Series*, in 1900, he likewise used the 12" Troughton and Simms theodolite, applying Eccles's method and measuring on 12 zeros, with 4 to 8 observations per zero.[35] He had to adapt to the nature of the terrain, which is why he began the framework with a quadrilateral, because the hills were too high for him to find adequate landmarks for a triangle. It was too cloudy in daytime for accurate measurements of vertical angles, so he took the readings at night. Combining knowledge and experience, he thus managed to get round the natural obstacles. The following season, fog and atmospheric errors plagued his measurements.[36] In November 1902, for the third season on this fundamental framework, the geodetic brigade sailed from Calcutta to Rangoon, then went by train to Sagaing, and from there were dispatched to Bhamo to begin building the stations for the principal series.[37] Meanwhile, Capitain Wood and

34 BL, IOR, V/24/3988, Report on the operations of the Survey of India, administered under the Governement of India, during 1889–1890, Calcutta, 1891, p. i.

35 BL, IOR, V/24/4041 Extracts from narrative reports of the Survey of India for the season 1900–1901, Calcutta, 1903.

36 BL, IOR, V/24/4042 Extracts from narrative reports of the Survey of India for the season 1901–1902, Calcutta, 1904.

37 BL, IOR, V/24/4043 Extracts from narrative reports of the Survey of India for the season 1902–1903, Calcutta, 1905.

Lieutenant Tillard simultaneously observed vertical angles from two separate stations, using a theodolite, in order to measure the difference in height between them. In addition to the 12" theodolite, the British used a new instrument here for the very first time, the Borda repeating zero circle, which they reported as giving very satisfactory results. Although French surveyors used the repeating circle extensively in other circumstances, for instance in Algeria, they preferred an instrument that was less accurate, but lighter, and easier to handle, for the triangulation of the mountainous terrain in Vietnam. After 1901, they settled for the field theodolite.[38]

In the course of the Franco-British Commission, it evolved that the British already used triangulation, whereas the French still based their topographical surveys on points whose positions were determined by astronomy. This lag on the part of the French applied only to the Indochinese Peninsula, and was most probably due to the fact that the British only had to extend a network already marked out in India, whereas the French had to create a geodetic system from scratch. Noting the differences in method, however, Colonel Woodthorpe invited Capitain Rivière and Lieutenant Seauve, the two French geodetic surveyors, to come with him to look at a triangulation station not far from their camp, along with Lieutenant Ryder and Doctor Lloyd. As Pavie reports it, it was a momentous discovery:

> The means employed by thc Colonel in the parts of the country where it was possible to take measurements consisted of laying bare prominent summits, and leaving only a few trees, so that the point could be seen from afar. Near these trees a signal was set up, in the form of a very high cone, made of bamboos split in such a way that the white was turned outwards, making them visible from very far away. They could last for a long time if not destroyed by accident or acts of malice.[39]

Not only does this quotation illustrate the training provided by the British for the French to overcome discrepancies in national surveying practices, it also broaches the strategies employed in building durable geodetic markers, which we shall look at next.

---

38 Édouard de Martonne, Jean Martin (eds), *La carte de l'empire colonial français* ..., *op. cit.*, p. 109.

39 A. Pavie, *Mission Pavie en Indo-Chine, 1879–1895*, *op. cit.*, p. 342–344.

2.3 *The Construction and Preservation of Geodetic Signals*

Geodetic surveyors tried hard to find natural features to serve as geodetic signals. In mountainous regions, the tops of trees or of pagodas were often used (which is doubtless why sketch maps sometimes mentioned a "remarkable tree", with the triangulation work in mind). Getting to the highest point was no easy task, and topographers' reports are full of dramatic descriptions of scaling heights and receding peaks:

> Over nearly the whole of Indo-China the finding of the highest point of a hill is not so easy as it looks. The forest is very heavy, and not even a glimpse of the surrounding country is to be had, unless one climb up a high tree at the summit of a lofty peak. [...] Often we found we had not reached the highest point, there being a higher in another direction, perhaps 2 miles off. Then hacking through jungle undergrowth, down and up steep slopes, wandering about and hesitating at half a dozen likely points, at last we gained the top.[40]

Tree-trunks were favourite landmarks, but the list provided by the Head of the *Service géographique* in 1902 shows a whole range of other solutions: the wall of a pagoda, a church or a fortification, a grave, a mass of stones, a small brick building – anything that could possibly be used was noted down [Figure 16].

Then a geodetic signal had to be erected, which was sometimes a complex process, especially in the delta regions where the land was predominantly flat. In 1899, for example, the construction of a high wooden structure on the dyke between Hanoi and Sơn Tây, on which a pylon with a geodetic signal was to be placed, caused two months of upheaval on the roads.[41] The French Resident of Sơn Tây was not against the work, but asked for the structure to be altered: "to achieve the desired result, it would be sufficient to alter the form of the structure by making an opening in it for rickshaws and pedestrians. I think it will be easy for the Artillery Department to make this change."[42] The geodetic stations thus altered the landscape, especially the signals of the primary and secondary networks, which were designed to last. The third-order network was a large sphere roughly one metre in diameter, in wood or in whitewashed stone from Biên Hòa, attached to the end of a long bamboo, and hoisted up to the top of

---

40 J. McCarthy, *Surveying and Exploring in Siam*, *op. cit.*, p. 132.

41 ANV1, RST 23607, Lettre du général de division Borgnies Desbordes, commandant en chef des troupes, au résident supérieur du Tonkin, Hanoi, 13 September 1899.

42 ANV1, RST 23607, Lettre du résident à Sontay au résident supérieur du Tonkin, Sontay, 27 September 1899.

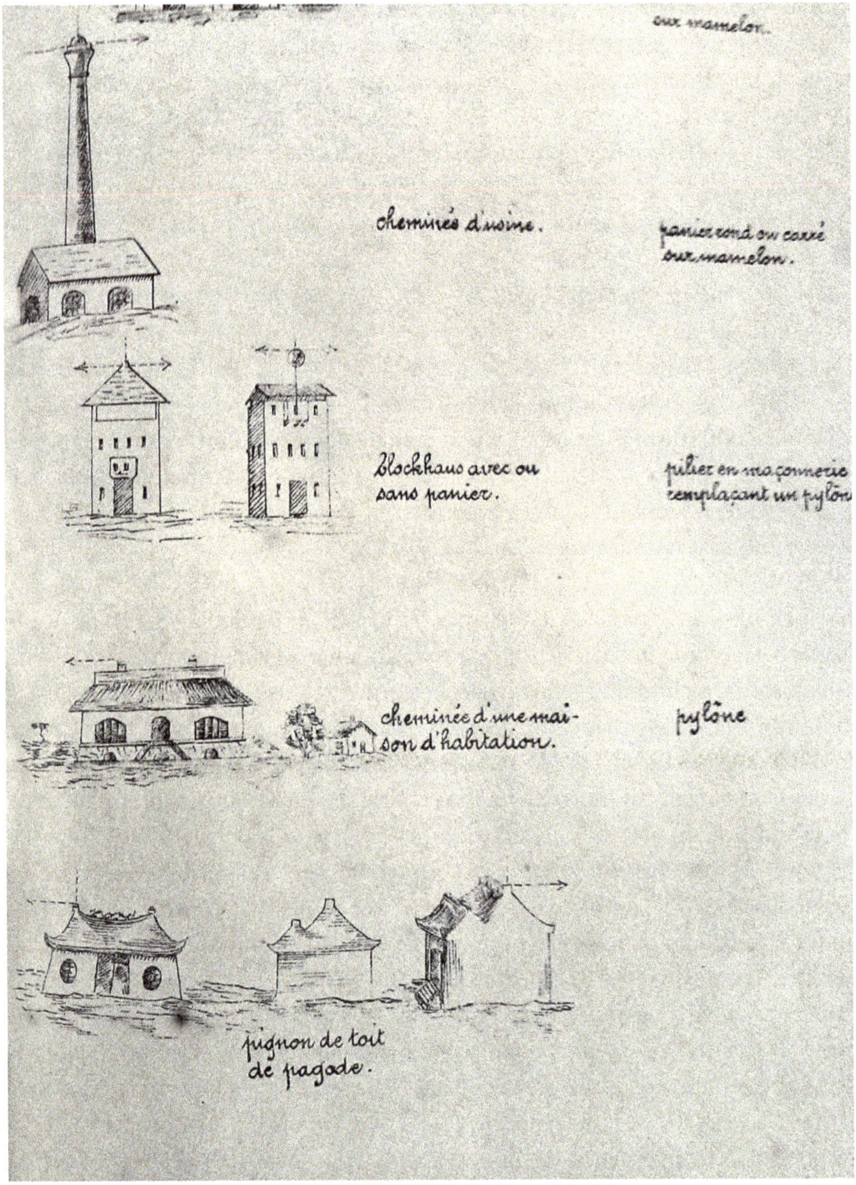

FIGURE 16 List of geodetic signals (detail)
SOURCE: ANV1, RST 73507, *Triangulation du delta* (1er fascicule), Hanoi, le 7 janvier 1902, par le lieutenant-colonel chef du Service géographique, Lubanski, p. 26

a tree. Alongside the signal was a geodetic mark set in tree trunks, small brick structures, or the walls of a building. In the French case, the mark consisted of a zinc plate bearing the following inscription: "*Service géographique de l'Indo-Chine*; triangulation of the delta; mark no…." [Figures 17 and 18]. Although these signals and marks were only required by the *Service géographique* for a couple of months, for mapping the terrain, they could be useful for other work later on, such as cadastral surveys or for planning the routes of roads or canals. This is why the question of their permanence arose.

The third-order signals were generally destroyed, either by the wind and the rain, or because they were plundered by the local population for their materials. For the authorities in Hanoi, this looked like acts of resistance, but the Resident of Bắc Ninh did not interpret it in this way, suggesting, rather, that the local population did not understand what the marks were for and the importance of preserving them in a good state.[43] Where the signals were destroyed with malicious intent, this seemed to occur only infrequently, but each rare incident gave rise to much toing and froing within the colonial administration. In April 1900, for instance, a geodetic officer from the *Service géographique* reported that a marker near Cầu Giấy had been taken and moved far from its original position.[44] Everyone assigned the responsibility for its conservation to someone else. In the ordinance Paul Doumer issued on 17 September 1900, co-signed by Guillemoto, his Director General of Public Works, he decreed that the costs for installing geodetic stations were to be borne by the *Service géographique*, but that their surveillance and maintenance were the responsibility of the Public Works Department. There were also attempts to make the local authorities responsible for the stations in order to reduce the amount of damage and theft. The Resident of Phủ Lý suggested that the local mayor and some dignitaries be present at the inauguration, and that they sign a document listing the sanctions incurred if the signal were destroyed.[45] A relatively simple operation could thus have manifold consequences.

The empirical methods employed – going up and down the land, surveying and observing the territory – ensured the scientificity of the topographers' results. They used much the same techniques as in Europe, but adapted these as necessary to the specificities of the Indochinese terrain. The fact that these

43 ANVI, RST 73614, Lettre du résident de Bac Ninh, M. Mahé, au résident supérieur du Tonkin, 5 December 1899.

44 This is where Francis Garnier in 1873, then Henri Rivière in 1883, was assassinated. Today it is a western suburb of Hanoi.

45 ANVI, RST 73614, Lettre du résident de Phu Ly au résident supérieur du Tonkin, 20 December 1900.

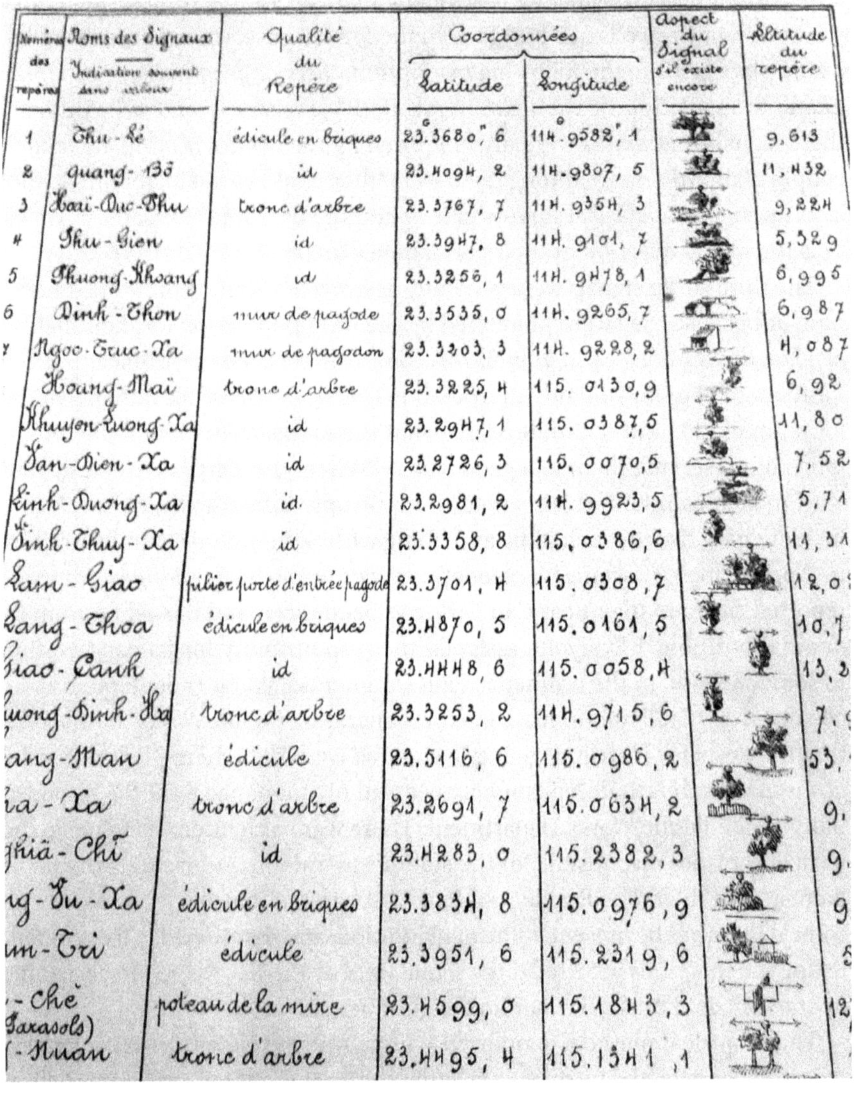

Numéros des repères	Noms des Signaux (Indication souvent sans valeur)	Qualité du Repère	Coordonnées: Latitude	Coordonnées: Longitude	Aspect du Signal s'il existe encore	Altitude du repère
1	Thu-Lê	édicule en briques	23°.3680″,6	114°.9582″,1		9,613
2	quang-Bô	id	23.4094,2	114.9807,5		11,432
3	Hoai-Duc-Phu	tronc d'arbre	23.3767,7	114.9354,3		9,224
4	Phu-Gien	id	23.3947,8	114.9101,7		5,329
5	Phuong-Khoang	id	23.3256,1	114.9478,1		6,995
6	Dinh-Thon	mur de pagode	23.3535,0	114.9265,7		6,987
7	Ngoc-Truc-Xa	mur de pagodon	23.3303,3	114.9228,2		4,087
	Hoang-Mai	tronc d'arbre	23.3225,4	115.0130,9		6,92
	Khuyen-Luong-Xa	id	23.2947,1	115.0387,5		11,80
	Yan-Dien-Xa	id	23.2726,3	115.0070,5		7,52
	Kinh-Duong-Xa	id	23.2981,2	114.9923,5		5,71
	Dinh-Thuy-Xa	id	23.3358,8	115.0386,6		11,31
	Lam-Giao	pilier porte d'entrée pagode	23.3701,4	115.0308,7		12,0
	Lang-Thoa	édicule en briques	23.4870,5	115.0161,5		10,7
	Giao-Canh	id	23.4448,6	115.0058,4		13,3
	Luong-Dinh-Xa	tronc d'arbre	23.3253,2	114.9715,6		7,
	ang-Man	édicule	23.5116,6	115.0986,2		53,
	ia-Xa	tronc d'arbre	23.2691,7	115.0634,2		9,
	hiã-Chi	id	23.4283,0	115.2382,3		9,
	ng-Su-Xa	édicule en Briques	23.3834,8	115.0976,9		9
	un-Tri	édicule	23.3951,6	115.2319,6		5
	-Chè (Parasols)	poteau de la mire	23.4599,0	115.1843,3		12
	-Nuan	tronc d'arbre	23.4495,4	115.1341,1		

FIGURE 17 Geodetic marks. An inventory of geodetic triangulation marks provides all the requisite information

SOURCE: ANVI, RST 73507, *Triangulation du delta* (1er fascicule), Hanoi, le 7 janvier 1902, par le lieutenant-colonel chef du Service géographique, Lubanski, p. 9

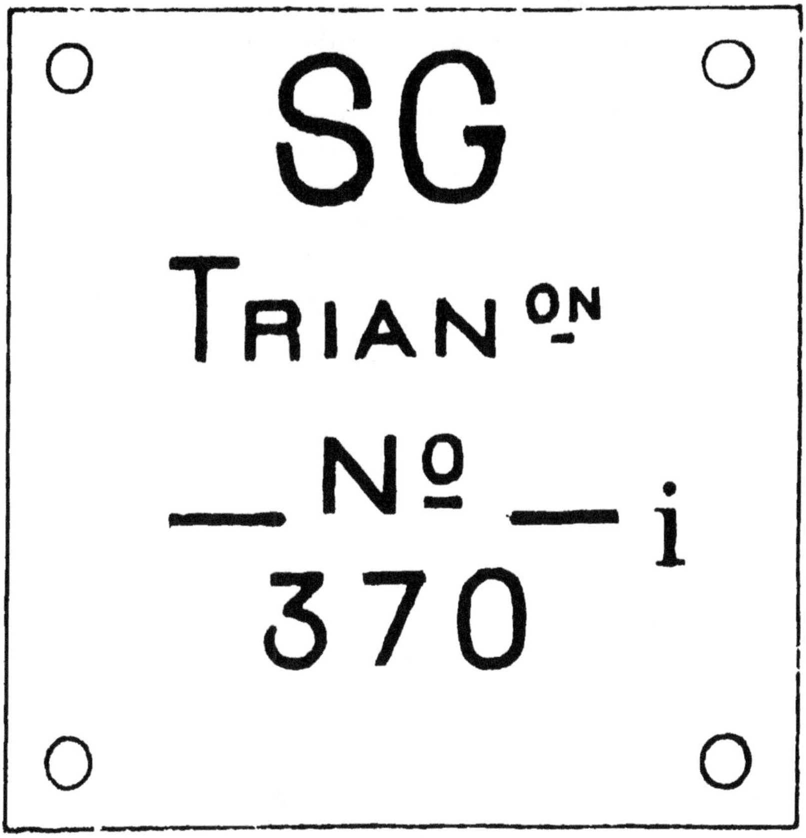

FIGURE 18 Geodetic marks engraved on a zinc plate

surveys took place in a colonial context does not seem to have occasioned radically different scientific techniques. However, the unfamiliarity of the territory to the geographer's trained eye could present difficulties for the mapping project.

3 Human Settlements and Dwellings: A Map-Maker's Headache

Human elements such as dwellings, and the distribution of populations across territories, were some of the best represented features on maps, but the topographers on the Indochinese Peninsula had difficulty mapping complex ways of life and identifying unfamiliar spatial patterns.

3.1 *The Problem of Mapping Ethnic Groups*

The early European explorers described the territory they were discovering in terms of the places they passed through, using the existing names of villages, rivers, and so forth. From these we are able to reconstruct their itineraries. However, in the mountainous regions which the explorers gradually reached, the diverse ethnic groups provided other ways of identifying localities, and even of naming places. Thus in 1898, the mining engineer Jean-Marc Bel mentioned "the central mountainous range, which could be called the *Halangs range*", from "the name of the group that lives there."[46] Here the human factor takes precedence, and the explorers describe their journey in terms of the populations they encounter. Just like the French cartographers in Algeria, who replaced the traditional mention of place names with ethnographic data, so those who penetrated the mountains of the Indochinese Peninsula realised how profoundly the territory was structured around human settlements.[47] However, they realised only slowly how very complex the ethnic groupings were.

Initially, explorers and cartographers had simply given general names to the populations in the high mountain areas, about whom no precise knowledge was available. On Yule's map, for example, many names appear in red capitals [see Figure 8], but often with a confusing note, such as "Unexplored Mass of Hills inhabited by a great many tribes of Indo-Chinese type in low stages of civilisation."[48] Garnier's map had similar features: the areas inhabited by ethnic groups were first left blank, but sometimes with a mention such as "Unexplored country", "Savages", or "*Moi*" (a generic Vietnamese term for the same notion). From this point of view, the Pavie Map did not add much to what was already known, but the expedition reports often addressed the complexity and diversity of the local populations. Exploring the Lai Châu region in 1890–1891, Pavie noted:

> Here is a detail which shows the population's variety in the area we crossed: in the 31 places where we stopped to spend the night, 5 times it was with some Thai from Lai, once with Northern Thai, 7 times with

46 BSG, septième série, XIX, 1898, p. 279.

47 Hélène Blais, "Les représentations cartographiques du territoire algérien au moment de la conquête. Le cas de la carte des officiers d'état-major (1830–1870)", in P. Singaravélou (ed.), *L'empire des géographes. Géographie, exploration et colonisation, XIXe–XXe siècle, op. cit.*, p. 124–134.

48 *New Map of Burma and the Regions Adjacent*, by Captain Henry Yule, 1857; MEP, K MYA 3.

> various Kha, twice with Yao, 6 times with Chinese, and 8 times with Lu peoples.[49]

Thus the Europeans were increasingly aware of the ethnic variety around them, even if the mention of "various Kha" points to the limits of their ethnographic knowledge, since "Kha" is a generic Laotian term meaning "savage" or "slave", and there is no further detail on the "various".

3.2 *A Terrain Irreducible to European Standards*

Identifying the ethnic groups was not the sole headache for the Europeans. The groups' distribution across the territory, and their population centres, received contradictory representations, showing the cartographers' difficulty in understanding spatial organisations so different from their own. However, place names helped the expeditions find their bearings on the ground, and, integrated into the resulting maps, showed how the territory, particularly its urban areas, was configured. Even the most modest sketch map or survey made during a reconnaissance included the names of villages and towns, or else had a note on the drawing, if the place name was not known. The early cartographers thus noted the presence of villages or dwellings, just as they did the vegetation and the locality's relief, as an integral part of the landscape and as a fundamental element of the map.

Sometimes the issue of how to represent this alien environment, its modes of settlement and spatial organisation, was discussed among the Europeans, particularly in the case of the very sparsely populated mountain regions.[50] The question of what a village is, and what one can call a village, in fact sparked a quarrel within the *Société de géographie* in 1881, and a similarly stormy debate erupted in the Royal Geographical Society in 1888, to which McCarthy contributed with a remark on the Siamese term for "village", *bān*, which could refer equally, he maintained, to a bamboo hut or to thirty houses – a position which understandably made consensus difficult.[51]

---

49 A. Pavie, *Mission Pavie en Indo-Chine, 1879–1895*, *op. cit.*, p. 125.

50 On the geographical distribution of settlements in the northern Indochinese Peninsula, see Rodolphe de Koninck, *L'Asie du Sud-Est*, Paris, Armand Colin, 2012; Michel Foucher (ed.), *Asies nouvelles*, Paris, Belin, 2002; for the rhythms of settlement, see Jean Michaud, Jan Ovesen, *Turbulent Times and Enduring Peoples: Mountain Minorities in the South-East Asian Massif*, Richmond, Routledge, 2000; Jean Michaud, *Historical Dictionary of the Peoples of the Southeast Asian Massif*, Lanham/Toronto/Oxford, Scarecrow Press, 2006.

51 BSG, sixième série, XIX, janvier-juin 1880; BSG, sixième série, XX, juillet-décembre 1880; BSG, septième série, II, juillet-décembre 1881; *Proceedings of the Royal Geographic Society*, nouvelle série, 10, March 1888, p. 117–134.

Both the British and the French sought to force the urban realities of India and the Indochinese Peninsula into the categories they already knew. For the census carried out in 1872, the British decided arbitrarily that a town was a place with more than 5,000 inhabitants, and this was enshrined in the 1876 Municipalities Act. This purely numerical approach seemed inadequate to the superintendent responsible for the census in Bengal, J. A. Bourdillon. He maintained, in 1881, that towns of less than 5,000 inhabitants could have the social characteristics of a town, while those of more than 5,000 inhabitants could well be agricultural centres lacking the social diversity characteristic of towns.

In the light of these problems, it was going to be no easy task to represent human settlements according to prevailing cartographic norms. In 1901, the Head of the *Service géographique de l'Indochine* issued new rules, recommending that stone or brick dwellings be visually distinguished from straw huts, and that stricter standards be applied to how place names were written (corresponding to the size of the place).[52] These changes reflected the colonial officials' vain attempts to depict how the habitations of Tonkin were hierarchised. After all, what objectively distinguished an area's administrative centre from a town or simply an outpost? Should the criterion be the number of inhabitants, the number of houses, or the presence of local authorities, which was a sign of the place's administrative importance? Certain criteria were clear (for example, the presence of military personnel defined an outpost), but others were less so. A "large town" in one province might not pass as such in another, and a military station, which was always represented by the French flag, might have two soldiers or several dozen. Not only were different categories used by different colonial civil servants in Tonkin, but they did not always refer to the same realities on the ground.

The terms that recurrently appeared on the maps, in italics and in brackets – town, village, abandoned town, ruins – point to the fact that the standard European cartographic categories were incapable of capturing the realities on the ground, and needed to be supplemented by written notes. And apart from these issues of categorisation, another problem was raised by temporary villages: since maps precisely attempt to fix or even freeze the geographical reality of a territory at a given time, how represent elements in flux?

52 ANV1, RST 73507, Carte de Phu Vinh au 1:25 000, édition de juillet 1901, dressé, héliogravé et publié par le Service géographique de l'Indo-Chine, étant chef de service, commandant Lubanski.

3.3 *Abandoned, Moved and Renamed Villages: The Joys and Woes of Cartography*

There are circumstantial reasons for territorial changes, such as the arrival of colonisers, the marking of a border, or a local war. But there are also structural reasons, related to traditions and customs. In the mountain regions, the majority of the populations were nomadic or semi-nomadic, and practised a slash and burn agriculture.[53] The clearing, and enrichment, of the land by setting it alight went together with a mobile population, and this was visible on the maps. The Europeans gradually become aware of this major distinction, between nomadic and sedentary populations in the Indochinese Peninsula. In 1889, Lieutenant Bythell, of the Survey of India Department, noted that:

> Another peculiarity is that from the Tao village eastwards, all the villages are known by a geographical name, whereas west of Tao the villages are named after the headman or chief. Probably this is due to the temporary nature of the villages to the west and permanence of those to the east.[54]

The lifespan of a village was often a key element in its description. This suggests that the Europeans closely observed the different ethnic groups, and were interested in their different lifestyles in this respect, adapting their methods to the reality of the terrain and attempting to understand the logics at work.[55] The frequency with which the different groups moved, and the form of their villages, influenced how the territory was represented: it is not unusual to read "(abandoned)" on French maps, next to a place name, when a village moved, or several identical place names may be found quite near each other, because the new village, built a few miles further on, was called by the same name. On a 1:120,000 colour sketch map (relief in bistre, roads in red, and rivers in blue) "made by the chief guard Quénel, commander of the Ai Lao military outpost", the movements of the population were represented, with the following

---

53 Charles Robequain and Pierre Gourou studied this method, which came to characterise tropical geography, and was used to explain the social and economic backwardness of the tropical countries. Charles Robequain, *L'Indochine française*, Paris, Armand Colin, 1935; Pierre Gourou, *Les pays tropicaux. Principes d'une géographie humaine et économique*, Paris, PUF, 1947; for an overview of the issue, Michel Bruneau, "Pierre Gourou (1900–1999)", *L'Homme. Revue française d'anthropologie*, 153, 2000, p. 7–26; Jacques Barrau, "Culture itinérante, culture sur brûlis, culture nomade, écobuage ou essartage?", *Études rurales*, 45, 1972.

54 BL, IOR, V/24/3988, Report on the Survey Operations with the Chittagong column, Chin-Lushai expedition force, by Lieutenant W. J. Bythell, season 1889–1890, p. lx.

55 BL, L/PS/20/83, Report on the Anglo-French Buffer State Commission, 1894–1895, by Captain Walker, p. 22–23.

explanation: "Ca-Dap, an evacuated Phutai village, the inhabitants built a new village in the mountains near Ai Lao (September 1892)", and not far from there, "a village built in September 1892 by the Phutai who had emigrated from Ho Sang and which they named after their old village".[56] This was depicted roughly, with red hatchings to show the village's overall surface area, and the mention of the number of houses written in – between three and twenty-five.

Although representing the number of dwellings in cartographic form seemed impossible, they still had to be counted, and written text was used for this. On many of the maps commissioned by the French administrators from the local elites in Tonkin, most probably dating from the 1890s, the number of houses, or *cagnas*, was added in pencil. The numbers varied considerably, suggesting that the definition of a "village" could not be based on the number of dwellings it comprised. On the map of the *huyện* (district) of Nguyên Bình, in the province of Cao Bằng, 10 *cagnas* were noted for a village, whereas on a map of the province of Phú Thọ, one "village" numbered 200 *cagnas*.[57] On the British side, only a few cartographic sources exist, but the reports from explorers and military personnel usually mentioned the number of houses in the villages, while the gazette at the end of the work, by listing the names of the places passed through, and their main features, gave some idea of the village's surface area, and the number of houses or inhabitants. For the Europeans, a village's size clearly seems to have been a key to understanding and defining the territory, even if they were unable to calculate it with precision.

The representation of organised inhabited areas is a fundamental feature of topographic maps, and the Indochinese Peninsula presented a particular challenge in this respect. All the categories applied proved inadequate, and even the definition of "village", a basic administrative and geographical unit, could not be stabilised. The Europeans put their best efforts into understanding all aspects of local spatial practices, in order to provide an up-to-date cartographic image of the territory, which had to accommodate the mobility of the populations. In the mountain areas, some villages existed only temporarily. The changes in place names gave rise to much confusion, which is visible on the maps, and conveys how alien this terrain appeared to the Europeans.

---

56 ANOM, GGI 14481, "Croquis fait par le garde principal Quénel commandant le poste d'Ai Lao," 1:120 000. 48. ANV1, Fonds du Service géographique, 371–09, Carte du huyện de Nguyên Bình, province de Cao Bằng; *ibid.*, 245, Carte de la province de Phú Thọ.

57 ANV1, Fonds du Service géographique, 371–09, Carte du huyện de Nguyên Bình, province de Cao Bằng; *ibid.*, 245, Carte de la province de Phú Thọ.

•••

In this enterprise of measuring and surveying the territory, calculating the geodetic coordinates and marking itineraries, the Europeans held to the methods they considered the most scientific: empirical observation and measurement. Their practices and instruments were the same as in the home country, which is why the notion of a "colonial science" must be re-examined. Did the first term influence the second? In other words, was the colonial context determinant on the programme of triangulation and topographical mapping?

Geodetic surveys were complex wherever they were carried out: in the Algerian desert it was difficult to find salient marks, whereas meridian arc measurement in the Andes encountered. The physical conditions on the Indochinese Peninsula, whether under colonial rule or not, hindered the scientific work, and the project of mapping the whole territory. Although the topographic party's security was of course affected by the context of European conquest and domination, it was, more generally, threatened by the hostility of local populations to foreigners. There are many examples outside of the colonial context which illustrate the dire consequences of misunderstandings: even Delambre and Méchain, measuring the distance between Dunkerque and Barcelona, encountered life-threatening situations when taken for Royalists or spies.

Lastly, as we have seen, the unfamiliarity of the territory made it difficult to map. Although this is often the case in colonial situations, it is not specific to them, since the land's alien nature is due, already, to its distance, its tropical climate, and so forth. The semi-nomadism of many ethnic groups in the north of the Peninsula, although found in other tropical countries, and in French colonies such as Madagascar, was unknown in France (and was also different from the nomadism practiced in the Saharan desert). It presented a major obstacle to any precise cartographic representation because the mobility of the villages, their changed names, and the lack of adequate definitions of types of settlement, prevented any faithful, standardised, image from being produced. Although observation and measurement remained the core tools of the topographers, they also consulted local people, to confirm their own findings and to improve their understanding of the situation. In the next chapter we shall explore who these local sources were, how the inquiries were made, and how the understanding it afforded was incorporated into the Europeans' constituted stock of knowledge.

CHAPTER 5

Consulting the Local Population

Whereas the principle of observation dominated European cartography, it could not always be applied in the northern Indochinese Peninsula because the cartographers were unable to have access to the territory. Instead of measuring it themselves they had to rely on second-hand information gathered from the local population for what they could not observe with the naked eye. The findings, for which, despite its inadequacy, the expression "indigenous knowledge" has been coined,[1] pertained to a precise area of territory, for large-scale mapping. We should bear in mind that "knowledge", understood as a body of information acquired in contact with a given reality, should be distinguished from "scientific knowledge", knowledge whose conditions can be formulated as laws, and whose findings can be verified by experimental methods. What role, then, did local sources play in the scientific cartographic project of the Europeans?

Relatedly, we shall ask how different interlocutors reacted to the Europeans' inquiries, how the first encounter was perceived, and what consequences this had for the cartographic projects. Historically speaking, the first relations between Westerners and local populations occurred in the context of exploration and topographic work, which took place before any form of colonisation, and thus outside of formal relations of domination.[2] During the period of conquest, however, the inquiries took place within the context of military reconnaissance expeditions that foreshadowed colonial annexation. Did this dominant position make the Europeans' information-gathering more difficult than had they been simple strangers? Who were their interlocutors, how was the process organised, and what different forms did it take?

The qualitative range of the information gathered can be attributed sometimes to the Europeans, sometimes to the local population, and sometimes to the contingent conditions of the meeting or the specificities of the Indochinese context.[3] Naming was a constant, and unresolved, problem throughout the

1 For an overview of this subject, see David Wade Chambers, Richard Gillespie, "Locality in the History of Science: Colonial Science, Technoscience, and Indigenous Knowledge", *Osiris* 15, no. 1 (1 January 2000), p. 221–40.

2 See Ch. 1.

3 On local populations' scope of action in the colonial encounter, see Gayatri Chakravorty Spivak, "Can the Subaltern speak?", in *Marxism and the Interpretation of Culture*, London, Macmillan, 1988, p. 24–28.

 | DOI:10.1163/9789004469853_007

period. The tendency, in colonial territories, was either to name things oneself, or to look for local names. Using one's own terms can be understood as a form of appropriation, by giving a name, according to criteria determined unilaterally, in an unfamiliar territory on which one acts. In the Indochinese Peninsula, both the British and the French clearly tried to use local place names, but they had to contend with extremely complex linguistic problems (as had the missionaries before them). The linguistic diversity of the territories, the diglossia of Vietnam, where Chinese was the written language and Vietnamese the spoken one, the Europeans' difficulties of understanding, and the absence of norms for transcription created countless problems.[4] These were present at every stage of the colonisation, from the exploration that preceded it, via the conquest and the boundary demarcations, up to the production of precise topographic maps. Moreover, the name chosen had to correspond to the map's objective: if it was to be consulted by the local population, then the local toponym was used, whereas if it was to allow Europeans to find their location or discuss an issue, then a simplified toponym was adequate. By examining the Europeans' attempts to find the right name – the criteria and the means employed, and the solutions proposed – we shall get a better understanding of the complexity of the reality faced by the topographers.

1 The Role of Intelligence: Local Information Sources during the Conquest

Informers were needed to supply the kind of information required by the topographers to map the unfamiliar territory they were encountering for the first time.[5] The Europeans initiated the process, with questions invariably concerning place names, roads, itineraries and distances. The replies were sometimes satisfactory, but sometimes inadequate, either because the interlocutors did not understand, or because they were reluctant to answer. The style of the

4 "Sometimes we give 'diglossia' the sense of a bilingual situation in which one of the two languages has an inferior socio-political status", Jean Dubois, *Dictionnaire de linguistique*, Paris, Larousse, 1974.

5 This sort of information can mostly be found scattered across travel writings and topographers' reports. The inquiries themselves rarely found their way into the archives. Exceptionally, between 1884 and 1886, the reports of heads of military detachments contain dozens of hand-written loose sheets entitled "Information provided by the Natives". These were records of the inquiries made in the villages as from the very beginning of the conquest. They provide a wealth of information concerning the relations between the French and the local populations, and what the French managed to find out from them. See SHD, 10H6, Dossier 15.

replies also varied, from a gesture indicating a direction, via speech to convey an exact place name, to a hastily executed sketch.[6] A detailed analysis of the logics underlying the sum of geographical information collected daily is essential for understanding how the colonial encounter operated, and what consequences this had for mapping the territory.

1.1 *A Methodical Approach?*

Collecting information was not the preserve of cartographers alone, but also of explorers and reconnaissance officers. In 1887, instructions were issued on this subject to the French troops:

> As a general principle, it is the duty of every officer and deputy officer, whether in transit or at the camp, and whether officially on such a mission or not, to make inquiries, and to gather information on anything that can help him orientate himself in the country; after his troops have pitched tent, he should make inquiries about the route, the neighbouring villages, and the number of men needed to get there, the name of the *huyen*,[7] of the province, etc.[8]

So it was not only the topographic corps that was enjoined to carry out systematic inquiries and to use local information, but military officers as well. However, since no advice was issued about the methods or sources to be used, the officer cartographers had a large degree of autonomy, and responded to the circumstances. The real conditions during the conquest were indeed so difficult that even if there were instructions, they were probably not followed in the field. In the conquest's opening phases, when the troops' progress actually depended on information gathered, the French military seem to have been quite insistent in their requests for information; in both the French and the British case, requests were addressed not only to the military officers, but also to the indigenous local authorities, who were enlisted for assistance. These often mentioned the imminent arrival of the geodetic surveyors and the topographers: "Moreover, I urge you to oblige the indigenous local authorities to supply any information that may be requested by the Officers responsible

---

6 This interweaving of writing, speech and gesture has been convincingly demonstrated, in the case of Africa, by C. Lefebvre, "Itinéraires de sable. Parole, geste et écrit au Soudan central au XIXe siècle", *art. cit.*

7 A *huyện* is a district.

8 SHD, 10H15, Dossier 3, sous-dossier 4, Instructions pour faire suite à l'ordre no 8, du général Brissaud.

for the topographic work."[9] What the topographers needed to know was simple enough: place names and distances, the state of the bridges, the number of rivers to be crossed, and the size of the roads, indeed all the practical details required in order to understand and represent the territory.

As such, they were on the lookout for individuals with a good knowledge of the local area, and they paid particular attention to the itineraries taken by the inhabitants, in order to have the most precise information on the roads, stopping places, and distances. This was fully part of the cartographer's work. During one of the operations with the military against the Chin-Lushai in 1889–1890, one topographer, Stotesbury, carried out the triangulation, while another, Bythell, "Devoted the whole of his time to plane-tabling and acquiring information as to the names of villages, &c., and the courses of the streams."[10] Reports from Border Commissions often described itineraries suggested by local sources.[11] The many revisions to the topographical maps drawn during this period also relied on local information. This situation raises the question of the extent to which the parties really cooperated, and who the cartographers' main sources were. The degree of precision in the answers to the topographers' questions seems to have depended largely on who was available for questioning; despite regular information requests, the results were at the mercy of chance meetings, and of how the fieldwork evolved.

1.2 *Collaborators Determined by Necessity not Choice*

These local inquiries involved a range of different protagonists – the political elites, dignitaries, traders, and ordinary inhabitants. The Europeans generally wanted their interlocutors to be indigenous to the area, but since they were looking for what they could not see with their own eyes, they tended to learn more from travellers and tradespeople.

During the conquest, the topographers' information sources were above all the local dignitaries and the local elites – the mayors of the villages the troops passed through, called *lý trưởng* in Vietnam – or, in the Burmese Shan States, the *Chao Pha*, the name given to local rulers holding a royal title. The first person questioned often indicated another, more knowledgeable person, and so a chain of varied sources developed. The army officer Durand reported that

9 ANV1, RST 37854, Lettre du résident supérieur du Tonkin au général Duchemin, commandant en chef les troupes de l'Indochine, Hanoi, 14 November 1894.

10 BL, IOR, V/24/3988, Report on the operations of the Survey of India, administered under the Government of India, during 1889–1890, Calcutta, 1891, p. 80.

11 BL, IOR, L/PS/20/D101, Report of Intelligence Officer attached to the Southern Party, Burma-Chinese Boundary Commission, 1897–1898, by Captain Evelyn William Norie, Simla, 1898, p. 21–28.

"I went up to the village of Dao-Quan, where I asked the head of the canton for information, and he put me in touch with a former caï[12] of the Annamite regular army, who knows the route very well."[13] Here, the veteran's knowledge and experience of the route seems to guarantee that his information will be precise, and trustworthy, a point which the officer makes in his report.

This case reveals that local elites, although the first to be contacted, could not always provide the information required because they generally only knew the areas under their jurisdiction. This explains the importance of travellers in the chain of information-providers: "All were committed to providing me with correct information, and various villagers were sent for, who had travelled in different parts of the country."[14] Traders and travellers provided useful practical information about areas which the topographers themselves were unable to access.

These particular sources were all the more valuable because the topographers' movements were sometimes restricted, and confined to an authorised zone. Town markets were a very important site of information collection, to further the Europeans' cartographic projects. Security issues were less acute there than on reconnaissance trips, and so inquiries were easier. In H. R. Davies's account of his journey to the Chinese-Shan States in 1894, he points out that his descriptions are based partly on what he observes, and partly on information gleaned from the indigenous populations of Bhamo, Sadon and other places.[15] He added:

> Much more information could be collected in Bhamo by an Intelligence Officer stationed there in the cold weather [...]. A great deal of information might be gathered about this country by an Intelligence Officer at Keng Tung, where there is a large population of Chinese-Shans and where many Chinese and Chinese-Shan traders arrive every cold weather.[16]

Kengtung was thus considered interesting for its mixed population, and especially as a crossroads for trade, traders being a prime information source for the Europeans. Traders travelled up and down the land, or moved between two

12 *Cai* is a Vietnamese term meaning corporal or guard.

13 SHD, 10H6, Dossier 14, Lettre de Durand, Phu Lang Thuong, à M. le lieutenant-colonel commandant supérieur à Bac Ninh, 7 June 1884.

14 Holt S. Hallett, *A Thousand Miles on an Elephant in the Shan States*, Edinburgh, W. Blackwood and Sons, 1890, p. 333.

15 BL, IOR, L/PS/20/94, Report on the part of Yunnan between the Bhamo frontier and the Salween, by Captain H. R. Davies, 1894–1895, Rangoon, September 1894, p. 1.

16 *Ibid.*

points, and were therefore familiar with the territory's routes and nodal points. They were liable to have up-to-date information on distances, and on the state of the roads.

As for who these traders were, those from Yunnan, called "Ho" or "Pathay", appear all over accounts by Western travellers. The British often mentioned their trade caravans of mules and ponies. They were exceptionally experienced in trade, which was expanding rapidly, benefitting the whole region, and for the Europeans, who tried to establish relations with them, they were the preferred information source for understanding the territory. They were Muslims, ethnically Hui, and very influential in these confines, although it is impossible to know whether their religion played a role in the Europeans' predilection for them.[17] In the Early Modern period already, they had a monopoly on the long-distance trade route between China and the northern Indochinese Peninsula.[18] The caravans left the commercial centres of China, especially Dali and Yunnanfu [Kunming] in the winter months between November and April, in order to sell their wares in the markets of the small towns of the frontier regions of the Chinese Empire – particularly in Burma, and somewhat less in Siam.[19] What role did they play in the local society, and how were they perceived? In the Kachin territories, they were treated as foreigners: they spoke a different language, dressed differently, and worshipped different gods.[20] Yet they were also familiar and well-known, because they were part of the local ethno-political systems, many of which paid tribute to the Burmese and Chinese courts. However, they were neither particularly admired, nor considered particularly knowledgeable by the local societies (unlike, for example, the Hausa merchants in Sudan).[21] Captain Walker's account of a trade

---

17 David G. Atwill, *The Chinese Sultanate. Islam, Ethnicity, and the Panthay Rebellion in Southwest China, 1856–1873*, Stanford, Stanford University Press, 2005; Andrew D. W. Forbes, David E. F. Henley, *The Haw. Traders of the Golden Triangle*, Chiang Mai, Asia Film House, 1997; Jianping Wang, *Concord and Conflict. The Hui Communities of Yunnan Society*, Stockholm, Almqvist & Wiksell, 1996.

18 C. Patterson Giersch, "Coton, Copper and Caravans: Trade and the Transformation of Southwest China", in Eric Tagliacozzo, Wen-Chin Chang (eds), *Chinese Circulations: Capital, Commodities, and Networks in Southeast Asia*, Durham, Duke University Press, 2011.

19 Ann Hill, *Merchants and Migrants: Ethnicity and Trade among Yunnanese Chinese in Southeast Asia*, New Haven, Yale University Press, 1998, p. 8; Wen-Chin Chang, *Beyond Borders. Stories of Yunnanese Chinese Migrants of Burma*, Ithaca, Cornell University Press, 2014.

20 Edmund Leach, *Les systèmes politiques des hautes terres de Birmanie. Analyse des structures sociales kachin*, translated by Anne Guérin, Paris, François Maspero, 1972.

21 C. Lefebvre, *Frontières de sable, frontières de papier. Histoire de territoires et de frontières, du jihad de Sokoto à la colonisation française du Niger, XIXe–XXe siècles, op. cit.*

caravan, during his work on the Anglo-French Boundary Commission of 1895, is rather negative:

> A caravan of Chinese of a very low-looking type were encamped on the small stream below my camp. They had come via Möng Hsai, four days' journey, from the upper-waters of the Nam Ou and were bound for Baw Hten. They described the route to Möng Hsai as being a good one and said there was a route from Möng Hsai to Luang Prabang, eight days' journey, but they had never travelled by it.[22]

A clear distinction can be made between long-distance traders, and the merchants who went from town to town, or from village to village. Once a caravan's goods had been unloaded in the market towns of the plain, local agents transported them to the sometimes far-flung mountain villages. These were often the only people ever seen by the villagers from beyond their immediate area.[23] The Europeans' cartographic work when they first arrived in the region was based on these routes, taken by the local people, and on the information gathered orally. This was a pragmatic decision, in the light of the very patchy information they had at their disposal. In fact, they made use of any information they could lay their hands on, while attempting to adhere to scientific principles.

1.3 *The Limits of the Exercise*

The topographers were relatively critical of the people they questioned, and especially of their own guides. The latter, who were chosen for their knowledge of places and languages, or who simply happened to be there, were the Europeans' first contacts, and thus they had a sometimes significant influence on exploration and mapping expeditions.[24] Recruited before the beginning of a campaign, or along the way, they were hired by the Europeans for their knowledge of the terrain. But Captain Walker, for instance, on the Anglo-French Boundary Commission of 1895, complained that his guide knew nothing about the course of a certain route in Laos: "More than this my guide could not tell

---

22 BL, IOR, L/PS/20/83, Report on the Anglo-French Buffer State Commission, by Captain Walker, Rangoon, November 1895, p. 20.

23 A. Hill, *Merchants and Migrants: Ethnicity and Trade among Yunnanese Chinese in Southeast Asia, op. cit.*, p. 46–47.

24 On the role of intermediaries in the history of exploration and travel, see the catalogue of the exhibition organised at the Royal Geographical Society in 2009: F. Driver, L. Jones, *Hidden Histories of Exploration: Researching the RGS-IBG Collections, op. cit.*

me. No one had come from Möng Hteng by this route for two years, and even about that he seemed very uncertain."[25]

The topographers tried as far as possible to check the information they received against the results of their own observations, which they considered to be more reliable than any oral reports.[26] A hierarchy existed between information and observation, and the cartographers always cited their sources, principally to preempt criticism. Sometimes this hierarchy was highlighted in the text: H. R. Davies explains at the beginning of his report that the information from local people is in small letters to distinguish it from reliable facts. This distinction appears in several British sources, but nowhere else,[27] and it not only appears in the text, but also on the sketch maps, through the choice of map symbols. This raises the question of the extent to which the information gathered was integrated into the Western maps of the area. A pencilled note on the "*Itinéraires de Langson par Kep et par* Chu"[28] states that "the dotted line going from one village to the next indicates the Annamite route," which probably derived from local inquiries, whereas alongside it an unbroken line in red pencil shows the route followed by the French column. The two lines, corresponding to two different information-gathering practices, are thus marked separately. A further guard against criticism surfaces with "they say that", or "claimed to be", or "probably" here and there on the map,[29] as a way for the cartographers to show that they are not liable for the possibly false information they were given. Their task was to make the inquiries as accurate as possible, and to cross-check the answers, but they ultimately did not have much room for manoeuvre, given the complexity of the exercise in a colonial situation.

Consulting the local population, in the context of conquest and pacification, was indeed not simple. The imbalance between the two parties could

25 BL, IOR, L/PS/20/83, Report on the Anglo-French Buffer State Commission, by Captain Walker, Rangoon, November 1895, p. 31.

26 "Having now personally visited some of the places which were described in the above report from native information, I have been able to make many additions and corrections to it. The following list is given, so as to make it quite clear how much of the information in that report stands good and how much of it is superseded by the 'Military Report on Western Yünnan' just published", BL, IOR, L/PS/20/94, Report on the part of Yunnan between the Bhamo frontier and the Salween, by Captain H. R. Davies, 1894–1895, Rangoon, September 1894.

27 *Ibid.*

28 SHD, 10H6, Dossier 13, Renseignements sur le Loch-Nam et les routes de Lang-son. Rapports de la place (Bac-Ninh), 1884–1885.

29 "Distances not known, but said to be [...]", BL, IOR, L/PS/20/D105, Report of the Intelligence Officer with the Southern Party, Burma-China Boundary Commission, 1898–1899, by Captain John Tod, Simla, 1899.

quickly make the cartographers' questions seem intrusive, or scare the inhabitants: these strangers, the Europeans, with their incomprehensible language, who gave nothing in return for the information provided, who did not say what it was for, and who did not follow local customs, were unlikely to inspire confidence. Silence could be the sign of a prudent reserve, and a wish not to get involved. Often, however, the Europeans' interlocutors simply did not know the answers to the questions asked. In 1890–1891, when Major Hobday was attempting to survey the territory up to the sources of the N'Mai Hka, in north-east Burma, he wrote:

> There was an absence of all information about the valley of the 'Nmaikha and the head-quarters of that river. "Nobody goes there" was the inevitable reply we received to our enquiries, and we could only elicit vague stories about the Marus or dog-eating Kachins, the Yawyins, and other wild tribes that inhabit those regions.[30]

He bemoaned the fact that the information gleaned was so vague that he was reduced to conjecture. The British ambitions were scaled down accordingly.

The local populations' fears essentially concerned the tax consequences of the cartographers' land surveys, and as such were not specific to a colonial context. Local hostility was a measure of the distrust of officials, as in Siam before 1895, in McCarthy's account, or in the French peasants' reactions to the engineer-cartographers responsible for drawing the Cassini map a century earlier. In 1910, when the Resident Létang was researching toponyms in his province, and turned to the municipal authorities, he was exasperated that:

> Some still do not understand the aim of the exercise: a certain lack of good will which I attribute to an unspoken fear – instinctive in the native – that their land taxes will increase. Hesitant or unclear answers on purpose.[31]

Not understanding, and mistrust of, European ways could thus explain the fact that local populations were often unforthcoming. The Resident also noted that some local officials, "still unfamiliar with map reading, and who for the most

30 BL, IOR, V/24/3989, Report on the Survey Operations in the tracts lying to the north and north-east of the Bhamo district, by Major J. R. Hobday, season 1890–1891, p. xliv.

31 ANV1, RST 73616, Lettre du résident de France à Thai Binh, Létang, au résident supérieur du Tonkin, 3 December, 1910.

part do not know *quôc ngu*,[32] did not always reply with the desired precision;"[33] similar shortcomings are reported for the area around Cao Bằng.[34] The countless question marks all over the French reports testify to acute problems of understanding, also on the part of the Europeans, who had difficulty understanding the Vietnamese terms, and making themselves understood by the local populations.[35]

We may wonder to what extent the European topographers knew the local languages. French officers were encouraged to learn Vietnamese and Chinese, but no training seems to have been given, so it was left up to the individual. Fernand Bernard, who later presided the Boundary Commission which demarcated Siam's frontiers, wrote during his first stay in Indochina to his brother Bernard Lazare (known for his support of Dreyfus):

> Would you believe it, dear brother, but I am beginning to Annamitise myself slowly [...] I bought three books, each of which contains a proven and rapid method for learning the Annamite language. The result is that I now know some 20 words, for having heard them pronounced, and for some 20 others I hesitate between three different meanings given by my three books, each with the same unshakeable assurance. [...] You understand that in such circumstances I shall have some difficulty learning enough of the language in two years to collect legends and traditions as you asked me to.[36]

The language barrier was real, and posed endless problems of pronunciation and spelling, since the French also had to learn Vietnamese in Latin script. Fernand Bernard's attitude changed and, in another letter to his brother six years later, his tepid enthusiasm had given way to something else: "If I was not learning Annamite, which now seems vital to me, if we want to be of any real help to this country, I would get bored, but I mostly hope that I will not

32 *Quốc ngữ* is the Vietnamese writing in Latin characters invented by missionaries in the 17th century.

33 ANV1, RST 73616, Lettre du résident de France à Thai Binh, Létang, au résident supérieur du Tonkin, 3 December 1910.

34 ANV1, RST 73616, Lettre du chef de bataillon Dehove commandant le 2e Territoire militaire au résident supérieur du Tonkin, Cao Bang, 11 June 1910.

35 SHD, 10H6, Dossier 15, Renseignements recueillis auprès du chef des Muong de Dong Lai et Dong Con, 10 December 1884.

36 AIP, Fonds Fernand Bernard, BEF1, Dossier 2, Lettre de Fernand Bernard à son frère Bernard Lazare, Hanoi, 18 September 1891.

be staying here too much longer."[37] He was pleased to be spending this second period in Indochina with more intelligent pursuits than during the first one, by learning the language, the history and the customs of Vietnam.[38] Not only did he learn the language, but he also intended to start learning Chinese, and he wrote to his parents to ask them where one of his cousins was, a certain Kahn "who is a well-respected China specialist, and must be a consul somewhere in China",[39] but we do not know what came of this. Again, learning the language was up to the individual, and although Fernand Bernard was not the only one to make such efforts, a great many French officers did not.

On the British side, there seem to have been more incentives for officers to learn Burmese or lesser-known dialects, to enable them to work more independently. Rules were issued in Burma after 1889 to encourage officers to learn the languages of the border areas, with a a reward of 1,000 rupees for any colonial civil servant who sat the standard exam for Chin or Kachin.[40] The decree added that there was no second reward for sitting two languages of the same group. The exam had three parts. First came conversation, to ensure that the candidate could converse with the local population, understand them, and be understood by them; then writing in English or Burmese, of sentences in dialect, to be explained in English; and, lastly, translation into the dialect. Changes to this system were constantly being proposed, with an updating of the list of languages concerned in 1903, presented by James George Scott, the Superintendant of the Southern Shan States, and C. C. Lowis, an ethnographer and the Superintendant of the Land Survey office.[41]

When there were too many discrepancies between oral and written forms, rectifying and correcting the topographic maps was discouraged. Thus in 1910 the Resident of Hòa Bình, criticising the number and the seriousness of the mistakes in the toponyms on the map, abandoned changing anything at all. He blamed the topographer's untrained ear, more attuned to Latin than to Vietnamese:

---

37 37. AIP, Fonds Fernand Bernard, BEF1, Dossier 2, Lettre de Fernand Bernard à son frère Bernard Lazare, Hué, 14 February 1897.

38 38. AIP, Fonds Fernand Bernard, BEF1, Dossier 2, Lettre de Fernand Bernard à son frère Bernard Lazare, Hué, 27 February 1897.

39 AIP, Fonds Fernand Bernard, BEF1, Dossier 3, Lettre de Fernand Bernard à ses parents, Hué, 12 March 1897.

40 BL, IOR, L/PJ/6/251/811, Rules for the encouragement of the study of languages of the frontier tribes of Burma, 23 April 1889.

41 BL, IOR, L/PJ/6/668/350, Gates, officiating Chief Secretary to the Government of Burma, to Risley, officiating secretary to the Government of India, Rangoon, 9 November 1903.

> The fact, for example, that Núi Hàng Ma comes out as "Núi Magna" shows that the topographer's ear was not familiar with the language of the country, which explains the very many imprecisions.[42]

So the quality of the inquiries carried out among the local populations depended not only on their availability, capacities, and good will, but also on the topographer's skills. Yet it would be naïve to think that a good mastery of the languages solved all the problems. The difficulty of naming correctly remained a persistent and even an insuperable obstacle:

> An officer-topographer, even with the finest attention to the phonetics, with the greatest tenacity, and the most ingenious methods for acquiring this information from the indigenous people, and even supposing he has the certificate in Annamite, and is capable of expressing himself fluently regarding his everyday needs in the field, will not always manage to name correctly all the places and orographic features over a largish area, along with their relative importance and connections.[43]

This situation was doubtless due to the complex reality with which the European topographers were confronted, quite apart from the problems of language and regardless of their own good intentions. To understand a territory that was unfamiliar to them, and to obtain information about what they could not see for themselves, they approached the local elites, and the traders who travelled the country. But their inquiries came up against their superiors' suspicion that the information obtained was untrustworthy, a taciturn local population, and the sheer complexity of what they were discovering. Names were central to these inquiries, and were often difficult to grasp orally, or else impossible to determine exhaustively: they remained a stumbling-block throughout the period.

2 Place Names, the Impossible Task

Naming villages, mountains and rivers is a fundamental aspect of cartographic work. The decision to favour one name over an other, when contradictory

42 ANV1, RST 73616, Lettre du résident de Hoa Binh au résident supérieur du Tonkin, 10 juin 1910. *Núi* signifie "la montagne".

43 ANV1, RST 73507, Lettre du général de division Dodds, commandant en chef les troupes de l'Indo-Chine à M. le résident supérieur au Tonkin, Hanoi, 15 June 1901.

representations of the same reality exist, can be considered to be a form of appropriation.[44] In the case of the European topographers working in a colonial context in the Indochinese Peninsula, the names given were not invented from scratch, but were gathered from available sources.[45] Two problems emerged here. The first was linked to the specificities of the territory, which baffled the Europeans, both the British and the French. Just like the 73 different terms for "bamboo" in Vietnamese, so rivers, mountains, and administrative units also had multiple names.

The second problem was the number of languages in the territory, with the result that a place name depended on the information sources used by the cartographers. These included existing maps, made by missionaries or other European explorers – which again multiplied the number of names and spellings – and above all the local populations, who alone could indicate the names habitually used. Every stage of an inquiry was problematic: understanding the local name, realising that there were generally several names for a single place, being able to transcribe it adequately – all these aspects made naming an arduous task.

We shall examine the changing methods used by the Europeans, from the period of exploration and conquest, when they discovered the linguistic complexity of the region, to the establishment of a colonial administration, at which point they had already acquired a certain experience of the terrain, and were looking for applicable solutions. Rather than supposing that practices developed in a linear fashion, we shall focus on how the Europeans adapted, or failed to adapt, their approach to the linguistic complexities of the region.

2.1 *A Multiplicity of Names for a Complex Reality*

The information sought by the British and the French for their maps were primarily the names of rivers, hills and mountains, and other elements of the relief; and administrative units and divisions. These were consequently the names which caused the most problems.

---

44 On this issue, see Philippe Pelletier, "La cartographie de l'Asie du Sud-Est et la dénomination de la "mer du Japon" (Mer de l'Est) aux XVIIe et XVIIIe siècles", *Péninsule*, 54, 2007, p. 58–97. What is true of the sea between Korea and Japan is also true of the sea between China and Vietnam, "South China Sea" for some, "East Sea" for others, names which are never neutral, as recent conflicts in the region attest.

45 This was not the case everywhere, even if finding the most accurate name was always the goal. See Hélène Blais, "Comment trouver le 'meilleur nom géographique'? Les voyageurs français et la question de la dénomination des îles océaniennes au XIXe siècle", *L'Espace géographique*, 30/4, 2001, p. 348–357.

For example, the different ways of defining a water course in Europe and in the Indochinese Peninsula were the cause of many a misunderstanding. In Vietnamese, the name depended on the context. On the French maps, the term most frequently used was *sông*, river. *Khúc sông* and *đoạn sông* were used to refer to a stretch of river, for instance the section of a river passing through a village, a locality, a district, or even a whole province. For tributaries, the terms were *nhánh sông* or *chi lưu*. Linguistic differences also emerged in relation to the size of the water course, and the region in which it flowed: generally, small- and medium-sized water courses such as streams, or the arroyos of the Delta, were designated by *lạch*, *mương*, *ngòi*, and *kênh*, whereas mountain water courses were *suối*. These many terms corresponded to the intricacies of the reality at hand.

There were yet further differentiations: whereas in England and France a river has the same name at its source and its mouth, this was not the case in the Indochinese Peninsula. When an official name was being chosen for the Mekong Exploration Commission, the marquis de Chasseloup-Laubat referred to "this River Laos, Meïcong, Song-Long – I really don't know what to call it, because its name varies, changes, and seems almost as unknown as its course."[46] The different sections of a river had different names, the name varied according to the languages spoken on its banks, and the proper name was often supplemented by a generic name for "large river". All these particularities bedevilled the work of the cartographers.

The British experienced similar problems in their exploration of northern Upper Burma to find the sources of the N'Mai Kha. Referring to Henry Yule's map, which gave the Tibetan and Chinese names for the river, Major Hobday pointed out that his own sources were in Kachin, and so none of the existing names could be retained.[47] This invalidated the use of earlier work, a situation which had become even more complex 15 years later when, during a joint survey of the Burma-Yunnan boundary, it evolved that five or six names could be used for a single river, depending on the different language groups and the Chinese officials, with the additional problem that the provenance of the information was not always given.[48] The same river could also be given different names depending on the scale considered, from the local village on the banks, where, within the same ethnic group, the name could vary over small

46 *BSG*, cinquième série, IX, January-June 1865, p. 385.

47 BL, IOR, V/24/3989, General report on the operations of the Survey of India, administered under the Governement of India, during 1890–1891, Calcutta, 1892, p. 75–76.

48 BL, IOR, L/PS/20/D239, Report on a joint survey of the Burma-Yunnan boundary, north of latitude 25° 35' north, by British and Chinese officials, March to May 1905, by Litton, p. 21.

distances, to the vast scale of the Irrawaddy itself. Streams or rivers sometimes took the name of the villages they passed through, and mountains had different names for different slopes. During the military expedition against the Chin in 1889–1890, Lieutenant Bythell mentioned the problem of naming geographical elements because often "the same river or hill is known by entirely different names, in two villages only a couple of miles apart."[49]

The same issues of language and scale beset the naming of relief. The term most often found on maps for a mountain was *sơn*, a Sino-Vietnamese word (*núi*, in Vietnamese). The term *sơn thủy* (*sơn* = *núi*; *thủy* = *nước*), the equivalent of *núi sông* or *non nước*, was used to refer to land or territory. Further variables were the size of the mountain (*đồi, núi, dãy núi, cánh cung,*), its aspect (the sunny side, the shaded side, etc.), or its substance, such as stone, or earth (*núi đất, núi đá, núi đá vôi, núi đá phiến, núi đất đỏ*, etc.). The European cartographers could not readily identify the realities to which these many names referred, and so were unable to represent or simplify the geographical features of mountains in a consistent manner. Their bewilderment and discouragement at being unable to name things correctly, especially given their principle of using local names – whereas no adequate method seemed to exist for collecting them – are easy to imagine. Moreover, the dilemmas were as acute at the end as at the beginning of the period, even if they were better understood later. In 1910, when the place names on the map of the province of Bắc Kạn were being revised, the Resident explained that:

> I also did not take into account the infinite number of special names by which the indigenous peoples choose to call each slope of a mountain, each rock whose shape has kindled the popular imagination, the string of little hillocks standing out against a mountain range, etc.[50]

The preferred solution seems to have been a pragmatic one. The Resident's justification for his choices here was to avoid complicating the work of revision by adding too many toponyms which would have made the map illegible. But the truth was that the topographers were overwhelmed by the complexities of the languages and the terrain.

The same applied to the nomenclature of administrative units: although the Europeans had no difficulty with the larger divisions, they often failed to grasp the smaller ones. The variety of names used to designate a village in Kachin

49 BL, IOR, V/24/3988, Report on the Survey Operations with the Chittagong column, Chin-Lushai expedition force, by lieutenant W. J. Bythell, season 1889–1890, p. lx.

50 ANV1, RST 73616, Lettre du résident à Bac-Kan au résident supérieur du Tonkin, March 1911.

was a major obstacle for the British.[51] In Vietnamese, the term *tỉnh* referred to the province, *phủ,* the major county town, *huyện*, the district, called *châu* in the Upper Region, and *xã* for the municipality. These were the terms used on the French maps. The municipality had many administrative subdivisions, and was itself sometimes called a "village" when there was only one. Even if the subdivisions were not entered on the maps, one can imagine that it did not make the Europeans' inquiries into place names any easier. *Thôn, xóm, ngõ, làng, bản, trại,* and *động* were terms referring to a village, "sub-village", hamlet and "sub-municipality." An inhabitant's reply to a French topographer who asked him the name of a village could go much like: "Here is is the *xóm* A of the *thôn* B of the *làng* C of the *xã* D, which is also called E …", insofar as Vietnamese villages often had two names, a scholarly name (in Chinese) and a common name (in Vietnamese). In addition to the details of naming were the problems of spelling the names correctly and entering them on the maps. This was perhaps the root cause of the problems.

2.2 *Transcription and Spelling*

The intricacies of Vietnamese spelling were not new to the French. Théophile Bigrel, a navy officer, Chief of Staff of the Governor of Cochinchina from 1870, and author in 1873 of a *Carte générale de la Cochinchine française* in 20 sheets, observed that the local population did not understand the maps made by the French, due to their faulty spellings, and that the map thus remained incomprehensible even to those whose territory it represented and whose language it used.[52] Dutreuil de Rhins devoted most of his *Foreword* to this very problem,[53] and one can say more generally that the two stumbling-blocks for the Europeans were the variety of languages used and their difficulty. Moreover, no standards had been set for transcription, and every cartographer thus had to choose the method he thought best.

Dutreuil de Rhins was against any idea of a universal writing system, which Francis Garnier advocated. He argued that there was no international standard, since neither the work of the French *Association française pour l'avancement des sciences*, nor the different systems devised by the Britons Max Müller (1855) and Lepsius (1863) provided an acceptable solution. There was no national

51 E. Leach, *Les systèmes politiques des hautes terres de Birmanie. Analyse des structures sociales kachin, op. cit.*

52 Théophile Bigrel, "Note sur une carte de la Cochinchine française", BSG, sixième série-tome VI, July-December 1873, p. 436–439. This map may be found at the BNF, Cartes et Plans, GE CC-826.

53 J.-L. Dutreuil de Rhins, *Avertissement géographique et orthographique sur la carte de l'Indo-Chine orientale, op. cit.*, p. 12.

standard either, and the French cartographers found it difficult to transcribe even the English names. So each cartographer decided individually, and there was no uniform approach. Francis Garnier, for example, used three different systems: he adopted the Jesuits' spellings for Chinese names, the missionaries' spelling for Vietnamese ones, and for all the other languages, he used Latin script, altering it slightly.[54]

The transcription of Chinese characters concerned not only maps of the Chinese zones adjoining Burma and French Indochina, but also all official Vietnamese place names, since they were written in classical Chinese characters. This choice was traditionally a mark of distinction, somewhat like the use of Latin in Europe, and local villagers and elites sometimes preferred to use the Chinese form rather than the indigenous name.[55] Adding Chinese names to the map was thus another hurdle for the European cartographers.

Dutreuil de Rhins consulted the Jesuits' maps for the place names in Chinese because, although they were far from ideal, they were the best available; the French cartographers could not borrow from the British because the latter had not managed to find a uniform system of transcription. When the Jesuit maps were imprecise, he consulted the British ones, but this only aggravated the problems of spelling:

> The literal translation produced an unusual spelling, and this is hardly surprising when one notices how greatly the English language disfigures, from our point of view, the pronunciation of Oriental languages.[56]

Since the British used the Jesuits' spelling, but also translated the Chinese characters directly, their maps had no uniformity either. Sometimes a single name given by the Jesuits resulted in several versions on the British maps, or an English term actually referred to several place names on the Jesuits' maps. Among the obstacles to achieving uniform spellings were the differences in pronunciation of the same Chinese characters in different provinces, the different transcriptions in French and English, and the lack of consistency even within one language (whether French or English). The consequences were sometimes serious, as when the 1897–1898 Boundary Commission could not start its work because of a problem of place names and finally missed the

---

54 E. Doudart de Lagrée, F. Garnier, L. Delaporte, *Voyage d'exploration en Indo-Chine, op. cit.*, p. 12.

55 Philippe Papin, *Histoire des territoires de Hà Nội. Quartiers, villages et sociétés urbaines du XIX*e *au début du XX*e *siècle*, Paris, Les Indes savantes, 2013.

56 J.-L.Dutreuil de Rhins, *Avertissement géographique et orthographique sur la carte de l'Indo-Chine orientale, op. cit.*, p. 13.

whole season.[57] The French had encountered the same problems ten years previously, when they had to name a place in consultation with the Chinese:

> We are here to agree the first official report of the demarcation, and for two days now we have been discussing, fruitlessly, how to refer to a position or place situated 50 metres from the Porte de Chine, on the road from the Porte to Dong-dang, at a spot where a stream crosses the road. We are wholly in agreement on the map, and this spot, but the Chinese commissioners refuse all our ways of referring to this place, while not proposing one themselves.[58]

As in the British case, the geographical location itself was not at issue, but naming it proved to be an insoluble problem: there was no agreement on an existing name, nor on a new one. Given the stalemates *in situ*, a system of Chinese transcription was gradually developed, spearheaded by Vissière, which the French Ministry for Foreign Affairs ratified.[59] Although still not perfect, it helped unify Chinese transcription practices.

The French attempted to achieve a uniform practice also for Vietnamese place names, and to a lesser extent for all others. In this, Dutreuil de Rhins found a useful source in Père Lesserteur's *Carte des missions de l'Indo-Chine*, from 1879. A former missionary, Père Lesserteur used only French letters which, when pronounced in French, came as close as possible to the Vietnamese sounds:

> Thus *ớ, ư, đ, d, x*, were replaced by their closest equivalent, *eu, u, d* (*dz-gi-zi*), *s* ... We decided against the German *ch*, and used a *k* or *gh'* instead; we adopted our *ch* and not the English *sh* for what the missionaries spelt as *s*. The Portuguese endings, *anh* and *inh*, were replaced by *agne, igne*.[60]

57 BL, IOR, L/PS/20/D101, Report of Intelligence Officer attached to the Southern Party, Burma-Chinese Boundary Commission, 1897–1898, by Captain Evelyn William Norie, Simla, 1898, p. 4.

58 P. Néis, "Sur les frontières du Tonkin", *art. cit.* 5.

59 ANOM, SGEOCOL 18, Compte rendu des travaux intérieurs exécutés par le Service géographique de l'Indochine pendant l'année 1903, Hanoi, 1 February 1904, par le lieutenant- colonel, chef du Service géographique, Claude de Mussy, p. 3. See the obituary of Arnold Vissière in the *BEFEO*, 30, 1930, p. 649–653.

60 J.-L. Dutreuil de Rhins, *Avertissement géographique et orthographique sur la carte de l'Indo-Chine orientale, op. cit.*, p. 20.

This third system for spelling place names was an improvement both on the erudite system of transcription, which was difficult to add to the maps, and not widely understood; and on the rough and incomplete simplifications, which were no real solution at all. However, it did not solve all the problems, especially as it removed the tonal marks. In 1886. the head of the Translation Department of the General Residency at Tonkin complained about "poor editions lacking detail, on which the names are so distorted that it is sometimes impossible to recognise the sounds of the Vietnamese characters."[61] And in 1901, the verdict was still unambiguous:

> Whatever care is taken establishing the 1:25,000 map of the Delta, however much one may guarantee the precision of its topography in the strict sense, there remain a difficulty and an uncertainty which those carrying out the surveys cannot ultimately overcome. I refer to the sensitive issue of the names, their accuracy, their relevance, and their correct spelling (in *quoc ngu*).[62]

In order to simplify work in the field, lists of place names of all administrative centres, from the smallest locality to the county town, were requested from provincial governors, to be supplied both in Chinese characters and in romanised Vietnamese. Yet even this did not prevent multiple errors. The head of the *Service géographique* accordingly guaranteed the "representational accuracy" of the map, which was indeed very precise and complete, but warned that it should be regarded as a "silent map" for the time being, until the Residents could help establish the right place names. Cartographers thus seemed always to be juggling too little and too much, trying to find a balance between maps that were too empty and those that were too crowded. Walking this line was particularly difficult for maps of the mountain regions.

For although official systems of transcription did exist for Chinese and Vietnamese, even if they might not be adequate for maps, there were no rules for transcribing the languages spoken by the different ethnic groups. This gave rise to untold problems:

> The correct way of writing "Chieng", according to the official way of spelling Shan, is "Kēng". The word is, however, pronounced "Kyēng" in Shan,

61 ANV1, RST 73611, Cabinet du résident général, service des traductions, à M. le résident général, Hanoi, 29 May 1886.

62 ANV1, RST 73507, Lettre du général de division Dodds, commandant en chef les troupes de l'Indo-Chine, à M. le résident supérieur au Tonkin, Hanoi, 15 June 1901.

> and "Cheng" or "Chieng" by the Laos, Khüns and Lüs. It is not pronounced "Kēng" by any one. Today it may be found in some maps and books written in all sorts of different ways, as "Kaing", or Kyaing (the Burmese pronunciation), Kiang, Kyang, Kyeng, Keng, Ching, Cheng, Chieng. In this report I have spelt it "Kēng" when it occurs in British territory, and Chieng in Siamese territory.[63]

In the light of the variety of transcriptions, which plagued both the British and the French, Captain Dussault proposed a simplified method for representing geographical names from Tonkin's Upper Region, in 1905. He advocated transcribing names phonetically, according to how they were pronounced, which had the advantage of being easier than writing romanised Vietnamese, which required specialist knowledge.[64] Dussault's table was submitted to the Residents of two provinces where these languages were used, for their approval. Scientific norms were thus tested against the realities on the ground. The terms collected were those used over a small area, for similarly pragmatic reasons. When the information sources were traders, the place names given could have a variety of origins, for example Chinese. This variety was not peculiar to Indochina, and in Algeria, for example, the cartographers gave a lot of thought to the use of Arabic or Berber names in regions where neither were dominant, especially the names of wells.[65] Returning to Indochina, different methods were used to standardise notation of the many native languages, including Dussault's, but each had its problems. A rational system of transcription of names was one of the key concerns of the successive heads of the *Service géographique*, and it was one of the fundamental difficulties encountered in establishing accurate maps. Maps quickly became illegible if overloaded with text, so an efficient solution seemed to be providing full information – glossaries, lists of place names, variants – on a separate sheet, and have the reader move back and forth between it and the map.

2.3 *Between Map and Text*

In order to overcome these linguistic obstacles, and help the maps be clearly understood, the British and French colonial authorities thus drew up long lists of place names separately. In this, they simply took over the pre-existing

63 BL, L/PS/20/16, Report on the Anglo-Siamese Boundary Commission, Northern Party, 1892–1893, by Captain Davies, Rangoon, 6 July 1893, p. 8.

64 ANV1, RST 76111, Note pour M. le résident supérieur, par l'administrateur chef du 2e Bureau, M. Bouzat, Hanoi, 11 January 1906.

65 H. Blais, *Mirages de la carte. L'invention de l'Algérie coloniale, op. cit.*

administrative documents produced by the Chinese, Burmese and Vietnamese administrations, where the presence of a name on a list was more important than a point on a map for determining to which administrative unit a certain village belonged, and hence what taxes would be levied.[66]

However, although the Europeans were likewise interested in taxation, they also used these lists to discover different versions of the place names, and so resolve the problems posed by multilingualism. In 1907–1907, the British transcribed toponyms into English for the first time, having battled with Burmese and Shan names.[67] They first requested the tax registers from the *Sawbwa* of Kengung in order to obtain the names of all the villages by circle. Then a British officer was assigned to transcribe them. Thus, on starting his work, each topographer was already in possession of a list in English of all the villages.

The French used a similar method, at about the same time, but the revision of the names on the map of Tonkin, which began in 1910–1911, encountered major difficulties for the Upper Region. The French wanted to write the official place names in Chinese characters on the map itself, along with their transcription into romanised Vietnamese – a customary procedure in the Residences – and the common name used by the local inhabitants, again in romanised Vietnamese, and in brackets.[68] This was highly ambitious, given the size of the maps and the number of place names. For the Delta regions, the only additions needed to meet these requirements were the local names, but in the Upper Region, which had many minority language groups, Chinese characters were not always used, except for the official names of administrative divisions, such as the province, district, county, or town. Instead, the local language was used (above all, for villages and hamlets). The rules were thus particularly difficult to apply, even if this remained an objective.

The regions of north-west Tonkin, where Thai was used, were also asked to add the name in Thai – bringing to four the number of names for one and the same place. Additionally, the most frequently used of the common names was to be underlined, if there were several of them, and all of them should be

66 For Burmese records, see Frank Trager, William Koenig, *Burmese Sit-tàns 1764–1826: Records of Rural Life and Administration*, Tucson, University of Arizona Press, 1979; Jacques Leider, "Taxation et groupes de service sous la royauté arakanaise. Un rapport d'enquête de 1803 dans un manuscrit de la Bibliothèque nationale de Paris", *Aséanie*, 1/1, 1998, p. 67–89; for Vietnamese records, see particularly Huy Lê Phan, "Địa bạ cổ ở Việt Nam", in *Tìm về cội nguồn*, Hà Nội, Thế Giới, 1998, vol.1, p. 231–242.

67 BL, IOR, V/24/4005, General Report on the operations of the Survey of India, administered under the Governement of India, during 1906–1907, Calcutta, 1908, p. 45.

68 This last norm thus ran counter to the method proposed in 1905 by Dussault, who discouraged the use of romanised Vietnamese, and favoured a phonetic transcription.

mentioned. These rules had to evolve in response to the realities on the ground, especially for places where Chinese characters and romanised Vietnamese were hardly known:

> The general principle to be followed: the name is the one used by the inhabitants of the locality in question; it must be written in Latin characters in accordance with the conventions of the administration in question […] Thus the name of a village inhabited by Laotians will be its name in the Thai language and not the name by which it may be known by the Annamites of the region. Nevertheless, in many cases, the name by which the majority of the Indigenous population of the region know a locality must also be written (for instance, in areas where there are many Thô and some Man tribes, the *mân* name of a locality inhabited by Man (*trai*) may not be known by the Thô, and if one asks a Thô the way to X …, using the map, (X … being the *mân* name), then he probably will not understand. So one must add this second name, and place it in brackets under the first one).[69]

The basic idea was therefore that the map should be useful in the field, when talking with the local people, and only secondarily for civil servants in their offices. The British also came to this conclusion, although the situation was simpler for them. The fact that the place names given by the ethnic groups were listed, and printed on the maps, not in brackets, was greeted as a real improvement by the political authorities, and made the maps immediately more functional.[70] Since, for the French territories, there were so many different names for the same place, the Residents sent in both the revised map and the list of the villages in their province. This list sometimes referred to other lists, because incorrectly transcribed place names would receive a number referring to lists where the correct transcription figured.[71] The links between map and text were therefore of vital importance, and one could not be used without the other. This was the case not only for the place names, but also for other terms on the maps. For these, the question arose afresh: in what language should they be written?

69 ANV1, RST 73616–01, Notice préparée par le chef du Service géographique, Salel, en vue de la révision méthodique des cartes du Tonkin, Hanoi, 20 April 1917.

70 BL, IOR, V/24/4005, General Report on the operations of the Survey of India, administered under the Governement of India, during 1906–1907, Calcutta, 1908, p. 45.

71 ANV1, RST 73616, Lettre du résident de Hoa Binh au résident supérieur du Tonkin, 10 June 1910.

Again, the area's linguistic diversity, and thus the different glossaries required, had to be weighed up against the Europeans' wish to ensure a homogeneous system of legible place names on their maps. Glossaries, showing correspondence between terms, became standard additions to reports, notes, and keys to the maps. They generally contained only a dozen words, those most used on the maps, but they could sometimes stretch to several pages. Without this scholarly apparatus, the maps were illegible and incomprehensible. Yet there seemed to be no rule as to which words were to be selected: should British and French terms be retained where they existed? Or should indigenous terms be privileged? Often the French maps showed Vietnamese terms for administrative divisions – *huyện* (district), *xã* (town) – and for natural features – *sông* for a river, *núi* for a mountain. *Trạm* – post houses – were also common.

In reality, a mixture of Asian and European terminology was used, without any over-riding rule. The British officer Tod used local terms on his "sketch map of the Valley of the Nam Kha", and provided their translation immediately after the Foreword to his Report.[72] On the map, the term "Nam" was thus not replaced by "river" – for example, "Nam Kha" – but the glossary mentioned "Nam = river, Man = village, Loi = mountain."[73] Or equally, "deep valley of Nam Mwi": the English term was used to describe the valley, and the Burmese term was retained for the river. Many toponyms began with "Man", and, in the mountainous areas, the term "Loi", followed by the place name: "Loi Mot Ting", "Loi Kawng Kwan". Certain terms appeared in English on the British maps, with the Burmese term placed in the glossary, but sometimes there was no glossary at all, and the same word was repeated in different languages.[74] All these examples suggest how difficult it was to name places, develop standards, and privilege a single language. There is even a French map on which Vietnamese place names are accompanied by common names in Latin, such as *flumen*, *rivus*, *aqua* and *lapis* – the work of a no doubt erudite, if sometimes careless,[75] French officer. [Figure 19] This mix of local and European words shows that the

72 BL, IOR, L/PS/20/D105, Report of the Intelligence Officer with the Southern Party, Burma-China Boundary Commission, 1898–1899, by Captain John Kelso Tod, map 3, "Valley of the Nam Kha above Pang Hsang and Mong Nga circle".

73 The Thai word *nặm* is close to the Vietnamese word *nước*, and is used for water courses, rivers and streams.

74 ANV1, SG 249, Carte du huyện de Văn Thắn (Tân Hóa Dạo) de la province de Yên Bái et du huyện de Yên Lạp de Phú Thọ, n. d.

75 "Cruditur [*sic*] duobus diebus", for "Creditur duobus diebus" ("we believe it takes two days"), ANV1, SG 374, Cartes de Cân Ấp (Vietnam) et Cam Mỏn (Laos) (a copy of 30 September 1982).

FIGURE 19 Map of Cân Ấp showing Latin words among Vietnamese ones
SOURCE: National Archives, Vietnam, Centre 1, SG 374, Carte de Cân Ấp (Vietnam) et Cam Mỏn (Laos)

maps were a middle ground between the indigenous sources, which provided the indigenouos terms, and the heritage of European cartography.

Overall, an overload and an unmasterable variety of information threatened all the maps, and to avoid this, choices had to be made, for which, on the eve of the First World War, no clear-cut standard had yet emerged. The plethora of names for a single place undermined the cartographers' grasp of the territory, and required decisions to be made on the most appropriate name to select. As such, the maps did not so much reflect a reality as register a selection made, with the remaining information being classed as insignificant, or abandoned because it was not in the preferred language. This brings us to the maps' functions, since the decisive criterion for inclusion was often its usefulness in the field, for the relations with the local population. Transcription, the passage from oral to written forms, constituted a further, unresolved, problem, which sometimes led to a place name's elimination, and in all cases was a complex and difficult process. Ultimately, it required a separation between map and text, and a to-and-fro between the two, to avoid the map becoming illegible due to the excess of information it sought to convey in order to make the territory comprehensible. Lists of place names, as well as a number of glossaries, were therefore linked to the map. These testify to the immense effort deployed to find the right name for a place, and to position it correctly on the map – an endless task, as a wealth of written sources suggest.

•••

We have seen that the Europeans carried out inquiries among the local population first to fill in the gaps in their knowledge of the territory, and later to draw up administrative maps, during the colonial conquest, and in the subsequent phases. From this colonial encounter we can learn much about the Europeans' objectives, the reactions of the local people, and the fate of the information gathered.

The outer parameters of these inquiries were determined by the topographers' mobility or forced immobility, which were in turn determined by the level of (in)security, the time available (often insufficient), and the extent of their means. Since the European cartographers could not go everywhere and reconnoitre all the terrain, they tended to ask for information either in the town where they were located, or in the villages they passed through. Although all sorts of people were interviewed – inhabitants, gossip-mongerers, traders – the Europeans had a clear preference for travelling tradespeople, because they too travelled up and down the territory, and could provide useful information for the Europeans about what lay beyond their horizon (unlike the local inhabitants, who saw more or less what the topographer saw). There was, naturally enough, no internal coherence to the information gleaned, not even retrospectively, since no one could claim to possess a general overview. The answers to the Europeans' questions – for instance, the route of a road, the presence or absence of a bridge, or the name of a neighbouring village – were based more on familiarity than on objective knowledge, and thus were also subject to suspicion, especially since this information was difficult to obtain and to confirm. The geographers' scepticism was a token of their efforts to arrive at scientific certainties, since acquired information was, for them, by its very nature less reliable than direct observation. However, even when the information gathered was subject to doubt, it was added to the maps and could be found on the finished products – the sign of a certain discrepancy between the slightly haughty discourses and the more pliable practice.

For both the British and the French, the goal of their inquiries was essentially, though not exclusively, to name places and things correctly. This was as laborious a task as it was unsuccessful, for a number of reasons. The lack of understanding between the two parties, due to linguistic difficulties, played a major role, but also the complexity of the reality encountered. Landscapes, administrative divisions and toponyms were not described in the same words as in Europe, and the many names for the same place, depending on the local languages and the informant's position, as well as the many uncodified transcriptions, or whose rules could not be applied on maps, occur frequently as problems in the archival material. The Europeans were sometimes clumsy in their requests, and the local people did not know how to reply or were unwilling to help, because they gained nothing in return for informing a stranger, whether the latter was a mandarin or a European topographer. Cartographers were not the only ones to encounter these difficulties, as the literature on the subject in other domains suggests. However, they had the task of staring an unintelligible reality in the face, identifying it, and writing it down, despite all the difficulties this entailed; from their initial bewilderment they had to develop workable solutions. They gradually devised suitable methods, one of which was to create lists to accompany their maps. The links between map and text thus became extremely important. Interestingly, while wishing to distance themselves from their Asian counterparts, the European cartographers actually returned to one of their strategies: appending an explanatory text, to be read alongside the map.

CHAPTER 6

Using Asian Maps: Borrowings and Reworkings

In addition to the information gathered orally, the Europeans used, and commissioned, maps made in Asia. These provided, at different periods of the colonisation, place names, practical information, or the path of a country's pre-existing borders. What place did these maps occupy in the production of cartographic knowledge, in the light of the Europeans' scientific methods? Were they treated in the same way as oral information, or were the British and the French inspired by the well-developed pre-existing cartographic traditions in the Indochinese Peninsula to use the Asian maps in new ways? Research on indigenous knowledge systems has focused particularly on oral traditions, because it centred on Africa.[1] It has limited relevance to the Indochinese Peninsula, and anyway does not discuss the uses of indigenous cartographic knowledge, except for ephemeral maps, such as those marked in the sand. The substantial indigenous map-making tradition in the Indochinese Peninsula stands out as an exception, with its wealth of hand-written cartographic documents produced in the field, mentioned in the sources, and found in the archives. We shall seek to understand what these maps are, their origin, in what context and with what objectives they were produced, and how they were used by the Europeans. We shall also establish whether the same types and number of documents were produced in different areas, whether the British and the French were interested in the same elements on these maps, and whether they used them in comparable ways.

Another, rarely asked, question arises here: did local mapmaking in these territories continue during the colonial period, or did it stop? Were the same maps made before and after the arrival of the Europeans? Where there are obvious changes, was this due to the more or less forced or freely chosen adoption of Western methods by local elites, or were there other reasons? Philippe Papin has shown that in Vietnam, in the Early Modern period, cultural transfers

1 Bruno Latour, "Comment redistribuer le Grand Partage?", *Revue de synthèse*, 110, April–June 1983, p. 203–236; Marie-Albane de Suremain, "Ambitions positivistes et savoirs autochtones: les traditions orales dans le Bulletin du CEHSAOF/IFAN", *Outre-Mers*, 2006, Sophie Dulucq, Colette Zytnicki (eds), *Savoirs autochtones, XIXe–XXe siècle*, p. 33–46; Id., "Coutumes, coutumiers", in Sophie Dulucq, Jean-François Klein, Benjamin Stora (eds), *Les mots de la colonisation*, Toulouse, Presses universitaires du Mirail, 2007, p. 37–38.

 DOI:10.1163/9789004469853_008

could take the form of copies and reproductions.[2] He suggests that other periods and fields be explored with this in mind. In the Indochinese Peninsula, then, during the colonial period, did traditional indigenous cartography gradually adopt European standards, or was there a fundamental incompatibility between the two, relating to the scientificity of the procedures, or to inherently different objectives? In order to establish whether one can speak of cultural transfer here, and if so, its extent, it is important to distinguish between what remains impervious to foreign influence, what is a form of adaptation to it, and what directly copies foreign procedures.

1 Checking Historical Borders

Since we rarely find in the archives both the map and the related text, it is impossible to match the stages in the map's production with the associated narrative.[3] However, it is possible to establish a typology. First, there were the Asian maps found *in situ* by the Europeans at the time of the conquest, or on another, unidentified occasion. Some of these were used to produce other, European, maps, or else they were translated, to extract the useful information. Then, there were the maps commissioned by the British and the French from their informants. The scale of these depended on how the Europeans' interlocutor represented the territory: King Đồng Khánh provided a map of the whole of Tonkin and Annam (a unique example), the provincial elites chose either local or province-wide coverage, according to their level of responsibility, travellers and traders tended to provide itineraries, and inhabitants generally concentrated on what they knew best, the very local area.

The Europeans sought out local chronicles and annals to supplement their knowledge of the power relations between territories, and their existing borders. These sources – plentiful at the Burmese and Vietnamese courts, present in the principalities of the northern Indochinese Peninsula, but rare in Siam – were not all of the same quality, nor were they made with the same goals in mind: many served to glorify the sovereign's exploits, or to record the legendary history of different kingdoms, and thus included religious references. In

2 Philippe Papin, "Un laboratoire pour l'étude des transferts", in Hoai Huong Aubert-Nguyen, Michel Espagne (eds), *Le Vietnam. Une histoire de transferts culturels*, Paris, Demopolis, 2015.

3 Clapperton's narrative of his explorations has often been used to elucidate the indigenous documents he brought back, reconstructing specific situations by linking word, gesture and writing. See C. Lefebvre, "Itinéraires de sable. Parole, geste et écrit au Soudan central au XIXe siècle", *art. cit.*

their historical and geographical emphasis, they went beyond any local scale. We shall ask whether these sources were actually used by the Europeans in practice, and what they made of these documents which looked so strange in their eyes.

Cartographic traditions varied across the Indochinese Peninsula. The quality and above all quantity of maps produced, as well as their nature, were not the same, and the numbers of documents to have come down to us were variable.[4] Vietnamese cartography stood out from all the others because it espoused the Chinese tradition. It was not only different, but also much larger than the Laotian or Burmese equivalents, which were almost unknown. For the French, was the abundance of Vietnamese maps of any help? And for the British, did the absence of maps pose a problem?

1.1 *Tracking Down Burmese and Siamese Chronicles*

Both the British and the French used official geographical documents from the different courts to gauge the extent of their new territories, particularly for boundary demarcation. The British took over the aspirations of the court of Ava, and the French Protectorate of Annam-Tonkin had the powers of the court of Hué delegated to France. This logic of political continuity influenced the form of the imperial territories, and explains the Europeans' efforts to find documentation concerning their extent and historical borders. If a territory or a province was marked on the ancient maps of the Empire of Annam, then French claims prevailed over those of Siam or China. What counted in this legitimation of the historical rights of France was simply the mention of the space on the map, not the precise surface area designated; if this mention went back many years, however precise or imprecise the represented location, then for the French this provided sufficient grounds for the claim to be considered valid.[5] Similarly when the British annexed Burmese territories, they claimed priority over the former lands of the Kingdom of Ava, against Siamese and Burmese interests. Geographical treatises were considered to provide the most substantial proof, but if these did not exist or could not be consulted, then the chronicles and annals of the territories concerned were accepted as second best. In general, the Europeans considered any historical or geographical document to be a legitimate source for a territorial claim.

---

4 These traditions have been the subject of several research publications, of which the most remarkable is John Brian Harley's and David Woodward's (eds), *History of Cartography*, Chicago, University of Chicago Press, 1987.

5 This argument was used by Gabriel Devéria, for example, whom we mention again later.

This approach was not new. In 1834, Richardson carried out an assignment for the British at Chiang Mai, in Siam, to fix the borders of the Tenasserim region, which was a Crown territory since the First Anglo-Burmese War (1824–1826). He studied the official Siamese sources to provide proof of the Burmese rights over the spaces to the east of the chosen border, the Salween River.[6] Historical documents were certainly used, but it can be difficult to identify which in particular. For example, the British consulted Siamese sources, but it is unclear whether they had access to the Royal maps found recently in the Grand Palace in Bangkok.[7] Sometimes a particular explorer mentioned that he had used a particular source, as when Robert Boileau Pemberton, who had joined the 44th Native Infantry Regiment of the British Indian Army in 1817, carried out surveying and exploration tasks at Manipur:[8]

> While I was living in Manipur, I established the existence of an ancient hand-written Shan chronicle, which proved to be indispensable for discussions on border issues with the court of Ava; and even if I could never obtain the document itself, the old Shan man who owned it nevertheless gave me permission to get it translated into the language of Manipur, under his close supervision, and on the basis of this translation I managed to obtain information on this kingdom's past political relations.[9]

This source most probably specified the tributes paid, from which the bonds of vassalage between the Burmese court of Ava and the neighbouring principalities could be deduced.

As for Burmese maps, they remained little known, but royal chronicles were in abundance, as Arthur Phayre (1812–1885) points out in his *History of Burma*, published in 1883. His work was based on his experience as assistant Commissioner of Tenasserim in 1846, then as Commissioner of the Arakan in 1849, of Pegu after its annexation in 1852, and then as Commissioner of all of British Burma, from 1862 to 1867. He was very impressed by the large number of historical sources available in the former Burmese imperial territories

6 T. Winichakul, *Siam Mapped: A History of the Geo-body of a Nation, op. cit.*, p. 68 (drawing on the Burney Papers).

7 Santanee Phasuk, Philip Stott, *Royal Siamese Maps: War and Trade in Nineteenth Century Thailand*, Bangkok, River Books, 2004.

8 Manipur was an independent kingdom situated between India and Burma. It was integrated into the Raj as a princely state only in 1891.

9 BL, V/27/64/160, Report on the eastern frontier of British India, with an appendix and maps, by Captain R. Boileau Pemberton, Calcutta, Baptist Mission Press, 1835, p. 108.

compared to the meagre number, or even total absence, of equivalent writings in the former Hindu kingdoms.[10] The British were able to consult the *Chronicles of The Crystal Palace* (*Hmannan Yazawin*), a work by the Royal Historical Commission of Burma, published in 1832 under the supervision of King Bagyidaw (1819–1837): it alternated historical and mythological tales, and covered events up to the eve of the First Anglo-Burmese war (1824–1826). The second volume was published in 1869, and went up to the second Anglo-Burmese war (1852). The third volume, the work of Maung Maung Tin, was published in 1905 but it went no further than 1885, the date of the third Anglo-Burmese war, and the abolition of the Burmese monarchy.

In addition to the Burmese chronicles, the princely courts that had preserved their own historiographers when they became vassal states also produced chronicles. For instance, the states of the Kingdom of Lanna (1259–1892) maintained their own sources, as shown by the different chronicles from Chiang Mai, Nan, and other principalities.[11] The Europeans supplemented the lack of documents on the Siamese side with these sources, since, as they pointed out, many historical documents had been destroyed when the Burmese army laid waste to the Siamese capital Ayutthaya in 1767, and only the major historical events could be reconstructed, leaving a large number of gaps.[12] But since the Siamese chronicles were mostly concerned with the mythological origins of the territories and local legends, they were anyway of lesser interest to the Europeans, who were looking for established borders. They seem to have found information of this sort for the territory of Luang Prabang, since a sketch map made during the Franco-Siamese border demarcation was described as "the historical frontier of the kingdom of Luang Prabang fixed some 150 years ago

---

10 Arthur P. Phayre, *History of Burma*, London, Trübner, 1883, p. xviii–ix.

11 The Chiang Mai annals were translated into French by the consul there: Camille Notton, *Annales du Siam*, 1, *Chroniques de: Suvaṇṇa khamdëng, suvaṇṇa k'ôm khăm Siṅhanavati*; 2, *Chronique de La:p'un*; 3, *Chronique de Xieng Mãi*, Paris/Limoges/Nancy, Imprimerie Lavauzelle, 1926–1932; other chronicles were also extensively translated and glossed See Sao Sāimöng Mangrāi, *The Pāḍaeng Chronicle and the Jengtung State Chronicle Translated*, Ann Arbor, University of Michigan, 1981; Sænluang Ratchasomphan, David K. Wyatt, *The Nan Chronicle*, Ithaca, Cornell University Press, 1994; David K. Wyatt, ʻarunrat Wichīankhīeo, *The Chiang Mai Chronicle*, Chiang Mai, Silkworm Books, 1995; Sithu Gamani Thingyan, *Zinme Yazawin: Chronicle of Chiang Mai*, Yangon, Universities Historical Research Centre, 2003; see also François Lagirarde, "Temps et lieux d'histoires bouddhiques: à propos de quelques "chroniques" inédites du Lanna", BEFEO, 94, 2010, p. 59–94.

12 Godfrey Eric Harvey, *History of Burma. From the Earliest Times to 10 March, 1824, The Beginning of the English Conquest*, London, Cass, 1925, p. 343.

during the Burmese invasion."[13] This date of 1767 regularly appears as the limit beyond which the Europeans cannot easily find Siamese sources.

In general, however, neither the British nor the French explicitly mentioned the annals they consulted. The exception was the Vietnamese cartographic tradition derived from the Chinese, to which the French referred quite precisely.

1.2 *The Rich Tradition of Map-Making in Vietnam*

The Vietnamese kings, concerned to master the territory over which they ruled and to defend its borders, particularly against Chinese threats, commissioned geographical treatises very early on.[14] Geographical knowledge thus circulated before the colonisation of Tonkin, and a few years before the conquest, Gabriel Devéria, the first interpreter, from 1873, to the French Legation in China, and correspondent of the *École spéciale des langues orientales vivantes* [Special School of Living Oriental Languages], wrote an *Histoire des relations de la Chine avec l'Annam-Vietnam du XVIe au XIXe siècle d'après des documents chinois traduits pour la première fois et annotés* (1880) [History of the relations between China and Annam-Vietnam from the 16th to the 19th century according to Chinese documents, translated for the first time and annotated].[15] In 1885, he published another work, on *La frontière sino-annamite. Description géographique et ethnographique d'après des documents officiels chinois traduits pour la première fois* [The China-Annnam border. A geographical and ethnographic description according to official Chinese documents, translated for the first time].[16] In this book, he first mentions the triangulation work carried out by the Jesuits at the request of the Chinese emperor, at the beginning of the eighteenth century. This work was the basis of the 31 maps drawn for the *Atlas*

13 ANOM, GGI 14491, *Délimitation de la frontière entre l'Indochine et le Siam. Royaume de Luang Prabang. Territoires contestés sur la rive droite du Mékong, Bo Sao et Ken Tao. Croquis 1 annexé à la lettre no 10 du 2 août 1898*, 1:100,000, Pak-Lay, 2 August 1898.

14 Tam Quach-Langlet, "La perception des frontières dans l'ancien Vietnam à travers quelques cartes vietnamiennes et occidentales", in P.-B. Lafont (ed.), *Les frontières du Vietnam. Histoire des frontières de la péninsule Indochinoise, op. cit.*, p. 25–62.

15 Gabriel Devéria, *Histoire des relations de la Chine avec l'Annam-Việtnam du XVIe au XIXe siècle, d'après des documents chinois traduits pour la première fois et annotés*, Paris, E. Leroux, 1880; for further information about the author, see Edmond Pottier, "Notice sur la vie et les travaux de M. Gabriel Devéria", *Comptes rendus des séances de l'Académie des Inscriptions et Belles-Lettres*, 44/2, 1900, p. 127–143.

16 Gabriel Devéria, *La frontière sino-annamite. Description géographique et ethnographique d'après des documents officiels chinois traduits pour la première fois*, Paris, E. Leroux, 1886.

of Kangxi, some thirty years before the Cassini enterprise.[17] They were little known in France, except for a copy made in 1729 by Jean-Baptiste d'Anville (1697–1782), the Geographer to the King, of the maps of the three Chinese provinces bordering on Tonkin. Noticing mistakes in these copies, Devéria went back to the original Jesuit maps and charted the whole Chinese-Vietnamese border from them.[18] He then set out to improve on them, adding further details from two official Chinese sources, the "Great Imperial Geography", and the "Great Administrative Encyclopedia".[19] The first of these, commissioned by the Emperor Qianlong and published in Peking in 1764, comprised 500 booklets containing a description for each administrative unit of the empire, with subdivisions from the province down to the town, including information on the population, taxation, the mountains and watercourses, the outposts and narrow passes, the distance in *li*, and the vegetation and agriculture.[20] The second source, the "Great Administrative Encyclopedia", was published in 1818, on the order of the Peking Court. It comprised 900 booklets dealing with institutions and official functions. Devéria consulted this Encyclopaedia's plates consisting of maps of the provinces, which principally showed watercourses. Considering the composition to be better than that of the "Great Imperial Geography", he decided to reproduce them, and so translated all the essential information on the provinces of Guangdong, Guangxi and Yunnan – the physical and administrative descriptions, and the descriptions of the populations – including a map for each. It is interesting to see how Devéria adapted the ancient Chinese maps, selecting for his version only those elements he considered relevant, that is, the water-courses and the mountains, and translating the place names and the territorial limits indicated [Figure 20].

17 On this Atlas and its connections with the European cartography of the time, see Laura Hostetler, *Qing Colonial Enterprise. Ethnography and Cartography in Early Modern China*, Chicago, The University of Chicago Press, 2001.

18 G. Devéria, *La frontière sino-annamite. Description géographique et ethnographique d'après des documents officiels chinois traduits pour la première fois, op. cit.*, p. 85.

19 The first of these is the *Da Qing yi tong zhi* [大清一統志], *Đại Thanh nhất thống chí* in Hán-Việt (Sino-Vietnamese), "Unified geography of the Qing Empire", which consists of a series of local monographs from the whole Qing Empire. It should be noted that the character 統 (*thống*) is written here by Devéria 銃 (*súng*), which is an error, since 銃 names an ancient weapon. The second, is the *Da-Qing huidian* [大清會典圖], *Đại Thanh hội điển đồ* in Han-Viet, "Inventory of the Qing Empire's maps".

20 On the official Chinese *Geographies*, see J. B. Harley and D. Woodward (eds), *History of Cartography, op. cit.*

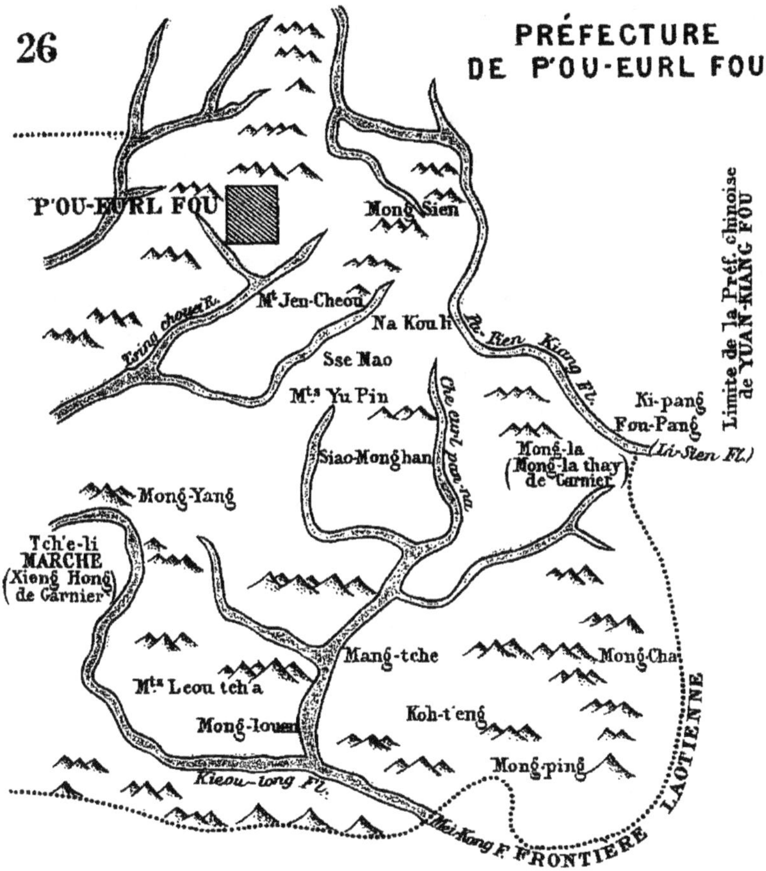

FIGURE 20 *Préfecture* of Pou-Eurl Fou, by Gabriel Devéria. It should be noted that François Garnier's presence in the area is mentioned (twice)
SOURCE: Gabriel Devéria, *La frontière sino-annamite. Description géographique et ethnographique d'après des documents officiels chinois traduits pour la première fois*, Paris, E. Leroux, 1886, p. 64 [http://gallica.bnf.fr/ark:/12148/bpt6k5438516v/f114.item]

Thanks to this work, Devéria could show that a border existed between China and Tonkin. The distances given in the "Great Imperial Geography" between the Chinese towns and the China-Vietnam border proved that it was a reality, and that a sufficient number of points existed to make reconnaissance possible. He observed:

> Even in the most recent editions of purely Chinese geographical works, the Peking court could find nothing to corroborate what some of its civil servants maintained, that *the border of China is not well known or*

> *demarcated*, and that between the two countries lies a vast region, and no one knows to which country it should be attributed.[21]

This is how he countered the idea of an ill-defined frontier region, while acknowledging that semi-independent and non-Chinese populations probably inhabited the area, and that dividing them up was made more difficult by the fact that the Chinese had gradually annexed their lands.[22] He also argued from the absence of any mention in the Chinese sources of the five most northern Tonkin provinces that China could not lay claim to them without contradicting itself. He also contested the route of the China-Vietnam frontier drawn by the Jesuits in the 18th century, showing that it left more territory to China than the official Chinese sources themselves, suggesting, therefore, that this route should be redefined. By combining European and Chinese sources, he hoped to provide the most precise information possible on these little-known frontier regions, and to base French territorial claims on Chinese sources.

In 1887, the French established the China-Vietnam border, using historical sources to prove to the Chinese that certain territories of the former Annam Empire should become French, in the name of political continuity. In this, they turned to the Vietnamese elites, and also to the Historiographical Office, attached to the Ministry for the Interior, for information on borders. The Office's Director, Hoàng Hữu Xứng, the author of the *Đại Nam quốc cương giới vựng biên* (*Treatise on the frontiers of the Đại Nam*), written on the order of the king in 1886, provided them with information from geographical treatises.[23] In his own work, he made a list of the place names, rivers and mountains that indicated the path of the border, a method later copied by the Europeans. The latter requested clarifications and explanations, when the toponyms or other information were unclear – a sign of the importance these sources had, in their view. It is worth noting that the French continued to attribute importance to the history of the area after military conquest: a new set of relations based on negotiation and diplomacy was being introduced, and the French made every

21 G. Devéria, *La frontière sino-annamite. Description géographique et ethnographique d'après des documents officiels chinois traduits pour la première fois, op. cit.*, p. vii–viii. Emphasis in the original.

22 These borderlands were inhabited already in the eighteenth century by a mixture of indigenous peoples, non-Han, as well as Chinese migrants. See Patterson C. Giersch, *Asian Borderlands. The Transformation of Qing China's Yunnan Frontier*, Cambridge, Harvard University Press, 2006.

23 ANOM, GGI 14451, *Renseignements sur les frontières de l'Annam, du Siam, du Laos, du Cambodge et de la Chine, extraits des livres géographiques composés par Huynh huu Xung, Ta tham, tri du ministère de l'Intérieur, bureau des historiographes*, 1887.

effort to provide a semblance of proof for their territorial claims, instead of imposing them by force. Hoàng Hữu Xứng's work was based on a large number of historical and geographical references:

> Given the loss of a number of books and manuscripts caused by the events of our Capital, I had no choice in my geographical treatise but to copy official documents concerning our borders, and to add those notes and opinions I found in the older geographies and in various books that I considered to have some importance for our limits. [...] These are the passages I position each in its due place and that I was able to assemble from ancient maps and various commentaries.[24]

It was not the first time that absent – lost, burnt or missing – sources had been mentioned. The Europeans leapt on these occasions in order to further their own designs, given the lack of available "historical proof". Documents could be lost due to invasions (the case evoked here), or else due to other unforeseeable events such as civil war or bad conditions of conservation.

The Europeans attempted to legitimate territorial claims through indigenous sources every time a new territory was annexed or borders were drawn. For example, when the French annexed Laos in 1893, the question of the historical rights of Annam over these principalities re-emerged. In order to add weight to their claims over the Laotian regions also coveted by Siam, the French invoked the notion of historic rights, and pointed to the original territorial situation. They positioned themselves within a history they were also writing, pointing to the sound basis of their political claim over spaces officially designated as independent. The British supported Siam's claims out of fear that France's influence would spread further over these Laotian lands. In the absence of documentation, all sorts of fantastical narratives flourished, which hindered diplomatic exchanges. This is why during the Anglo-French Border Commission set up to establish the borders and determine the fate of the three contested frontier provinces, the Governor General of Indochina ordered his second in command in Hué to "ask Van Minh to have all the documents relating to the rights of Annam over Xieng-Hong and Xieng-Kheng found as a matter of urgency."[25] It was again in Hué – Vietnam's centre of power – that the French looked for documents on Laotian territories. This may have been

24 *Ibid.*

25 ANOM, GGI 26665, *Mission Pavie. Commission franco-anglaise pour constituer un État tampon entre la Birmanie et l'Indochine. État de Xieng Kheng. Lettre du gouverneur général de l'Indochine au vice-résident supérieur à Hué*, Hanoi, 25 August 1894.

because the French were unable to gain access to the sources locally, or else because they wanted to compare viewpoints.[26] Whatever the reason, they adopted the Vietnamese viewpoint, and used the royal geographies and other available documents to base their territorial claims on historical data, and so to legitimate their demands by taking over the position of the powers they were replacing, in this case the court of Hué.

The Europeans thus used the official geographies of the Indochinese Peninsula with a very precise objective in mind, namely to determine the limits of their territories. They were interested neither in how the maps looked, nor in the symbols used, but only in the place names that marked the border route. Given this, exchanges and borrowings were naturally limited. Hence it would be overstating it to say that the official Asian geographies contributed to fashioning the European representation of these spaces, because the Europeans paid little attention to the course of the rivers and the symbols for the mountains. At most, the geographies gave them some idea of the history of the territories, which they could exploit as they wished. And these sources were used to varying degrees. Either translations existed already, as was the case with Gabriel Devéria's translation of the Chinese sources, or they were made on the spot when the borders were being demarcated, at the request of the Europeans. The meagre number and low informative quality of the Siamese and Laotian sources were a sad contrast to the many informative Burmese and Vietnamese archives – in the opinion of the Europeans. For this is the real issue here: the Europeans used the historical documents with a particular end in view, namely to find out where borders lay, in order to know which territories they could legitimately claim on the basis of historical precedent. Both the British and the French sent the information gathered to their Border Commissions for use in the field in their negotiations with Siam or with China.[27] Again, what interested the British and the French were not the myths of the country's foundation, nor praise of the Sovereign's greatness, but the territorial losses and gains over the years, insofar as these could support their present territorial claims.

Did this tradition of official geographers continue after colonisation? In the case of Burma, there could be no further royal cartographies after the sovereign's deposition in 1885. In Vietnam, however, there was the case of the *Geography* undertaken during the reign of Đồng Khánh (1886–1888).

---

26 See the work of translation and explanation of the chronicles of Xieng Kheng by Volker Grabowsky and Rēnū Wichāsin (eds), *Chronicles of Chiang Khaeng: A Tai Lü Principality of the Upper Mekong*, Honolulu, University of Hawaii Press, 2008.

27 See Chapter 8.

1.3 *Đồng Khánh's Geography*

Đồng Khánh's *Geography* was the swan song of royal Vietnamese geographies, the last work of its type. It comprised 23 booklets bound together, devoted to the Vietnamese northern and central provinces.[28] Officially commissioned by the King, in reality the work was coordinated by the National Office of the Nguyễn Annals.[29] Nothing is known of its authors, except that they were provincial mandarins – provincial governors, and heads of counties or districts. It is impossible to identify them precisely, because the reports were not dated and these posts were held for short periods. The key architect of this vast project might have been the chief advisor, Hoàng Hữu Xứng, but his name does not figure anywhere. It was a collective work, based on a vast network of collaborators from every district of the country. The officials in charge could dispatch envoys to obtain first-hand information from the inhabitants themselves, concerning place names and distances, thus guaranteeing that the information was precise and trustworthy. Each mandarin probably received the same standard form requesting information, which was then distributed to heads of counties and districts.[30] These forms provided the fundamental information on which the reports, sent back to the Royal Court, were based, and from which Đồng Khánh's *Geography* was compiled. This method seemed both efficient and reliable as a way of gathering information from all the four corners of the kingdom. The project probably originated in Tự Đức's wish to update the *Encyclopaedia of the Empire of Dai Nam* [*Đại Nam nhât thông chi*], in 1882. Although some mandarins worked faster than others, and the distance from the capital was also a factor, it is probable that all the reports had arrived at the Office of the Nguyễn Annals by the end of the reign of Tự Đức (1847–1883). However, corrections were made subsequent to Đồng Khánh's reign, particularly of place names, which adds further unknowns to the history of this *Geography*, although the most probable scenario is that the French were behind the project.[31] Whether or not this was the case, the French certainly followed it closely.

28 See the impressive volume published on the subject, from which this information is taken: Gô Đức Thọ, Philippe Papin, Nguyễn Van Nguyên (eds), *Đồng Khánh địa dư chí*, Hanoi, Institut Hán Nôm/EFEO, 2003.

29 This Office had more in common with the Chinese High Command than it did with the colonial geographical departments, even if it was indeed the place where different imperial *Geographies* were developed. The French turned to this institution to verify the historical rights of Annam over the territories they coveted.

30 Ngô Đ. T., P. Papin, Nguyễn V. N. (eds), *Đồng Khánh địa dư chí, op. cit.*, p. lxxiii.

31 This is the hypothesis argued in *ibid.* It is also supported by Nguyễn Đình Đầu (an interview carried out in Ho Chi Minh City (Saigon), February 2015).

The map-making process used for Đồng Khánh's *Geography* was certainly its most striking innovation. It borrowed from European scientific techniques in its use of direct inquiries in the field, and it was revolutionary in introducing district-level maps. Despite the difficulty in assessing whether European practices were imposed, it is possible to examine what was borrowed and what was not, and to evaluate whether the Vietnamese reproduced European ways, or whether they transformed these.[32] As with the British and French maps, one of the fundamental goals of these Vietnamese productions was to establish place names and the most detailed notes on them. The instructions given by the Director of the *Service géographique* to the Residents whenever the maps were revised were very clear, and the British issued similar forms, requesting all the information necessary for making the map, including the name of the local chief, that of the ethnic group, and its location. In this respect, the mandarins operated in exactly the same way as the Europeans, and borrowed the model questionnaires, which were the building blocks of the final map. However, the real conditions in which the *Geography* was composed remain mysterious, and data such as distances, the number of tax-liable individuals, and the amounts of tax payable seem to have been derived from earlier practices, and were passed up the hierarchy of officials in the traditional way. This obliges us to refine our notion of 'borrowing', which in this case applies to the practices but not to the finished product. Whereas in general the Siam adopted European cartographic methods, and produced topographic maps that entirely abandoned the indigenous cartographic tradition, Đồng Khánh's *Geography* remained, in its execution and in its choice of symbols, in the tradition of official Vietnamese geographies. What was new, and showed the adoption of European techniques by the Vietnamese, was the fieldwork, and the fact that the same instructions were given to all the mandarins, such that the information gathered was standardised and systematised; at the same time, indigenous techniques were used for making the map itself.

Sometimes indigenous maps were indeed used to provide an historical precedent for European claims concerning the path of a border, but at other

---

32 We agree with the terms of Philippe Papin's position: "We shall now add a clarification, in order to dispel any ambiguity: there is no value judgement implied in this way of seeing cultural transfer, no connotation of inferiority for a culture that imports something unchanged, and no methodological regression. It is worth noting that the opposite position runs no less risk of malevolent interpretation. To say that transfers are constantly being adapted, reinterpreted, ingested, to appropriate the foreign, could signify an overestimation of the exogenous culture, considered so attractive that it can only seduce and be appreciated." P. Papin, "Un laboratoire pour l'étude des transferts", *loc. cit. paragraph 18* (https://books.openedition.org/demopolis/468).

times they were used by the Europeans simply to understand the area and be able to govern it, especially since the spaces in question were almost entirely unknown.

1.4 *Collecting Ephemeral Maps*

Interest in indigenous geographical practices, in the wealth of knowledge they contain, and the way this is passed on, has given rise to research on a whole range of spaces.[33] It is common to read of a local inhabitant sketching a rough map for a stranger, and this is an almost obligatory scene in travel writing.[34] Yet although frequently mentioned, these sketches are rarely reproduced by explorers and have only recently begun to interest historians. Bruno Latour's work has helped us begin to formulate how Europe capitalised on indigenous knowledge. He refers to the episode recounted by La Pérouse when some Chinese drew a map for him in the sand, on the island of Sakhalin.[35] This map was simply an indefinitely reproducible drawing for the locals, but for La Pérouse it meant that the sole goal of his journey had been achieved, namely to acquire written proof that Sakhalin was an island, evidence that he could send back to the Court of Versailles. One could point out, however, that La Pérouse already knew that Sakhalin was an island – he was well-placed to confirm this – and that the map in the sand did not so much prove that Sakhalin was an island as prove that the local population knew this. Be that as it may, the fact remains that the map in the sand was conclusive for La Pérouse, and so one can speak of a capitalisation of geographical knowledge. Using the same example, Michel Bravo shifts the focus from the visual trace to the verbal exchange that took place afterwards between the local inhabitants and the explorer.[36] In order to illustrate how much more powerful than writing are gesture and

33 For example, Joël Bonnemaison, *Gens de pirogue et gens de terre*, Paris, Orstom, 1996; Béatrice Collignon, *Les Inuit. Ce qu'ils savent du territoire*, Paris, L'Harmattan, 1996; for North Africa, see Dominique Casajus, "Las, le temps non, mais nous nous en allons …", *Textuel*, 56, 2008, p. 185–210; for Sub-Saharan Africa, see the article cited earlier: C. Lefebvre, "Itinéraires de sable. Parole, geste et écrit au Soudan central au XIXe siècle", *art. cit.*

34 C. Jacob, *L'empire des cartes. Approche théorique de la cartographie à travers l'histoire*, *op. cit.*, p. 58.

35 B. Latour, "Comment redistribuer le Grand Partage?", *art. cit.*; Id., "Les "vues" de l'esprit: une introduction à l'anthropologie des sciences et des techniques", *Culture technique*, 14, 1985, p. 4–29.

36 Michael T. Bravo, "Ethnographic Navigation and the Geographical Gift", in David N. Livingstone, Charles W. J. Withers (eds), *Geography and Enlightenment*, Chicago, The University of Chicago Press, 1999, p. 199–235.

speech in the acquisition of geographical knowledge, he emphasises the signs of respect exchanged between the parties.

There are some rare examples of ephemeral maps of this sort being made for European explorers in the northern Indochinese Peninsula. Hallett recounts that after calculating the position of the hills around him, near Kiang Dow, north of Chiang Mai, he began to make a sketch, but the curiosity he aroused from the villagers was such that he asked their chiefs to draw a map for him:

> Whilst I was sketching the hills and fixing their positions, the villagers came crowding round me, and peeped from all directions at the picture I was making. When I had finished, I made the most of the opportunity by getting the head-men into the temple to make a map of the country with matches and bamboo strips on the floor. All were very good-natured, and I learned from them the position of the sources of the Meh Ping, Meh Teng, Meh Hang, Meh Pai, Meh Nium, Meh Pam, and other streams.[37]

Thus all means were considered good to achieve the most prized result, namely, in this case, a graphic geographical representation. One can imagine how the exchange between Hallett and the local chiefs must have proceeded, with Hallett trying to convey the information he wanted to know, and the headmen reacting to him by positioning the river sources directly on the improvised map. Regardless of whether the information was true or false – which only a subsequent inquiry could ascertain – we have here an instance of capitalisation in Bruno Latour's sense. Likewise for the French explorer Jules Harmand, who returned from his journey in 1877 with a sketch made for him by a Laotian of the course of a tributary of the Mekong – called, according to the names probably written by Harmand on the copy of the sketch included in his account of his travels, "Great River" [Figure 21]. Harmand gives details not of the document itself, but of how it was made, thus revealing contemporary cartographic practices in this space:

> It was drawn in white chalk on a slab of black lacquer, the same as those used to make drafts of letters. Then the sketch map was etched onto a palm leaf with a stylus, and by rubbing engraved lines with a mixture of oil and soot from a lamp, they were penetrated so that they stood out in black, distinctly and indelibly.[38]

---

37 H. S. Hallett, *A Thousand Miles on an Elephant in the Shan States, op. cit.*, p. 332.

38 "[…] die mit weisser Kreide auf eine schwarzlakirte Tafel gezeichnet war, wie man sie dort für den Entwurf von Briefen benutzt. Dieselben werden später mit einem spitzen Griffel

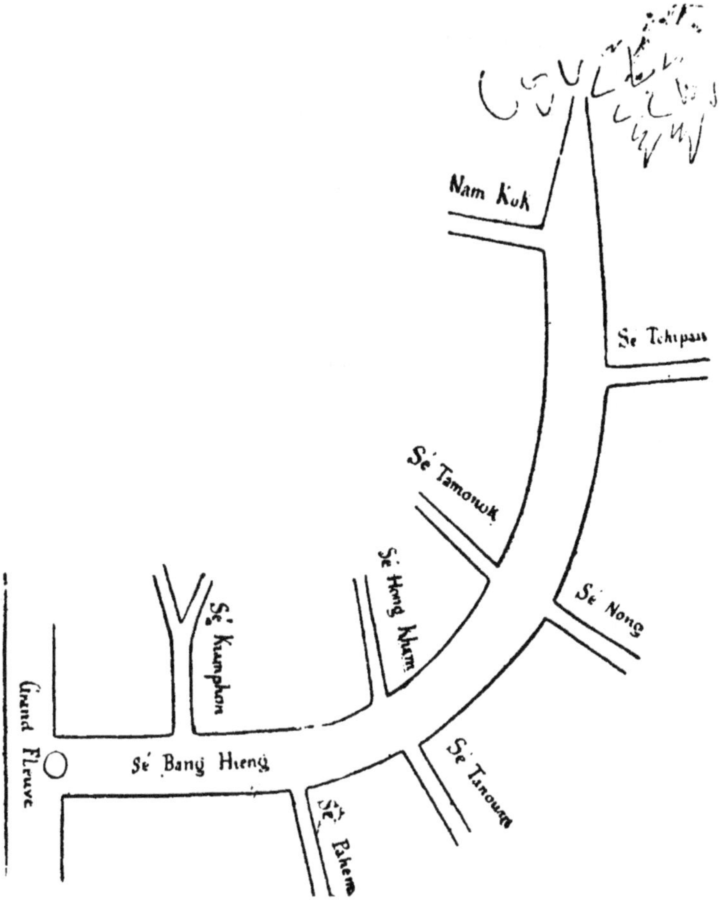

FIGURE 21 Copy of a Laotian map
SOURCE: "Im Innern von Hinterindien (nach dem Französischen des Dr. Harmand)", Globus, 38/14, 1880, p. 212

Harmand seems more interested here in the map-making process than in the information conveyed, and his scientific curiosity is roused more by a different technical procedure than by new knowledge of the territory. So this dimension should also be taken into account in the encounter between different cartographic traditions. A further scenario was that the local document was

auf ein Palmblatt eingerisst, und dann die Schriftzüge mit einer Mischung aus Del und Lampenruss eingerieben, worauf sie deutlich und unverlöschbar in Schwarz hervortreten", "Im Innern von Hinterindien (nach dem Französischen des Dr. Harmands)", *Globus* 38/14, 1880, p. 213.

considered unworthy of interest. For instance, when Jules-Léon Dutreuil de Rhins, employed as Navy Captain by Tự Đức between 1876 and 1877, made a "map of the river of Hué",[39] he mentioned a map made by the captain of a trading ship: "this map is regrettably nothing more than a simple sketch, and this is a pity especially because we intended to compare it with our own recent work."[40] In this case, no capitalisation is possible. It is even more difficult to analyse these productions when the local maps come down to us without any precise context.

1.5 *Compensating for Scarce Information*

The French made use of indigenous maps before the colonial expansion, as a collection of 16 Vietnamese maps sent to the *Dépôt de la Marine* between 1880 and 1881 testifies.[41] [Figures 22–26] They were maps of the provinces, and closely resembled the official Vietnamese geographies from which they had doubtlessly been taken, with traditional map symbols, the customary focus on rivers and mountains, and the text written in black Indian ink in Chinese characters. The translations into romanised Vietnamese or French, with corrections suggesting that interpreters had been used, corroborate John Whitmore's contention that they were read by the French.[42] After consulting 13 Vietnamese maps at the Bibliothèque National de France, and at the Bayerische Staatsbibliothek, Whitmore concluded that these local maps, consulted by the French, had helped them in the military conquest of Tonkin. At the time, there was indeed so little information that the question arises of how the military managed to get their bearings in this territory they did not really know. What documents did they consult? Translations on the maps generally concerned itineraries, journey times, and the routes to take. On some documents, the timing, added in pencil, suggests how the military used these maps, for example, on the bottom left of sheet 8 [Figure 27] we find:

> From *hai ninh* via *laon gian* and *luc ngang gia hiên* village arriving at *chau bien yen* 9 d. From *quan yên* via *hoanh bo vi Lai* village 5 days on foot; this route is difficult.[43]

39 Jules-Léon Dutreuil de Rhins, "Carte de la rivière de Hué, levée en 1876 et 1877", BSG, sixième série, XV, January–June 1878, at the end.

40 BSG, sixième série, XIX, January–June 1880, p. 15.

41 BNF, Cartes et Plans, Service hydrographique de la marine, portefeuille 180, division 2. 22.

42 J. B. Harley and D. Woodward (eds), *History of Cartography, op. cit.*, p. 506–507.

43 BNF, Cartes et Plans, Service hydrographique de la marine, portefeuille 180, division 2. 22, feuille 8.

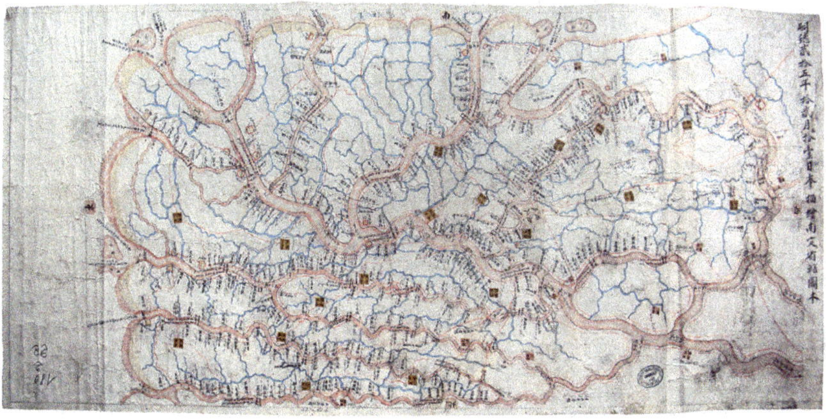

FIGURE 22 Map of Nam Định province, 1872
SOURCE: BNF, Cartes et plans, Service hydrographique de la marine, portefeuille 180, division 2.22, feuille 1

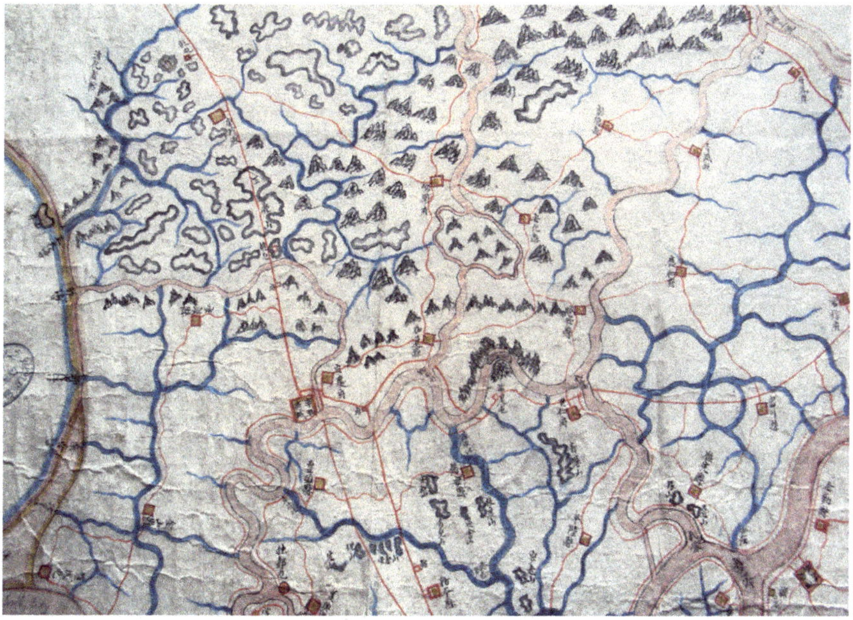

FIGURE 23 Map of Ninh Bình province (detail)
SOURCE: BNF, Cartes et plans, Service hydrographique de la marine, portefeuille 180, division 2.22, feuille 3

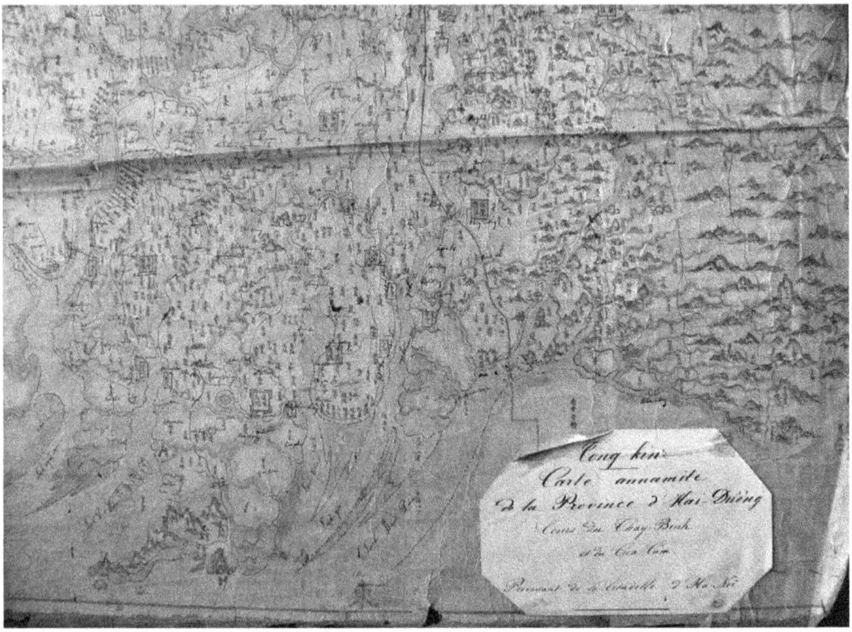

FIGURE 24 Map of Hải Dương province (detail). Explanation in French: "Tong-kin. Carte annamite de la province d' Hải Dương. Cours du Thay-Binh et du Cua Cam. Provenant de la citadelle d'Ha-Noï"
SOURCE: BNF, Cartes et plans, Service hydrographique de la marine, portefeuille 180, division 2.22, feuille 4

This annotation was undoubtedly meant for troops on the march. It shows that although the scale of the maps was too small to be useful for military purposes, Vietnamese maps were used in advance of the conquest.

The French and the British also made use of local knowledge during the period of conquest itself. At several points, the European officers' sketch maps referred to indigenous maps copied or consulted. A map called "*Routes de la région de Lang Son d'après les cartes annamites*" ["Roads in the region of Lang Son according to Vietnamese maps"] [Figure 28], notes in its key that indigenous sources were used, and it specifies that the map follows "the information provided by a Chinese map taken at Nuï-Bop." Here, a Chinese map has replaced a Vietnamese one, as though the French made no distinction between them. It should be said that it was not always easy to tell the two apart insofar as Vietnamese cartography used Chinese map symbols, as well as Chinese characters. The only sign that a map was made by the Vietnamese was when demotic script was used, that is, when the Chinese characters expressed Vietnamese. The French simply took the elements useful for understanding

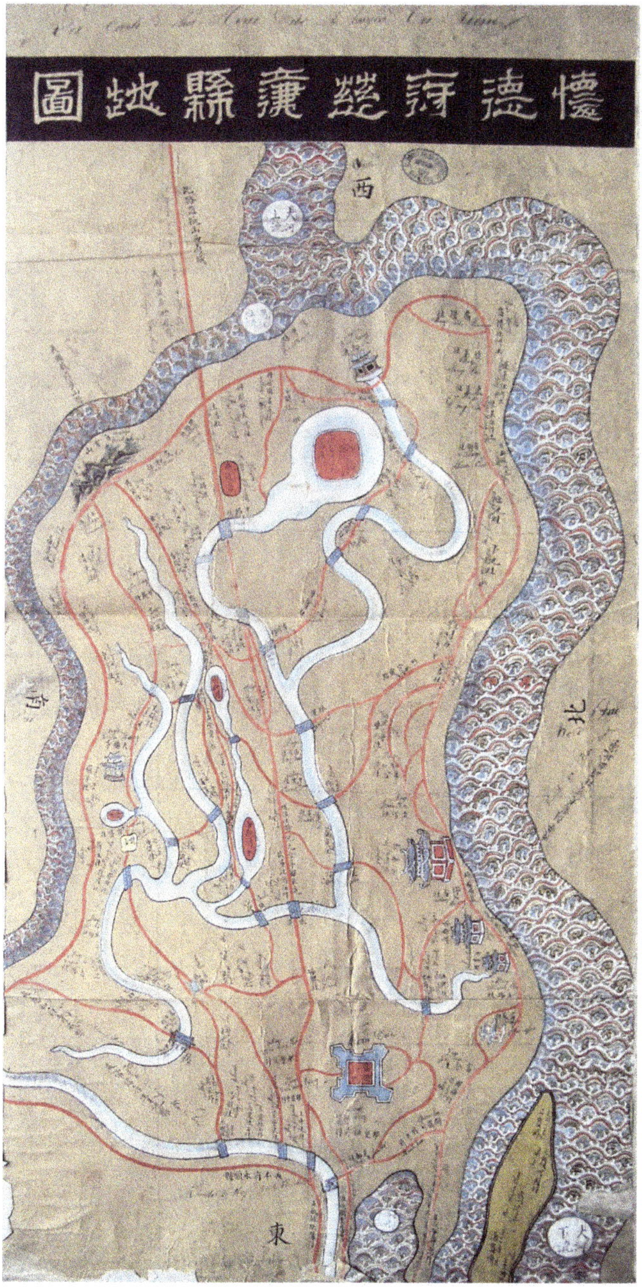

FIGURE 25 Map of Từ Liêm district, Hoài Đức phủ, Hanoi province
SOURCE: BNF, Cartes et plans, Service hydrographique de la marine, portefeuille 180, division 2.22, feuille 6

FIGURE 26 Map of Thái Nguyên province (detail)
SOURCE: BNF, Cartes et plans, Service hydrographique de la marine, portefeuille 180, division 2.22, feuille 13

FIGURE 27 Royal Geography of the Vietnamese province of Quảng Yên
SOURCE: BNF, Cartes et plans, Service hydrographique de la marine, portefeuille 180, division 2.22, feuille 8

the territory, in this case adding a remark – "battle of 23 and 24 June 84" – that was not present on the indigenous maps, but that appeared on the final "*Routes de la région de Lang*".[44] There also appeared a list of a few towns and villages, with the distances between them in kilometres, and additions and corrections made in pencil, doubtless by the officers – for example, 96 km instead of 72 km. They must have thought the original map was inaccurate, perhaps because they themselves had covered the same territory. Although it is difficult to know which information was provided by the indigenous maps, and which was subsequently established by the French, the final document is clearly a hybrid scientific product resulting from contacts between the two parties.

The first campaigns of the Survey of India in Upper Burma employed similar methods. For the 1886–1887 campaign, the British maps were based on "The information obtained from maps of the country prepared by Burmese

44 On military cartography, and on maps used to commemorate events and glorify battles, see V. Pansini, *L'œil du topographe et la science de la guerre. Travail scientifique et perception militaire (1760–1820)*, doctoral thesis supervised by Jacques Revel, EHESS, Paris, 2002.

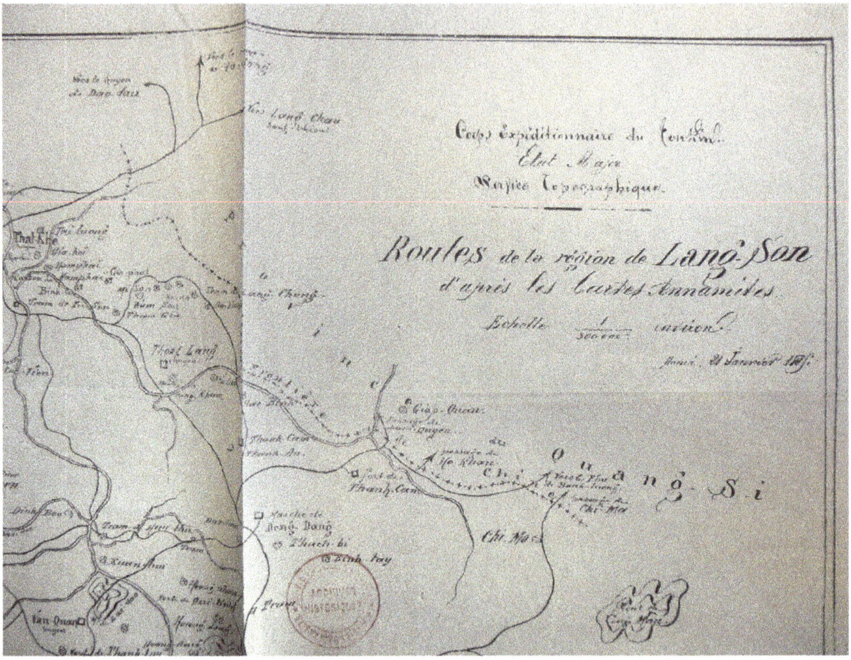

FIGURE 28 Roads in the region of Lang Son according to the Vietnamese maps, by the *Service topographique de l'état-major du Corps expéditionnaire du Tonkin*, Hanoi, 21 January 1885
SOURCE: SHD, 10H3, dossier 9, sketch no. 53 or 61, scale roughly 1:300 000. Dimensions: 43,5 × 39,5 cm

draftsmen and [...] the sketches and reports of trained Military Intelligencers belonging to the Native Army who have been trained at Roorkee."[45] These maps were therefore made by indigenous cartographers who were on the staff of the British administration in India. They were the first to provide the British with a sense of what the territory looked like. However, sometimes the indigenous maps discovered locally were mentioned not so much because they enabled gaps to be filled in the Europeans' cartographic knowledge, but because they featured something of geographical interest.

In April 1888, the Vice-Resident of Hải Dương sent a "Vietnamese map of the *huyen* of Vinh Bao" to the Resident General, adding that:

> it covers precisely the region for which you requested topographical information. The names of the villages are written in characters and in

45 BL, IOR, V/24/3985, Report on the operations of the Survey of India, administered under the Government of India, during 1886–1887, Calcutta, 1888, p. 67.

> *quo ngu* [*sic*], as well as the names of the cantons. It will be easy to recognise and distinguish them on the map, where the name of each village is followed by the word *sà* (village) and the name of each canton by the word *tông* (canton). Moreover, the names of the cantons are placed in a little rectangular-shaped frame.[46]

With the map, he thus sent an explanation, as a kind of substitute for a map key, to make the document understandable. It stressed the map's conformity to standards. The complexity of the real territory is once again palpable, with its administrative subdivisions such as the town and the canton represented in distinctive ways on the map. Sometimes indigenous maps were expressly commissioned by the Europeans, when they were not found by chance on the territory. This was already the practice with previous explorations, before the conquest, and it continued after it. As colonisation began in the region, and European officials began to settle, indigenous sketches and maps were commissioned more frequently, in order to obtain information about the territory the civil servants would have to get to know and govern.

The Asian maps were often used by the Europeans for the measurements they contained. Two sketch maps show this.[47] On the first, a route painted in white, running between two parallel lines in black ink, is interrupted by five framed place names, which represent post houses.[48] Commentaries on the itinerary taken are written in the margin of the map, in Chinese characters translated into romanised Vietnamese, with many spelling mistakes, for instance, "From Lạng Sơn province to Bắć Ninh province, there are 36 small waterways one can cross"[49] or "small craft can get as far as the Tram̆ Pier."[50] [Figure 29 and Figure 30] Again, experience and observation are the privileged instruments of this cartography which only represents itineraries. The map is designed for use, highlighting the itinerary and the stages, with comments on the points of departure and arrival, and information on the state of the road. On the second sketch map, the river represented in orange takes up the whole

46 ANV1, RST 73618–5, Lettre du vice-résident au résident général, Hai Duong, 9 April 1888.

47 They were found at the SHD, 10H6. They have no title, date, or author, but their presence in the military archives suggests that they were requested or used by the French Army in the first years of the conquest. The first one has been analysed in depth in my article "Se repérer au Tonkin. Usage d'un croquis vietnamien par les Français lors de la conquête, 1883–1886", *Revue d'histoire du XIXe siècle*, 56/1, 2018, 173–185.

48 The five names, in Chinese characters and romanised Vietnamese writing, are "trạm Bắc Mĩ, trạm Bắc Ngàn, trạm Bắc Lê, trạm Lạo Quan, trạm Ninh Nhân."

49 "Từ tỉnh Lạng Sơn đến Bắc Ninh có 36 khe suối đều nhỏ hẹp có thể lội được."

50 "Sâ này cứ mùa này thì thuyền ván bé đến bến trăm mà thôi."

FIGURE 29 Itinerary in the province of Lạng Sơn
SOURCE: SHD, 10H6, 75 × 50 cm. By comparing a report, the places on the map, and the text "here lives the cai Vinh", we can conclude that the map was drawn by the Vietnamese soldier asked by the French about the road, see chapter 4

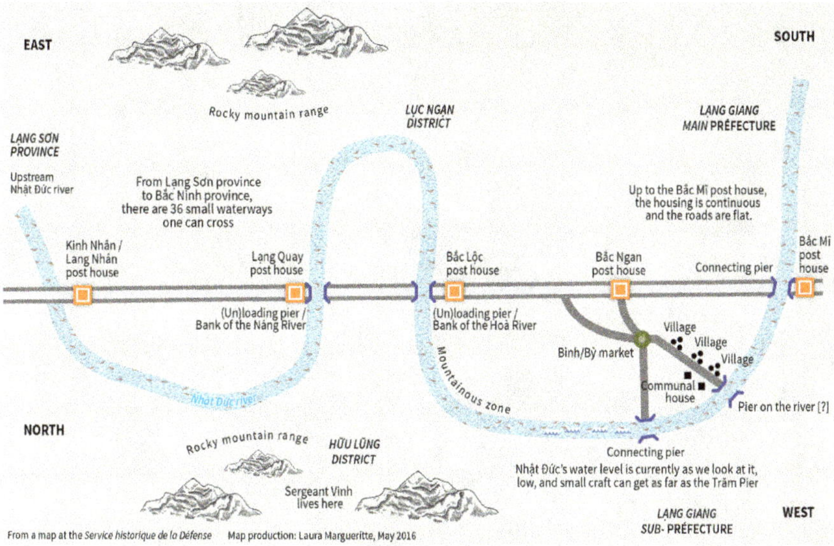

FIGURE 30 Explanatory sketch

map [Figure 31]. The distances between the different villages mentioned, and the measurement of the river for the whole length of the route taken are written in Chinese characters.

> From the village of Phác to that of Trú Hựu, a breadth of about 60 feet; Up to the waterfall, an average of roughly 3.5 ft deep; From the village of Phác to that of Cẩm Lý, an average of more than 15 ft deep; From Cẩm Lý to the junction with Phố, a breadth of about 200 feet; From Cẩm Lý to Trạm Điền, a breadth of about 200 feet; an average depth of more than 14 feet.[51]

The characters were translated into French on the map in pink pencil, often quite carelessly and with mistakes. There seems to have been some confusion between the units of measurement, mixing feet and metres:[52] "More than 200 m wide, 14 m deep", "200 Vietnamese m [*sic*] roughly – 15 m deep", "120 m 15 m deep", "60 m wide 3 m d[eep]". An annotation was added in French that does not correspond to anything in the original: "Beyond points A and B, the width

51 自 樸至 住佑 横 濶 約陸 拾尺零;以 上四 平常 約深 叁尺 半;自 樸以下至 錦里 平常 約水 深倣 拾五 尺零;自 錦里 以上 至我 巴铺 横闊約貳 百尺 零;自 錦里 以下 至湛 田横 闊約 貳百 尺零 平常 水深 倣拾 肆尺 零. I thank Nguyễn Thu Hoài for his help with the translation.

52 One foot = 0,04 metres.

FIGURE 31 Vietnamese sketch with French annotations
SOURCE: SHD, 10H6, 38 × 30 cm

is too narrow to cross in anything but a small sampan". So the indigenous map was used for the measurements it provided, translated and supplemented with the same end in view. On the British side, although we do not know how the Siamese maps discovered in the Grand Palace, Bangkok, were used, nor whether the Europeans had access to them, one cannot but notice the importance given to the dimensions of the transport routes represented. Of the 16 topographic maps, only two have no measurements of routes, and ten show distances and times, as was the custom in the Siamese literary tradition.[53] On six of the maps, regardless of whether there were distance and time markings or not, Siamese units of measurement were used.[54] On the map of Muang Thalang (Phuket), which represents the south of the Indochinese Peninsula, a text in Siamese notes that "the journey from Muang Lakh on to Muang Thalang

53 S. Phasuk, P. Stott, *Royal Siamese Maps: War and Trade in Nineteenth Century Thailand*, *op. cit.*, p. 22–23.

54 1 *sen* = 40 metres and 1 *wa* = 20 metres. An older unit of measurement is also used, the *yod*, which is equivalent to approximately 16 kilometres.

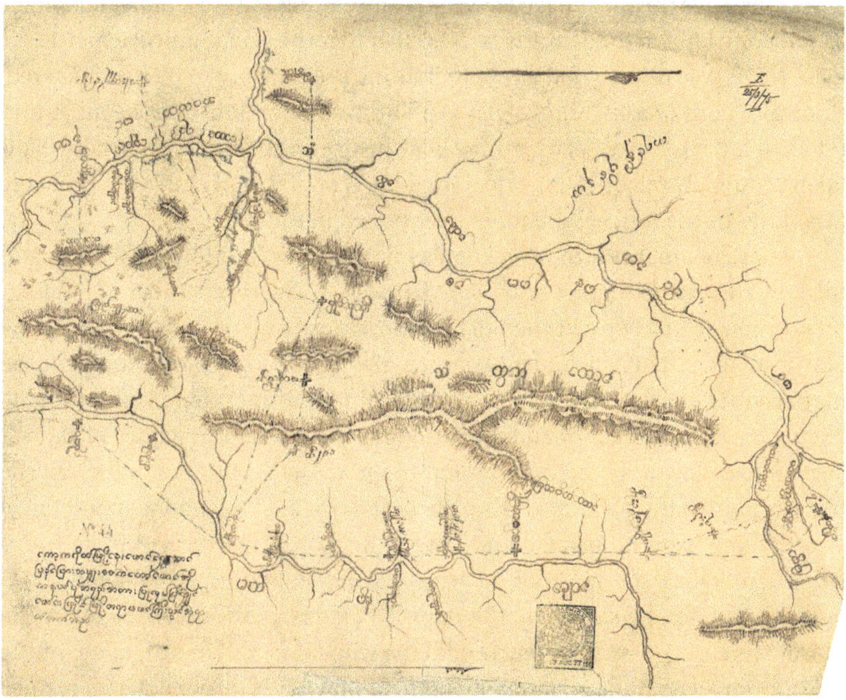

FIGURE 32 Burmese map of the region of Chiang Mai
SOURCE: RGS, mr Burma S.39, manuscrit no 44

takes 18 days."[55] The comments supplying additional information thus focused primarily on the distances, and the time taken to cover them, a sign that the Europeans and the local elites shared a common interest in these aspects.

1.6 *Commissioning Maps to Get One's Bearings*

Apart from turning to the local population as information sources, the Europeans also asked the local elites, heads of provinces and heads of villages to make maps for them, mostly for use by the colonial administration. A text accompanying a Burmese map explains the map production process [Figure 32]:

> Everything repeated by Maung Shwe Aung of the town of Kokareit was transmitted to the Lord Master Head of Justice in Moulmein after the territory had been delimited by the commander in chief Maung Oh.[56]

55 S. Phasuk, P. Stott (eds), *Royal Siamese Maps: War and Trade in Nineteenth Century Thailand, op. cit.*, p. 83.

56 I thank François Tainturier, U Htun Yi, and Tin Naing Win for this translation.

Although we do not know who Maung Shwe Aung was – a simple inhabitant or someone known to the British – it is clear that the geographical expertise of local chiefs was put to good use in delimiting the territory to be administered, since the "Lord Master" (*thakin paya* in Burmese) mentioned in the quotation refers to a British civil servant, the Chief Justice in the former Burmese capital of Moulmein (1827–1852). The dotted lines, added in pencil, were probably itineraries or the territorial borders mentioned in the text.

French documents also reveal details of the role played by local elites and the how their cartographic skills were harnessed. The commander of the military station at Núi Bóp attached to his report "a map made and brought by the Mayor of Yen Chau",[57] who had come to give him information. Although the map itself is not among the records, a local map is mentioned, and a indigenous practice. The commander continues: "I have had the names of the villages noted down, dictated by an interpreter. The blue pencil indicates the point occupied by the pirates." Knowledge is here adapted and recycled, since clearly there were no place names on the map, or else they had to be translated. This raises the question of map-making standards: what should, and what should not, be shown on the map? Details not considered useful by the Mayor were added by the French military, who had a different reading of the territory and who pursued different strategic objectives through the map, since for them the names were as important as the position of the enemy. As with the consultation of local inhabitants, the role of the interpreter was important in rationalising information for the map and creating a scientific object out of a collaboration between local protagonists and French officers.

The local elites provided maps at different scales, from the province down to the village. When local inhabitants were approached, it was generally for large-scale and precise representations of their village area, since this was the terrain they knew intimately. Thus Slymm, a Conservator in Korat, mentioned in the key of the map he was compiling that "reliable sketches by indigenous foresters and inhabitants of Mhinelongyee, Maihonthoon, and other parts of the foreign states" had been used, a remark that illustrates the role of local inhabitants in the Europeans' acquisition of a large-scale symbolic representations of the area.[58] We can note in passing the author's stress on the information's

57 SHD, 10H6, Dossier 16, Noui Bop, 12 March 1885. Le capitaine Falcon commandant le poste de Noui Bop à Monsieur le colonel commandant la circonscription Phù.

58 RGS, mr Burma G.13, "Map of the teak localities in British Burma, Burma Proper and the Shan and Karennee States, Scale: 1 inch = 16 miles; signed M. Slymm, Conservator of Korat, Salween Division, 19 October 1871, True copy, Revenue Surveyor, Amherst District, ms no 49".

FIGURE 33 "Reconnaissance carried out by the Né Nguyen Thu Lâm from 18 to 21 September [1884]"
SOURCE: SHD, 10H6, Dossier 10. Cartes, levés et croquis, 1884

reliability, to justify himself, and the priority given to the roads to take and the main communication routes.

On this latter point, it is noteworthy that already in 1795 Francis Buchanan-Hamilton asked a Shan slave who was making a map of Chiang Mai for him to change it and add distances.[59] If we compare the two maps we can find the trace of British influence in the representation of the territory. Discussions with inhabitants who had travelled in neighbouring territories or with trade caravans were sometimes accompanied by drawings, some of which have come down to us. For instance, a "Reconnaissance carried out by Nhaqué Nguyen Thu Lâm from 18 to 21 September" included a sketch map featuring a mixture of traditional local symbols and practical information probably requested by the army[Figure 33]. The relief was represented in the manner of traditional Chinese cartography, as was the fort. The size of the roads and the paths was mentioned each time, and the many bridges represented were accompanied by

59 Joseph Schwartzberg, "Southeast Asian Geographical Maps", in J. B. Harley and D. Woodward (eds), *History of Cartography, op. cit.*, vol. 2, p. 741–827, here p. 747.

their name or type. This information was probably requested by the military, given their interest in routes and crossings. One can infer from these elements that an encounter really did take place between two types of knowledge, which coexist on the final sketch map. The latter displays the practical experience of the local inhabitant combined with the adaptations required by the Europeans for the map to be communicable and usable. There are few examples of these types of maps on the French side, many more with the British. The most prized information remains the distances covered, the itinerary followed, and the towns passed through. For example, the map "brought by a Shan trader from Kaing Tong", Duce Tseng Beng, on Sunday 15 October 1871, represents the Shan States [Figure 34]. The rivers – the Salween and the Mekong – are represented in red, and the route taken, in black dots, is placed at the centre of the map, crossed by place names framed in circles or squares. The map's primary and essential objective seems to have been to represent the route taken. Since itineraries were based on information obtained from travellers and traders, they were often accompanied by drawings, on which local knowledge and European annotations and translations coexisted. Although few documents of this sort exist, suggesting that they were not systematically requested from the local populations, or were not preserved, the Scott collection has quite a few examples,[60] which show that for regional maps, distances and the territory covered were indeed the focus. These Shan maps depict the luxuriant vegetation, the roads in double red lines, the pagodas as landmarks, the towns and the bridges. The military stations and the stages are often represented, and on at least one map the distances are provided in English units or else in terms of the number of nights spent getting there.

•••

Thus Europeans used Asian maps for a range of precise goals: to understand the territory during the period of conquest, to ascertain the historical borders between the territories in preparation for the work of the boundary commissions, and to learn the place names, in order to facilitate the post-conquest administration of the area. These maps were of all sorts, and all contributed to the British and French cartographic projects: full-coverage, general maps, maps of the provinces, of itineraries, large-scale, and even ephemeral maps. Information-gathering was an oral activity, but the maps are the written trace, preserved or not, of the exchanges between European cartographers

---

60 *Ibid.*

FIGURE 34 Map of the Shan States by a merchant, 1871
SOURCE: RGS, mr Burma S.39

and local populations. No equivalent traces exist for French colonisation in Algeria or Sub-Saharan Africa, nor in such large quantities. They are thus a particularly valuable source, specific to Asia. Their distinctive characteristics include an absence of orientation marks, and of a key, and if they have a title it tends to have been added afterwards, in pencil. As for the nature and scale of these maps, they varied according to the needs of the Europeans who requested them.

It is difficult to measure the extent to which the information gathered and the knowledge acquired were incorporated into the Western maps. The Asian maps provided some detailed information, but, overall, they were considered insignificant and of no transferrable value. Apart from place names, distances, and the best itineraries, which remained much sought-after information, what ultimately figured on the topographic maps? Not a single Chinese or vernacular character, for a start, from which we can deduce that multi-lingualism was an intermediary stage to help the colonial government, and that the Asian terms were later removed. Thus the Asian cartographic documents obtained at the time of the conquest or to prepare the work of the boundary commissions, before the Europeans started their own mapping, were considered obsolete once the Europeans had produced their own maps, and no longer needed to refer to the indigenous ones.

The final sketch map did of course display two types of knowledge, testifying to the local inhabitants' lived experience and their attempts to make this comprehensible to the Europeans. But hybridisation was more in evidence in the drawings, the working drafts, and in the replies to specific requests from the colonisers. In these cases, local practices blended with European elements. There are few examples of these sorts of documents, and they are restricted to sketches and intermediate maps, so although they may be interesting in themselves, they should not mask the fundamental discrepancies between the two approaches to cartography. On the final versions of the British and French maps, no sign of these hybrid productions remains.

The presence of the Europeans in the Indochinese Peninsula put an end to the official geographies written there. The King of Siam paved the way for this, followed immediately by Burma, after the deposition of its king, then, progressively, in Vietnam. A Royal Geography was still published in 1886 by the Hué court, but it was the very last one, presumably because such an enterprise appeared futile subsequently, given that in producing maps and geographical descriptions Vietnamese mandarins and European topographers pursued the same goals, if they pursued any: to build a State administration.

This included the administering and control of the population, of defence, and of taxation. Transfers of cartographic practice were perhaps one sign of the fusion of interests, in which case the division would not be between Asian and European cartography, but between the administrators and those they administer. Defending and administering a territory, and extracting its resources, require knowledge and potentially also transformation of a country's spatial organisation, the path of its borders and its administrative divisions.

CHAPTER 7

Colonial Roads and Territorial Reconfigurations

> Their idiocy is on a par with their ignominy. We came to Tonkin to open up the Chinese provinces of Yu Nam and Quay Si to trade. It might have seemed natural to open up that famous Red River route immediately, after all the talk about it. A simple path two to three metres wide leading up the valleys of the Red River, the Clear River and of Lạng Sơn seemed an obvious choice; but Monsieur de Lanessan preferred to have roads of 11 metres wide built in the Delta region, costing more than half a million, inaugurated with much pomp and ceremony and expense, and which reverted to rice paddies with the first rains. Luckily the old Vietnamese roads of 7 metres wide still exist, and they will do. 200 kilometres north of Tuyen Quang is a post called Ha Yan. There is no road leading to it except a dreadful path you can sometimes take in winter if you do not have baggage. So all the convoys take the Clear River route, and at high water you can make three kilometres a day. The last convoy took 72 days to get there. The troops at Han Yan had asked for a full ration of bread as an extra for 14th July. You can imagine the splendid health conditions at this post. Well, all of this is gloomy, my dear brother, and there are some awful scoundrels in this world.[1]

∴

There are several levels of criticism here: the different treatment received by the Delta and the Upper Region, the mismanagement, and the poor adaptation of the colonial roads to the Indochinese terrain. Road and rail construction were grist to the mill of the advocates of colonisation, who stressed the infrastructures developed thanks to the colonial powers, whereas its opponents highlighted the exploitation of the local labour force entailed by building transport routes, with serious human consequences, for instance the roughly

1 AIP, Fonds Fernand Bernard, BEF1, Dossier 2, Lettre de Fernand Bernard à son frère Lazare Bernard, Tuyen Quang, 12 September 1892, p. 2–3.

 | DOI:10.1163/9789004469853_009

12,000 individuals who died building the Yunnan railway.[2] This colonial violence did indeed exist, but we shall focus here on the road as a geographical object rather than as an infrastructure with human and social implications.

Descriptions of the itineraries of roads, or maps of the different stages of a journey, are very common among the geographical documents produced between 1885 and 1914. This suggests that territories were defined above all through lines, and that they were interpreted in terms of roads, rivers and other communication routes, such as railways and telegraph lines. These gradually built up a network, and structured the spatial organisation of the empire.[3] The British and the French needed to control their extensive new territories for three essential reasons: military (to ensure the circulation of information, orders and troops), economic (exploitation of resources, and the search for outlets for their colonial production), and political.[4] Distances became obstacles to be overcome in order to link up spaces. This is why the colonial powers considered measuring and mapping the land to be vital to the conquest, and then government, of the new spaces. Journeys, explorations, and reconnaissance missions were thus of central importance because they provided information on distances, junctions, crossing-points, towns, and camps.[5] The abundant descriptions of roads in the reports show that the map's preeminent feature was the line: space was represented first and foremost as crossed and linked to other spaces. We shall examine how the primacy of linearity is conveyed in the representations of the territories, and whether road travel is described in order to inform about the journey undertaken, or in order to guide others who may take the same route in the future.

In the light of how important maps of itineraries and roads were to the Europeans, we shall address how communication networks affected the

---

2 This has become a central argument in the critique of colonial empires, for example in Rithy Panh's documentary, *La France est notre patrie* ["France is our Homeland"], broadcast in 2015, which relies on archival footage, particularly of huge road-building projects in Indochina.

3 A network can be defined as a set of lines or relations interconnected in more or less complex ways. A physical network shapes space, underpinning movements orientated by main routes and junctions. See J. Lévy, M. Lussault, *Dictionnaire de la géographie et de l'espace des sociétés*, Paris, Belin, 2003.

4 Claude Nicolet demonstrates these features for the Roman Empire in *L'inventaire du monde. Géographie et politique aux origines de l'Empire romain*, Paris, Fayard, 1988, p. 9.

5 The Europeans thus completed an "inventory of the world", not only by pushing back the frontiers of their known world (with reference to the Greek *oikumene*), but also by filling in the spaces left blank. On the notion of inventory, see Claude Nicolet, *L'inventaire du monde. Géographie et politique aux origines de l'Empire romain, op. cit.*; Emmanuelle Tixier du Mesnil, *Géographes d'Al-Andalus, de l'inventaire d'un territoire à la construction d'une mémoire*, Paris, Publications de la Sorbonne, 2014.

concrete shape taken by the imperial territories over time. Many pre-existing thoroughfares were maintained, like the Mandarin Road or the old trade routes, some of which were enlarged. Others were created from scratch, from the mountain path to the international road.[6] Who decided on their route, according to what criteria, and to what extent were decisions delegated to the local authorities? Moreover, roads do not simply structure territories, they also transform and reorganise them. As Chad Haines has shown for the north of British India, the path of major roads generates spatial transformations, by integrating spaces which had previously been too distant or enclosed, or else by orientating a region towards one centre rather than an other.[7] So which territories did the British and the French privilege or neglect? And what role did roads play in the way these territories developed? These questions are particularly relevant in the case of Yunnan, which was the focus of intense competition between colonial powers in the Indochinese Peninsula. The different British and French proposals for the development of transport routes, and their chronology, express this rivalry clearly enough, although this competition needs to be qualified. The arguments put forward, and the degree of execution of the plans, as well as the disagreements between those who were for, and against, Yunnan's incorporation into the Indochinese Peninsula, reveal the Europeans' goals in the area, and, more generally, raise the question of whether it was the transport routes that constructed the spatial identity of these marginal regions.

1 Travelling across the Territory

From the itineraries of explorers marked in red on the final maps to the geographical documents focusing specifically on transport routes, via the many reconnaissance sketches, which mostly show only the path taken, many European sources represented the land in linear form. The territory took shape for the Europeans first as lines. The course of major roads, and the nodal points of population centres – from the smallest village to the city – were the main territorial markers. Can this mode of representation be attributed to a modern,

6 The Emperor Gia Long (1762–1820) had the 1730-kilometre long Mandarin Road built along a north-south axis, to link the main Vietnamese towns from Lạng Sơn to Cà Mau in the Mekong Delta. Its name comes from the mandarins on their administrative missions, who travelled the country along this road.

7 Chad Haines, "Colonial Routes: Reorienting the Northern Frontier of British India", *Ethnohistory*, 51/3, 2004, p. 535–565; Id., *Nation, Territory, and Globalization in Pakistan: Traversing the Margins*, London, Routledge, 2012.

Western, scientific practice, or does it draw on earlier models and other spaces? Do the descriptions of the spatial organisation of the imperial territories show French and British practices to be identical?

1.1 *Working Drafts: The Road at the Centre*

Exploratory expeditions and military reconnaissance from the early years of the conquest and during the period of pacification focused on land and river routes, with a view to obtaining the largest amount of information possible concerning how easy they were, the quality of the bridges, and the possible fording points. This particular geography was not concerned with coverage of the territory, but with giving a linear vision of it in terms of its road network. The sometimes lengthy comments generally gave details about the route's practical aspects, and contained considerations on the itinerary taken. The experience narrated was an incomplete one, comprising only information considered important, namely the particular path taken, in the interests of a future journey, and to show what had been achieved. For example, the comment "Came from Chiang Mai, we took tea leaves"[8] informs us about the resources of the country, but also, and above all, brings the singular moment of the journey back to life and gives it a dimension of universality by being written on the map.

The quality of the transport routes was a key piece of information provided. The "Itinerary from Trai Lam to the ford of Thac Lo", described in January 1885 by a French officer of the Tonkin Second Infantry Regiment, mentioned four difficult sections, a "first set of very dangerous rapids because of the rocks", and a "fourth occurrence of rapids which the boat could not pass because of the current" [Figure 35]. The sketch focuses on the river route and its navigability.[9] Another type of information concerned the stages in the journey. The military dimension comes to the fore here, with the importance of places for camps and bivouacs, but this division of the journey into daily segments had greater importance than simply the passage of troops. In the "Map composed jointly by Tsaya Pay and Ko Shoay Kho of the district between Moulmein and Zimmay",

8 RGS, mr Burma S.39, manuscript no. 45, "Copy drawn by Maung Kaung, Land Survey, 24 June 1871".

9 These sorts of remarks, on the navigability of rivers, can also be found in Siamese cartography. The map of Muang Lakhon/Ko Mak/Thalang Sai, for instance, which represents the region of Penang and Phuket, bears the inscription "only small boats can take this route". S. Phasuk, P. Stott, *Royal Siamese Maps: War and Trade in Nineteenth Century Thailand, op. cit.*, p. 88.

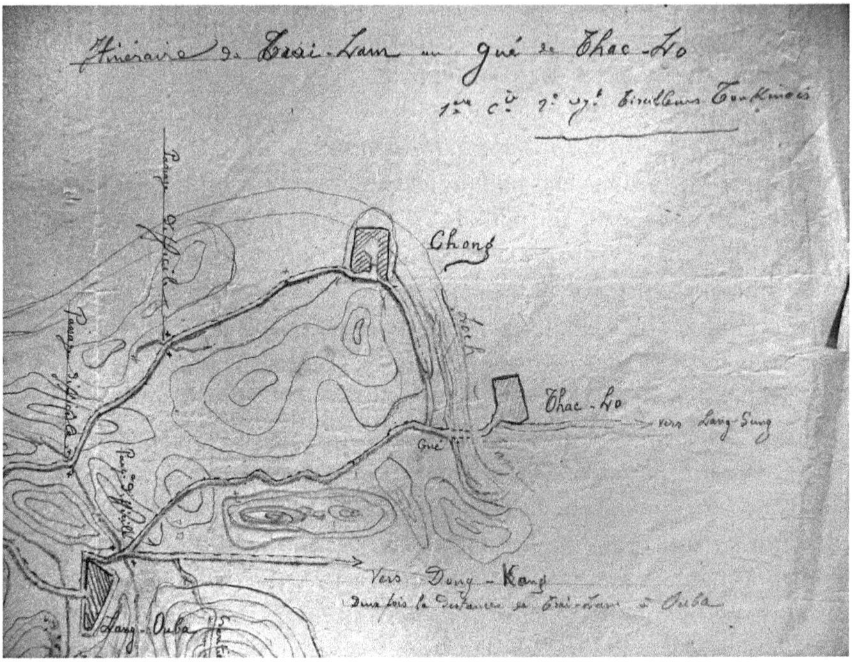

FIGURE 35 Itinerary from Traï-Lam to the ford of Thac-Lo
SOURCE: SHD, 10H3, dossier 9, a sketch attached to the "Compte rendu de la reconnaissance exécutée par la première Compagnie du 2e Tonkinois, de Treï-lam au gué de Thac-lo en passant par Lang-Ouba" ["Report on the reconnaissance carried out by the First Company of the Second Tonkin Regiment, from Traï-Lam to the ford of Thac-Lo via Lang-Ouba"], 22 January 1885

the places chosen to camp appeared at least twice.[10] They also often appeared on the topographic maps, as though they had been transferred directly from the itineraries to these more sophisticated representations, thus importing a trace of the experience of past travels. The importance of dividing up the journey and segmenting the space seems clear: the aim is to make the journey comprehensible by adapting its representation for other Europeans.

The third type of information was the distance covered, which was on almost all sketches and reports. The territory was conceived and represented first and foremost through the measured length of the journey, and the time it took. The unit of measurement was usually temporal – the number of hours, or days on foot – but it could also be spatial – kilometres or miles – without

10 RGS, mr Burma S.35, "Map composed jointly by Tsaya Pay and Ko Shoay Kho of the district between Moulmein and Zimmay", Indigenous map with English lettering, Tracing no. 32.

any very clear explanation for the choice.[11] The distances were rarely precise, however, either because there were no accurate measuring instruments in the early days of the conquest, or because there was no time to use them during the reconnaissance. In the document entitled "Reconnaissance carried out by Né Nguyen Thu Lâm from 18 to 21 September", the approximate distances in kilometres, which may or may not have been measured, were given just below the title.[12] The Vietnamese author seems to have had to make do with few means and no measuring instruments, except for his stride, which might explain the lack of precision. The distance between different points was often written out in full in the margin of the drawn itineraries. More generally, although mention of the time taken and the distance in kilometres or miles was standard, the unit of measurement – a cultural parameter – varied. A fish merchant, for example, noted in his account of a journey that he climbed a mountain "four times the height of a pagoda".[13] Here, a local landmark served as an informal and pragmatic unit of measurement, something which the Europeans were unable to absorb into their own cartographic standards, with the result that the distance quoted gave no more than an order of magnitude. Asian units of measurements were also cited, and they were no less difficult to integrate: "The Chinese measure distances in *li*, but the *li* varies according to the nature of the road taken."[14] This is why measurements in terms of time on foot were preferred. However, there was a further variable here, because an explorer or a merchant, for example, did not spend as much time as a scientific mission, a military detachment or a caravan of traders. The British captain Walker wrote:

> Letter-carriers come through by this route from Den Ben Pu in 17 days to Chieng Kawng, but for troops or ordinary parties of travellers it would be necessary to allow 25 days for the journey.[15]

---

11 Distance, in the geographical sense, is the number of metres or kilometres between two points. This is the metric distance. Yet "distance" can also be measured by other indicators, which are sometimes more relevant to the separation in question, for example, distance-cost or distance-time. See J. Lévy, M. Lussault, *Dictionnaire de la géographie et de l'espace des sociétés, op. cit.*

12 SHD, 10H6, Dossier 10, Cartes, levés et croquis, 1884.

13 SHD, 10H6, Dossier 13, Voyage d'un marchand de poisson de Luc-Ngan à Langson.

14 BL, IOR, L/PS/18/B116, Letter from James Turner to John Halliday, of the Arracan Company, Talifu, Yunnan, 30 October 1897.

15 BL, IOR, L/PS/20/83, Report on the Anglo-French Buffer State Commission, 1894–1895, by Captain Walker, Rangoon, November 1895, p. 24.

Within a single reconnaissance unit, a distinction was sometimes made between the time taken by the men on foot, and the time taken by the animals. In one instance, Walker clarified that fourteen days' march corresponded to the cattle, while the donkeys could do the same journey in six days. He also mentioned that one particular stretch was calculated according the pace of a pedlar, averaging 12 to 15 miles.[16] Measuring and dividing up the territory was the Europeans' overriding concern. Transport routes, divided into sections in order to measure the time taken or the distance covered as precisely as possible, and duly described in terms of their form, their surface quality and their ease of use, could therefore be called cartographic subjects in the fullest sense.

The surveyors went further than simply representing the road. They also drew what they saw on either side, and so presented a particular (and partial) vision of the territory, depending on what they were able to observe. This is apparent in the reports, where the road remains the point of reference, but also on the sketch maps, where only the relief and the villages lying near the path, and the vegetation, were drawn. The rest of the page was usually left blank, because it was considered to be too far from the path of the road. The sole features of the landscape, relief and vegetation to be noted were what could be seen by an observer at any particular time, as he advanced on his journey.

The working sketches thus gave shape to the territory around transport routes and their surroundings, from the perspective of the traveller, as though only these structures existed. Yet they also bore the traces of lived experience, through the date of the reconnaissance (always mentioned), and sometimes other features of the journey. The details reveal the singularity of a particular itinerary taken at a particular time, although other elements were included to help subsequent expeditions: the mention of distances, and places to strike camp. These connected the singular mission report to the cartographer's goal of a scientifically valid, hence universal, representation. However, only certain sections of the roads, and scattered pieces of territory, were treated on the sketches, without any overall framework. Did a change in scale, on the general maps of the colonies, produce less partial representations of the territory?

1.2 *Route Books, Road Maps and Maps of Communications*

The key on the general map of Indochina drawn by Dutreuil de Rhins shows a hierarchy between the "main road in Lower Cochinchina" and "the approximate

16 BL, IOR, L/PS/20/1, Report on the Keng Tung Keng Cheng mission for 1893–1894, by Captain Walker, p. 47–48.

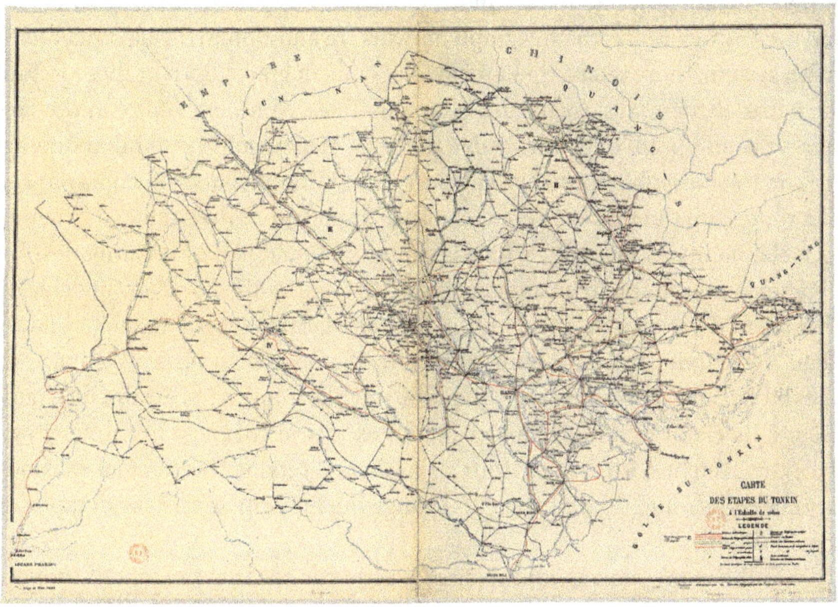

FIGURE 36 Map of stages for Tonkin, May 1894
SOURCE: BNF, GE C-1920

course of the road, according to the information obtained by Europeans or from the travel narratives of locals."[17] Below it is information on the roads in Laos, taken from Dr Néis's explorations in the region, namely a list of ten itineraries, and the number of days' march for each. The descriptions were transposed directly onto the map, without changes, but selected for their usefulness in their new format. The itineraries, also transposed unaltered, completed the picture and clarified it. In this process, what information disappeared and what was retained? Were the maps, at the end of the day, anything more than a collection of particular itineraries, and did they enable a qualitative leap in the representation of how the space of these territories was organised? Although the British and the French favoured different media – text, maps – in their attempts to master the terrain, both were updated at roughly the same time and pace.

A *Map of Stages for Tonkin*, the *Carte des étapes du Tonkin*, drawn at the scale of 1:500,000, was made by the *Bureau topographique des troupes de l'Indochine*

17 Jules-Léon Dutreuil de Rhins, *Carte de l'Indo-Chine orientale*, Paris, Dépôt de la guerre, 1886. BNF, Cartes et Plans, GE C-756 (1–4).

in 1894 [Figure 36]. The roads, and the distances in kilometres, are in black, and the key focuses on the telegraph network. In addition to the electrical telegraph system, there is also the optical telegraph, in green. Both of these follow the paths of the roads, rivers and boundary lines. Another symbol in the key marks the end-point of the navigable route (by steam), and yet another one, an anchor, the navigable route by sailboat (junk). All transport and communication routes are therefore covered. Another import from the sketched itineraries is the mention "bivouac", which is sometimes qualified by a Vietnamese or a French name after it in brackets, such as " bivouac of the Rock", or "bivouac of Thai Nien." The very specific mention of a "small cave", or "the hamlet of Ranh Chau", were most probably transferred directly to the map from a sketch, and, like the encampments, bear the mark of singular itineraries. So the full map, despite its claim to a more universal and extensive coverage of the territory than the sketched itineraries, thus nevertheless retains the trace of particular trajectories, and does nothing to change the linearity of the cartographer's transpositions.

With the map's second edition, in 1899, one can see that the representation of the territory has changed [Figure 37]. On this map, in colour, the rivers are shown in blue, with their names, the place names are in black ink, and the distances, in kilometres, are written in red above the roads, also in red, and represented by lines. The map's key represents nodal points, for example, a black dot in three concentric circles for a territory's capital, in two circles for the capital of a province or an administrative *cercle*, in one circle for a *secteur* or a delegation, all alone for a military post, and a smaller version for a village. Sometimes village markers are replaced by the mention "bivouac", to show the itinerary taken by the reconnaissance team, with its different stages. The distances noted are of 20 or 30 kilometres on average, with extremes of one and 330, west of Lai Châu. As such, they generally correspond to the average daily distance covered by the Europeans. The directions noted on the eastern side of the map and towards China, show the time required, in number of hours or days, to reach the indicated destinations. A distinction is thus made, within the map, between distances in kilometres obtained through observation, and time-distances that use second-hand information about the imperial confines, especially in the case of Chinese towns. When a charted itinerary is transformed into a map, this dual system of measurement is retained, as is the linear conception of the territory. Blanks remain, and a road will suddenly stop – corresponding to the end of a reconnaissance trip – and then start again at a neighbouring village, where another itinerary began or ended. The differences are visible to the naked eye, with dense, connected lines in the eastern

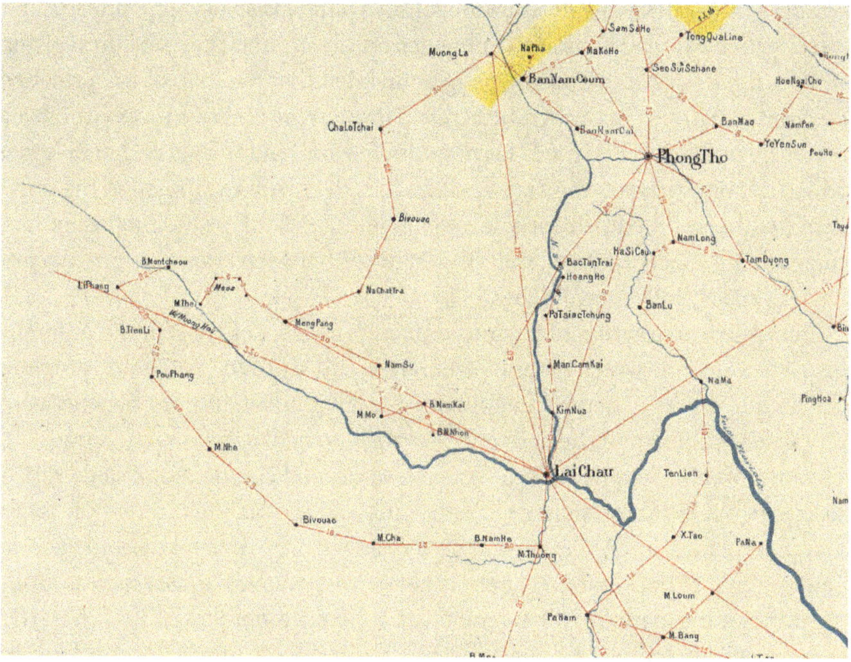

FIGURE 37 Map of stages for Tonkin, May 1899 (détail)
SOURCE: BNF, GE AA-2074

Red River delta, and blanks further west. However, overall, this *Carte des étapes* manages to represent the major spatial features of Tonkin, unlike the British road maps.

The British too tried hard to keep their itineraries up to date. In 1904, the *Burma Route Book*, the road map of Burma, replaced an earlier edition of 1894, just when the first edition of the French *Carte des étapes du Tonkin* was being published.[18] It was a compilation made by Captain Nangle at the request of the Intelligence Branch, organised into four parts, arranged geographically: the roads of the north of Burma, then the north-east, the west and, lastly, the south. The new edition had been required due to the wealth of information collected on the routes between 1894 and 1904, and because several roads had been built over the period, making the first edition obsolete. As in the case of the maps, it was important to update the material regularly, even if this was so difficult to achieve that some maps and descriptions of itineraries were out

18 BL, IOR, L/PS/20/D141/1, The Burma Route Book, Calcutta, 1904.

of date before they were even printed. Cartography is a very long process and maps will always lag behind the changes on the ground they seek to register. In 1901, the Resident of Tuyên Quang demanded that the map of the province be entirely redrawn because it bore no relation to the reality. In support of this claim, he pointed to two roads leaving Chiêm Hóa, built in 1897 and 1898–1899, which did not figure on the map at all. In fact, the regular mapping of the province had been scheduled for a few years later, and the provisional map had therefore not been updated. But discrepancies between map and territory are anyway structurally unavoidable.

In order to obtain the most up-to-date information, officials in the different districts were sent sketch maps of itineraries, and asked to check and complete the details. It was generally the province-level colonial authorities who were mobilised, although in the case of the British no actual maps were requested. The information supplied was always the same, and closely resembled French diaries of operations: the point of departure and arrival, the number of stages and their names, intermediate and total distances, the time taken, and general comments. The information generally concerned practical issues such as available stocks, fuel supplies, or the quality of the encampments. The itineraries were drawn according to a specific code, to produce uniform descriptions, in which roads were divided into stages, and the journey was segmented into days of marching; the day of marching was the standard unit of measurement. As stated, the *Burma Route Book* was a compilation of all the routes taken by the military in the years 1886–1888 or in the early 1890s, supplemented by information from local sources. Each British mission report contained a mini route book, which was then integrated into the general *Route Book* in 1894, and again in 1904. The British military personnel were pragmatic in their approach, referring back to previous reports rather than repeating descriptions of itineraries already mentioned.

So the 1904 *Burma Route Book* brought together all the itineraries undertaken, and remained up to date, thanks to the input of the provincial authorities. The British seemed satisfied with a compilation of itineraries, combined with detailed descriptions of the routes, rather than opting for a general map specifically produced to show the transport and communications network. The latter would have shown the overall spatial organisation of Burma, but this was anyway available in the form of the general topographic maps.

The French used the same methods as the British to keep their information on routes up to date, but they added maps. At the beginning of the 20th century, they undertook the huge task of replacing the earlier *Carte des étapes* with a new *Carte des communications*. The borderlands, under military administration, were the first areas to be represented in booklets of travel stages, with distances in kilometres, the time required for each leg of the journey, the means of transport used, and details concerning marching columns and detachments. The booklets, which were similar to the British route books, were used for two military 1:500,000-scale maps published in April 1894 and in 1903 (close to the publication dates of the *Burma Route Book*).[19] The idea was to replicate this method for the non-military provinces. In November 1902, the Governor General of Indochina requested information from the Resident of Tonkin concerning transport and communication routes in the provinces under civil administration.[20] The priority was to represent the road links between territories, and the other lines of communication such as the telegraph, much like the *Carte des étapes* before it. The Resident Superior of Tonkin gave clear instructions to the Residents to measure distances between the Resident's headquarters and the main sites of his province such as the European towns, plantations and other productive operations. The instructions were issued on 6 December 1902, and exactly 13 days later, on 19 December 1902, the head of the *Service géographique* asked the Superior Resident of Tonkin for:

> The assistance of the Resident administrators of the different provinces of Tonkin in updating the fragment of the map concerning their territory as regards the marching stages, the river routes, the post and telegraph offices, and the limits of their provinces, according to the attached instructions and table of conventional symbols.[21]

On 30 December 1902, all the Residents received a memo on the production of the map, and the table of the conventional signs to be used, devised by Mussy, the head of the *Service géographique*. There were railways, in use and under construction, tramways, with the same distinction, roads – metalled, suitable for motor vehicles, or simple mule tracks – rivers, navigable by dinghy all the time, or at high water only, navigable by sampan or not navigable, the administrative centres, the limits of the province, in green, and of military territories, in red, the post or telegraph offices, or both [Figure 38]. The memo explained the table and defined the terms used in it.

In early May 1903, the *Service géographique* declared that the information collected by the Resident Superior of Tonkin on the existing transport

---

19 BNF, Carte et Plans, GE C-1919 (1–4) et GE B-8235. These maps are available on: gallica.bnf.fr.

20 ANOM, GGI 66185 (C6).

21 ANV1, RST 73507, 25, Lettre du général de division Coronnat au résident supérieur du Tonkin, Hanoi, 19 December 1902. The idea was to take the *Carte des étapes* drawn at a scale of 1:500,000, and to add to it railways, tram lines, and telegraph lines.

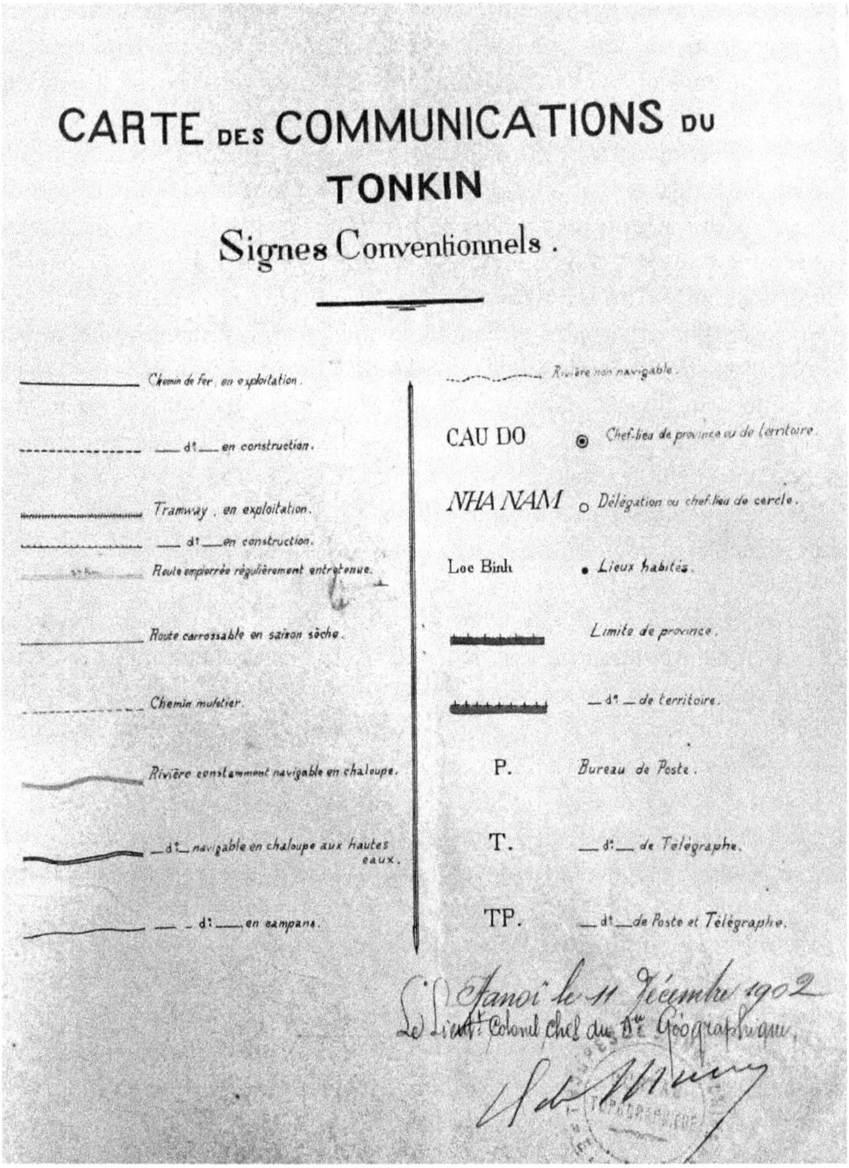

FIGURE 38 Conventional signs used for the *Carte des communications du Tonkin* [*Map of Communications of Tonkin*]
SOURCE: ANVI, RST 73507, Annexe signée par Claude de Mussy, chef du Service géographique, à la lettre du général de division Coronnat au résident supérieur du Tonkin ["Letter from the Major General Coronnat to the Resident Superior of Tonkin, Annex, signed by Claude de Mussy, Head of the *Service géographique*"], Hanoi, 19 December 1902

and communication routes was inadequate. The overall situation was comprehensible, but too imprecise to be of use. Paul Beau therefore asked the Residents to:

> [...] redo their work and complete it, taking into account the precise indications contained in the attached memo and model, devised by the *Service géographique*. Apart from the schematic map of his circumscription, on a scale of 1:500,000, every head of province must provide a booklet of stages similar to the one that exists in the military Territories.[22]

The sketch maps submitted in late 1902 and early 1903 were indeed of very variable quality, from a bare outline to very careful work. It was rare for a key to be supplied because the report or table of distances generally gave explanations for each province. The map itself was insufficient, and only made sense as an illustration of the roads described in the report.[23] The desire to be exhaustive was again in evidence, with all the distances marked from one point to the next, as stipulated. The most important administrative centres, the border points, and the population centres were marked. But the main roads traversed mostly uncharted space, such that the map of Tonkin was essentially a representation of its abstract spatial structure. Bridges were the focus of much comment, and concern, with many Residents noting the presence of "makeshift bridges", requiring "prior repairs, since they are in very bad condition." Crossing-points were already mentioned as a major preoccupation in the first sketches from the 1880s, and little change seems to have taken place since then.

The colonial powers thus described the territory of the Indochinese Peninsula first in terms of communication routes, in line with the European tradition of road maps. The French tended to represent the territory visually, through sketches and maps, whereas the British favoured written accounts. Although potentially more precise than maps, written accounts were less able to convert a prior experience into a range of possible future ones, that is, to convert a particular itinerary into something universally accessible. British Route Books were compilations, and, for Burma, they did not provide a clear

22 ANOM, GGI 17541, Lettre du gouverneur général de l'Indochine, Paul Beau, au résident supérieur au Tonkin, 6 May 1903.

23 The 23 reports accompanying the sketch maps mirrored the 25 booklets that formed the basis of the reports later sent to the Court of Đồng Khánh. They were shorter and less precise, but they used a similar method, in which text and image were mutually reinforcing, and both were produced by the provincial authorities. Pride of place was given to roads, highways and other routes. See Ngô Đ. T., P. Papin, Nguyễn V. N. (eds), *Đồng Khánh địa dư chí, op. cit.*, p. xvi.

image, comprehensible at a single glance, of the territory's spatial structure. The French *Carte des communications du Tonkin* went further in this direction, even though one had to refer to the reports to obtain practical details about the roads. The latter were always foregrounded, regardless of the nature of the geographical document. So why were the Europeans so interested in roads, and in what ways did they contribute to empire-building?

2 The Margins Reconfigured by Roads

Roads were not only a primary concern in cartography, but their planning and building were the object of much debate since, whatever their scale – a much-used main thoroughfare or the mountain paths linking isolated villages – roads organised the movement of people and of goods. In the light of the many roads built during the colonial period by the colonial administration, we shall ask in what ways the local elites were involved, and the effects of the resulting territorial reconfigurations on the local population in their daily life and movements. Building a new transport route can produce spatial homogeneity, but equally, spatial differentiation, depending on the density of the territorial communications coverage. Being located close to lines of communication can bring a competitive advantage, but also make outlying places even more distant.[24] Which territories were given precedence, and why, in these road building projects? What were the routes ultimately chosen, and with what territorial consequences? On this issue, unrealised projects are as illuminating as completed ones, in that they reveal divergent viewpoints on how the territory should be conceived.

2.1 *Why Build Roads?*

Although roads had a central role to play in organising the imperial territories, they did not all have the same strategic importance for the Europeans. Cross-border roads were a major military issue for the British and the French. On the one hand, before frontiers were demarcated between empires or kingdoms, thoroughfares were viewed as providing access for armed attackers, and hence represented a threat of invasion by a neighbouring country.[25] On the other hand, roads in the border regions enabled the military to keep an eye on

24 See Jean-Marc Offner, *s. v.* "Réseau technique", in See J. Lévy, M. Lussault, *Dictionnaire de la géographie et de l'espace des sociétés*, *op. cit.* p. 872.

25 BL, IOR, L/PS/20/83, Report on the Anglo-French Buffer State Commission, 1894–1895, by Captain Walker, Rangoon, November 1895, p. 67.

movements there, and gave rapid and easy access to the frontier, on which all effective control and repression depended.[26] In addition to the road network, there were also railways and the telegraph network, with its interconnections and extensions to strengthen control by the colonial administration. In 1892, Monsieur Massie, the French manager of the post at Luang Prabang, reported that the English were beginning to install an optical and electrical telegraph system between Chiang Mai and Luang Prabang, that is, between northern Siam and a territory which at the time was coveted by both the British and the French, and where it was therefore quite sensitive to implant communications materials.[27] Priority was given to linking the margins to the centre of the empire, and connecting up its different areas. But who decided on the routes of the road, rail and telegraph networks? Can one speak of a politics of road planning?

The Chinese, Siamese, Europeans, and the local populations were each responsible for building and maintaining certain roads, depending on their size, and on the way they shaped and linked different spaces. The colonial administration, which was responsible for public works projects, took its orders from central government. The presence of roads consequently gave visibility to the nation state in the furthest reaches of its territories.[28] British and French road-building in the borderlands of the Chinese provinces thus not only provided much sought-after economic outlets, but it additionally expressed the imperial power of the metropole. The Yunnan Railway, built by the French, is a good example of this, as is the road from Burma.

Although Engineering and Public Works Departments piloted many of the road-building projects, local governors often had delegated powers in this field. On 14 November 1894, Captain Walker noted that the route to Möng Pawn had been improved and enlarged to form a 6-foot-wide road, probably on the orders of, and monitored by, the *Sawbwa* himself, who had already made improvements to many roads in his State, right across to its Eastern frontiers. Walker was similarly impressed two days later when he took another road, renovated and broadened to a width of 12 feet by order of the *Sawbwa* of Monè, and with milestones every quarter mile.[29] Local rulers thus also took

26 Eric Tagliacozzo, *Secret Trades, porous Borders: Smuggling and States along a Southeast Asian Frontier, 1865–1915*, New Haven, Yale University Press, 2005, p. 77–78.

27 ANOM, GGI 14470, Extraits du rapport de M. Massie, gérant du poste de Luang Prabang, pour la période du 11 au 31 mai 1892.

28 C. Haines, *Nation, Territory, and Globalization in Pakistan: Traversing the Margins*, *op. cit.*, p. 6.

29 BL, IOR, L/PS/20/83, Report on the Anglo-French Buffer State Commission, 1894–1895, by Captain Walker, Rangoon, November 1895, p. 2–3.

responsibility for maintaining and enlarging roads. Walker hoped that the Burmese, or Indian, colonial government would compensate them for this.

The situation was similar in the French territories, in the frontier zones where roads were first built by the military and where the Chinese were also active, during the conflicts of 1885. The French tried increasingly to delegate road-building and maintenance to the local rulers, for example in the territory of Đèo Văn Trì, which received funds from the Government General for this.[30] Sometimes it was the colonial administration itself that delegated powers, as when Captain Tod, a member of the Boundary Commission demarcating the China-Burma border, mentioned in 1899 that significant efforts had been made in the State of Hsi Paw to improve transport routes, officially under the authority of Lieutenant Drage, but in fact thanks to "a very competent Burmese man, who had been assigned responsibility for road-building."[31] Whether carried out by Europeans or delegated to local officials, road-building played a major part in the exercise of sovereignty, which explains why the French and the British were wary of the Chinese encroaching upon their own territories, yet did not hesitate to extend their own major road systems towards China. The rivalry between the two powers for ascendancy in Yunnan, and the different plans for incorporating this province of south-west China into their respective imperial territories, generated some singular territorial reconfigurations.

2.2 *The Competition for Yunnan*

The representations of Yunnan changed considerably from the 1880s to the turn of the century. In December 1886, a Chinese mandarin described the roads from Yunnan to Burma and to Tonkin as dangerous, with attacks by "barbarians" and "savages", as well as tigers and panthers. Foreign travellers were strongly discouraged from taking them.[32] In 1895, during a British reconnaissance in Yunnan, Captain Davies wrote that the peripheries of Mong Lem and Yaunnanfu [Kunming] had been left out of all explorations and geographical inquiries, not only by the Europeans, but also by the Chinese. He deplored this

30 See P. Le Failler, *La rivière Noire. L'intégration d'une marche frontière au Vietnam, op. cit.*

31 BL, IOR, L/PS/20/D105, Report of the Intelligence Officer with the Southern Party, Burma-China Boundary Commission, 1898–1899, by Captain John Kelso Tod, Simla, 1899.

32 ANOM, GGI 14146, Dépêche du Tsong Li Yamen à M. Constans, envoyé extraordinaire de la République française à Pékin, le 22e jour du 11e mois de la 12e année Kouang-Sin, 17 December 1886. On the representation of the ethnic minorities of Yunnan by the Chinese officials of the Qing dynasty, see C. Patterson Giersch, *Asian Borderlands. The Transformation of Qing China's Yunnan Frontier, op. cit.*

geography of ignorance.[33] Yet the provincial authorities were obliged to make and update maps on a regular basis, including maps of Yunnan.[34]

Both the French and the British had expansionist designs on Yunnan, due to its geographical situation at a cross-roads, the wealth of resources it was supposed to possess, and the access it would give to Imperial China. Expeditions to study the roads between India and China started already in the mid nineteenth century.[35] The two powers sent out explorers and trade expeditions, such as the Colquhoun and Hallett expeditions, and the *Mission lyonnaise* (1895–1897), each camp keeping an anxious eye on its rival.[36] Travel writings and reports kept alive the myth of Yunnan, but it was deflated time and again by the facts gathered: the exploration of the Mekong, the Red River and the Salween demonstrated their limited navigability, and the expeditions into China, particularly the *Mission lyonnaise*, downscaled the expectations of wealth to be gained from Yunnan and its neighbouring provinces, due to the difficulty of the terrain.

Yet despite the expeditions' findings, business entrepreneurs went on believing that Yunnan had vast mineral resources, and potential for opium production, and that the trade routes from the rich province of Sichuan could be redirected towards Tonkin, or equally Burma.[37] The crux of the rivalry and fears between France and Britain was, indeed, trade, and trade routes – new and ancient, to be diverted or created. Apart from, but sometimes associated with, these economic considerations were other, national and imperial, aspirations, such as a "Greater France" or a "Greater Britain", as the Prince d'Orléans explicitly states in his account of his travels in the region:

If we are successful, then first we will be able to cross a territory from East to West that is particularly important for our possessions in Indochina, since it forms *their natural zone for peaceful commercial expansion* [...]. Secondly, we will be able to pursue the *very French* endeavour started 30 years ago by Francis Garnier, namely the exploration of the river that, from the point at which

33 BL, IOR/L/PS/20/94, Report on the part of Yunnan between the Bhamo frontier and the Salween, by Captain H.R. Davies, 1894–1895, Rangoon, September 1894.

34 Nanny Kim, *Mountain Rivers, Mountain Roads: Transport in the Southwest of Qing China, 1700–1850*, Leiden, Brill, 2012.

35 See Chapter 1.

36 Archibald Ross Colquhoun, Holt Samuel Hallett, *Report on the Railway Connexion of Burmah and China*, Allen, Scott & Company, 1888; Chambre de Commerce de Lyon, *La mission lyonnaise d'exploration commerciale en Chine, 1895–1897*, Lyon, A. Rey, 1898.

37 A text added to the 1895 Gérard Agreement, and signed by him on 20 June 1895, arranged for three towns in south-west China to be opened up to trade. French industrialists accordingly began mining operations there. Charles Fourniau, *Vietnam. Domination coloniale et résistance nationale, 1858–1914*, Paris, Les Indes savantes, 2002, p. 525.

it enters Indochina until its estuary, *is geographically, if not yet politically, entirely ours*.[38]

So the province of Yunnan is here considered attractive only insofar as it is a continuation of Tonkin, a "natural" extension of French Indochina. In 1899–1900, the Governor General Paul Doumer attempted to extend the political reach of French Indochina over Yunnan, and he was not the only one. Several article titles in *La Géographie*, the review of the *Société de géographie de Paris*, likewise presented Yunnan as the next step after Tonkin.[39] At the Universal Exhibition of 1900, the provinces of South-West China were included in the pavilions of French Indochina, and when a new edition of the map of the Pavie Mission was published in 1902, a note mentioned "the enlarged frame, to allow the inclusion of the Chinese province of Yun-Nan, which lies in the zone of French influence". This map also showed the path of the future railway from Haï Phong, the capital of French Indochina, to Kunming, the capital of Yunnan, thus reinforcing the idea of a close link between access to China and economic outlets.

Doumer's plans came up against those of Delcassé, the Minister for Foreign Affairs, who was assisted by Paul Cambon, the French ambassador to London. They were committed to reconciliation between the French and the British.[40] Doumer consequently abandoned the idea of formal political control of Yunnan, but maintained the project of the railway.[41] Unlike the British who, in the person of Lord Curzon, the Governor General of India, announced in 1901 that the project of a railway line between Bhamo and Yunnansen [Xiangcheng] was to be abandoned, the French kept working on the Yunnan line, section by section: from Hải Phòng to Lào Cai, completed in 1906, reaching Yunnansen in 1910.[42]

38 *BSG*, septième série, XVI, 1895, p. 391. Our underlining.

39 *La Géographie*, huitième série, I, janvier-juin 1900, p. 267–288, and II, juillet-décembre 1900, p. 294–296.

40 C. Fourniau, *Vietnam. Domination coloniale et résistance nationale, 1858–1914, op. cit.*, p. 549.

41 Rang-Ri Park-Barjot, *La Société de construction des Batignolles, des origines à la première guerre mondiale (1846–1914)*, Paris, Presses de l'université Paris-Sorbonne, 2005; Désirée Lenoir, *Le consul qui en savait trop. Les ambitions secrètes de la France en Chine (1886–1904)*, Paris, Nouveau Monde éditions, 2011.

42 On 10 December, 1901, in Rangoon, the Viceroy of India declared that no railway would be built between Burma and Sichuan, via Yunnan, because it would be sheer madness (Entry on "Lord Curzon et le chemin de fer du Yunnan", *Bulletin du Comité de l'Asie française*, Paris, December 1901, 1st year, 9, p. 376).

In the late nineteenth century, the penetration of the railways in the Indochinese Peninsula and elsewhere was enthusiastically followed by entrepreneurs, reporters and diplomats. Like the steam ship and the telegraph, the locomotive was conceived as an instrument of competition between Empires, full of strategic potential, since it reduced the time needed to cross and link different territories. Transcontinental railways were built, such as the Trans-Siberian railroad or the rail network of India. Delimiting spheres of influence for the construction of railways became a way of managing the antagonisms between the powers, as well as a way of defending imperial territories such as India, when under threat from Russia and Afghanistan in Central Asia.[43] Other transcontinental railway projects were proposed, sometimes for years on end, only to be later abandoned. The idea of the Trans-Saharan railway, for example, was pursued from 1870 to the 1950s, in five phases, but it was never executed.[44]

In the case of the Indochinese Peninsula, the British constructed a railway network in Burma, where the North-South routes from Rangoon to Pyay, and from Rangoon to Myitkyina via Mandalay were operative by the end of the century, and were the main thoroughfares organising the Burmese territory. The project of a Burma-Siam-China railway, advocated by Hallett and Colquhoun up to 1895, was abandoned as too complex and expensive. The most important project was the route from Mandalay to Lashio, eastwards, which could end up at Yunnan if it was extended, but it was finally abandoned at the beginning of the 20th century.[45] The railway line between Saigon and Mỹ Tho was built by the French in 1885, the first in Cochinchina, followed in 1890–1894 by the line between Hanoi and the *Porte de Chine* (north of Lang Son) to the North-East. For the French, the most important project was the railway route to Yunnan, which was completed in 1910, whereas a similar project by the British was abandoned.

43 John Fisher, "Lord Curzon and British Strategic Railways in Central Asia before, during and after World War I", in Thomas John Fisher, "Lord Curzon and British Strategic Railways in Central Asia before, during and after World War I", in Thomas G. Otte, Keith Neilson (eds), *Railways and International Politics. Paths of Empire, 1848–1945*, London, Routledge, 2006, p. 137–155.

44 Michael Heffernan, "Shifting Sands: The Trans-Saharan Railway", in Stanley Brunn (ed.), *Engineering Earth: The Impacts of Megaengineering Projects*, New York, Springer, 2011, p. 617–626.

45 For a summary of the different projects, see Nigel J. Brailey, "The Railway-Oceanic Era, the India-China and India-Singapore Railway Schemes, and Siam", in Thomas G. Otte, Keith Neilson (eds), *Railways and International Politics. Paths of Empire, 1848–1945, op. cit.*, p. 94–111.

The railways were instrumental in reshaping the territories they crossed, creating new movements and restricting others. In a note on the "Perspectives for trade with Yunnan and extending the railways", written on 19 November 1895, Edward Neel, the Head of the Burmese Public Works Department, took up Archer's conclusions of 1894 to the effect that only limited trade existed between Yunnan and Chiang Mai (Siam): only 9 caravans, or a total of 400 animals, had plied that route from China in 1894, with 800 the previous year. One of the explanations could be the loss of pack animals due to illness, but it was not the only one:

> Another reason is doubtless to be found in the improvement of the communications in the Shan States, and the extension of railways in Upper Burma, which will tend to drive all the trade of the Eastern Shan States to Upper Burma, unless communications with Lower Burma are improved.[46]

What transformations did these thoroughfares bring about, and how were trade flows affected by the railway? Although the railway did not cover enough territory to improve internal trade, it boosted exports significantly, such that in Indochina their value doubled between 1913 and 1926.[47]

A new geography of trade emerged with the railways. In the colonial period in North Africa, an "inversion of space" occurred when certain internal, mainly north-south, trade routes such as the Trans-Saharan routes, badly hit by the change, were replaced by routes leading to the coast.[48] Were similar displacements produced in the Indochinese Peninsula? Trade between the Indochinese Peninsula and the Chinese Empire went back a very long way, and the centuries-old routes taken by the caravans aroused the curiosity of the Europeans.[49] In the 18th and 19th centuries, under the Qing dynasty, trade

---

46 In BL, IOR, L/PS/18//B72, Prospects of Trade with Yunan, and Railway extension, E. N[eel], 19 November 1895.

47 India is an exception in this respect because the railway system consisted of more than a few access lines, and opened up the whole territory, and with it, a national market. See Dharma Kumar (ed.), *Cambridge Economic History of India, Volume II (1757–1970)*, Cambridge, Cambridge University Press, 1983.

48 This expression was coined by Marc Côte, *L'Algérie ou l'espace retourné*, Paris, Flammarion, 1988.

49 Paul Pelliot (1878–1945) was a member of the EFEO from 1899 to 1911, and had a brilliant career between Paris, Indochina and China. He decided to collect documents fundamental to the history and geography of the countries of the Indochinese Peninsula, particularly the Chinese sources. In 1904 he published "Deux itinéraires de Chine en Inde à la fin du viiie siècle" ["Two itineraries from China to India at the end of the eighth century"]

increased, using four main routes from and to Yunnan, particularly the one via Mengzi which follows the valley of the Red River to Hanoi and Hai Phong.[50] The Chinese merchants imported copper, rice, sugar and pepper from Tonkin, but above all cotton from the markets of Ava, Bhamo and Theinnee in Burma. They also looked for new routes to buy cotton in Siam, in Chiang Mai and Kengtung, and to expand their trade in mules. The mid nineteenth-century Taiping and Panthay Rebellions halted trade, which started again in the colonial period, but reorientated: whereas previously regional or international trade had been organised along north-south lines, it was gradually shifted to an east-west axis, towards the two ports of Saigon and Hai Phong.[51]

Although trade was expanded mostly in the south of the Peninsula during the colonial period, with maritime connections to Hong Kong, Singapore and Canton, it also continued in the north. Some of the trade routes used previously, such as the caravan routes, were adopted by the Europeans. For instance, a journey from Tonkin to Burma in June and July 1901 followed the path taken by the trade caravans between Dali and Bhamo.[52] The British were also interested in this route, especially since at the time they were developing the route between Dali and Bhamo, and were contemplating the need for a railway line between Dali and Yunnanfu.[53] Both the British and the French tried to find caravan routes, and mentioned them in their reports.

Another factor in the geographical impact of communications networks were the changes relating to the construction phase itself. Large-scale projects require significant resources. At the inauguration of the Yunnan Railway in 1910, the project's director, Monsieur Getten, evoked the "difficult and expensive recruitment of thousands of coolies not only in Yunnan, but also in Kweichow, Kuang Xi, Kwangtung, Tonkin, in Northern Annam and as far as

---

(BEFEO, 4, 1904, p. 131–143), in which he mentioned the route of the Red River and that of the Lô River, which was more difficult but considered by the Chinese to be more direct. The article was accompanied by a commentary on the geographical naming system, which was a cornerstone for future histories of the region. On Pelliot, see the proceedings of the conference *Paul Pelliot. De l'histoire à la légende*, Michel Zink, Jean-Pierre Drège (eds), Paris, Académie des Inscriptions et Belles-Lettres, 2013.

50 C. Patterson Giersch, "Coton, Copper and Caravans: Trade and the Transformation of Southwest China", in Eric Tagliacozzo, Wen-Chin Chang (eds), *Chinese Circulations: Capital, Commodities, and Networks in Southeast Asia*, Durham, Duke University Press, 2011, p. 41.

51 P. Brocheux, D. Hémery, *Indochine. La colonisation ambiguë (1858–1954)*, *op. cit.*

52 "Un voyage du Tonkin en Birmanie", Bulletin du Comité de l'Asie française, 9, December 1901, p. 378.

53 BL, IOR, L/PS/10/13. Memorandum by Litton (the British consul at Tengyue) and reports by Wilkinson (the British consul at Yunnanfu) on the best route, 1904.

Fuzhou, Ning Po and Tien-Tsin".[54] Once the itinerary of the railway had been decided, the project also had to mobilise over 60,000 workers, in an area which stretched well beyond the line of the Red River. It transformed the territory crossed, as huge numbers of people flocked to the region, where "whole villages sprang up along the railway line, a busy, teeming crowd of workers and traders, always on the move."[55] Whatever the bias and limited vision of this representation, it shows that road construction produced new territories, insofar as communities were created as the line extended. This raises the further question of whether the communities disappeared after the construction work moved on, or whether they became permanent.

As we have seen, the two powers were thus thwarted in their ambitions for Yunnan – the French did not manage to integrate the province into French Indochina, and the British decided not to build a connecting railway – but the episode had considerable territorial impact. The rivalry between the French and the British, and their different approaches, have been amply treated in the scholarly literature. What a geographical focus has been able to show, through examining the plans proposed, the arguments advanced, and the altered territorial dynamics generated by the implementation of the selected project, is the importance of this rivalry in concretely shaping the imperial territories of both the British and the French.

2.3 *Redefining Territories*

As from the first decades of colonisation, other territories, apart from the hotly contested province of Yunnan, were reshaped by the construction of communication routes. Their influence on the territories should be considered over a long time period, and particularly between the two Wars, when infrastructures were developed intensively in the colonies. However, we shall look at the beginning of this process, in order to understand how transport and communication routes reshaped territories, and the extent to which the colonial powers were key agents in this.[56]

Spaces were reshaped not only by the lines marked by roads or railways, but also by altered nodal points and crossroads. The colonisers tended to keep the former administrative capitals as centres, and to preserve the pre-existing

54 BL, IOR, L/PS/10/13, Dossier 1032 (1910), discours de M. Getten, directeur général des chemins de fer du Yunnan.

55 *Ibid.*

56 This issue has been addressed for the colonial period in India, Ian J. Kerr, *Building the railways of the Raj, 1850–1900*, Delhi, Oxford University Press, 1995; and also for a longer time-span: Ian J. Kerr, *Engines of Change: The Railroads that Made India*, Greenwood, 2007; John Hurd, Ian J. Kerr, *India's Railway History: A Research Handbook*, Leiden, Brill, 2012.

territorial hierarchies. Where a strong administrative system existed prior to colonisation, as in Tonkin, with a hierarchisation of centres and a network covering the whole territory, enabling spatial control, the colonial powers tended to maintain the spatial organisation already in place.[57] However, over time, the hierarchy between the centres could change, with some gaining in importance and others being abandoned. Transformations were not only caused by colonisation, however, but also by local situations, such as the conflicts in the second half of the 19th century at the Chinese frontier. H. R. Davies, a member of the Boundary Commission for the frontiers with Siam, noted that he did not meet any caravan on the route which had previously been the main highway between Mandalay and Kengtung:

> Its desertion is due partly to the war between Möng Nai (Monè) and Laikha (Legya) in 1886–87, during which the town of Laikha was burnt and the country almost depopulated, and partly to the construction of the railway. The object of the traders now is not so much to get to Mandalay itself by road, as to get to anywhere on the railway, whence conveyance to Mandalay is easy.[58]

Sometimes the Europeans tried to revive a hierarchy which existed before the conflict, and to reinstate an older spatial organisation. Roads were important instruments in this process, for example the one between Fort Stedman [Maing Thauk] and Mong Nai. Mong Nai was a fortified town and a state capital, which therefore had a certain importance. It boasted 3,000 houses some thirty years before the conquest; the *Sawbwa* of Mong Nai was a highly prestigious position in the hierarchy. But conflicts erupted in 1887, and by 1893 less than 50 houses remained. Walker thought that opening up a road from Fort Stedman would help the town back on its feet.[59] The end point of the line, at the limits of British territorial possessions, was thus envisaged as becoming a centre again, and regaining its former political and strategic functions. However,

---

57 Pierre Singaravélou (ed.), *L'empire des géographes. Géographie, exploration et colonisation, XIXe–XXe siècle, op. cit.* See particularly the article by Paul Claval, "Réflexions sur la géographie de la découverte, la géographie coloniale et la géographie tropicale", p. 7–26.

58 BL, IOR, L/PS/20/16, Report on the Anglo-Siamese Boundary Commission, Northern Party, 1892–1893, by Captain H. R. Davies, Rangoon, 1893, p. 13.

59 "When the Fort Stedman-Monè carroad is opened, Monè will rapidly increase in importance, and in the event of a railway through the Southern Shan States, would become both its centre and its terminus eastward." BL, IOR, L/PS/20/92, Anglo-Siamese Boundary Commission, Southern Party, 1892–1893, by Captain Walker, Rangoon, September 1893, p. 2.

this redesign of territories got no further than the idea, because the railway was finally not built, and the British administration, located at Fort Stedman, was moved to Taunggyi, which was chosen as the capital of the Southern Shan States in 1894. The French consul pointed out that the town was "at the head of a very good road leading to Xieng Tong, along which there would be nothing simpler than to build a railway".[60] Preparatory work was carried out by the Department of Public Works in 1905, to link this road by train to the main North-South Rangoon-Mandalay-Myitkyina road.[61]

The specific case of the Shan States raises the question of the links between road and territory. Among the British conquests, certain territories remained marginal, yet this did not prevent them playing a central role in the flow of trade across them.[62] On the French side, the growth of the town of Điện Biên Phủ, in Tonkin, over the period considered, was due to the military and administrative functions it was allotted, and the construction of roads to enable access. One need only examine the size of the dot and the letters for Điện Biên Phủ on topographic maps of this frontier region to understand that this town expanded, a fact that is confirmed by the population figures.

Through road building, the colonial powers could thus reinforce existing hierarchies or create others, by privileging access to a particular place, for defence purposes or trade. Apart from changing the balance between towns in this way, roads could also affect the degree of integration of borderland areas into the empire.

The major thoroughfares seem to have been designed to orientate the border regions towards the empire's centre, and away from whatever lay beyond the frontier. Chad Haines argues this point on the basis of how the territories lying outside the British Empire were gradually incorporated into it through the building of the road network. This model applied particularly to the territory of the Gilgit Agency.[63] As access roads, they originated in territory

---

60 ANOM, GGI 250170, Lettre de M. Dautremer, consul de France à Rangoon, à M. Rouvier, président du Conseil et ministre des Affaires étrangères, Rangoon, 22 July 1905.

61 Among the black market trade routes between Burma, Thailand and China, the one linking Mandalay to Siam by Taunggyi, Lai Hka and Keng Tung, was the royal road for jade and opium, the products which constituted the largest volume of trade between Burma and Siam. The jade route, taken by the thirteenth-century Mongol invaders, was revived at the end of the seventeenth century by Chinese traders from Yunnan, and became the route to Burma. The Kachins were central to this trade, which was still active under British rule, but had ceased by the time of the Chinese Communist Revolution. P.-A. Chouvy, *Les territoires de l'opium. Conflits et trafics du Triangle d'or et du Croissant d'or. Birmanie, Laos, Thaïlande et Afghanistan, Iran, Pakistan*, Geneva, Olizane, 2002, p. 174.

62 *Ibid.*, p. 325.

63 C. Haines, "Colonial Routes", *art. cit.*

already secured, and were driven through to spaces to be connected with, or incorporated into the empire, in order, additionally, to cut them off from neighbouring areas.

This is how the French proceeded at the western frontier of French Indochina, seeking to draw Upper Laos and particularly the plateau of Trán Ninh[64] towards their own possessions in order to pry them away from Siamese influence. In 1904, a topographical brigade was sent onto the Plateau together with an expedition to study the path of the railway.[65] Here, the *Service géographique*, the Public Works Department, and the French colonial government joined forces, and in 1912, the government commissioner Barthélemy, located in Xieng Khuang, published a memo entitled "*Le Tran Ninh. Comment le tirer de son isolement*" [The *Tran Ninh*. How to bring an end to its isolation].[66] He provided precise information concerning the geographical characteristics of the plateau, its geology, and the natural resources to be found there. He criticised the dreadful state of the roads, and provided a costed outline of a road connection from the plateau to Tonkin. The road was thus the primary protagonist in integrating outlying territories into the empire and separating them away from their traditional partners.

In the case of non-demarcated frontier zones, where no clear line marked the territory's limits, roads defined the frontier. They began in stable, central areas, and led out to spaces where imperial power was not yet secured. Access to these borderlands was vital, not solely for military, strategic and trade purposes, but also because the construction of a road actually *produces* physical territories, and enables populations from the same imagined community to mingle.[67]

Where a border was already marked, roads made it a reality because it generated social transformations along its path. Imperial subjects experienced the border through the roads they had the right to travel on or not – the roads were thus sites of control. Moreover, the border's path itself depended on access to these frontier zones, without which the empire's limits could not be fixed or mapped. The description of transport routes holds pride of place in Boundary Commission reports, suggesting how important these were in defining the final line.

---

64 This was the Vietnamese name for today's Xieng Khouang plateau, in north-eastern Laos.

65 ANOM, SGEOCOL 18, Compte rendu des travaux sur le terrain effectués par le Service géographique de l'Indo-Chine pendant la campagne 1903–1904, Hanoi, 25 July 1904, par Friquegnon, chef de bataillon, chef du Service, p. 5.

66 ANOM, GGI 15700, Laos, Tran Ninh, 1912.

67 Benedict Anderson: *Imagined Communities: Reflections on the Origin and Spread of Nationalism*, London: Verso, 1983, new edition 1991, 2006.

Roads, however large or small, thus made the border a reality. The major trade routes crossed it, while the local paths created links between military posts, and thus reinforced the line physically, like the boundary markers. However, transport and communication routes, although necessary, were not the only requirement for accessing certain towns and frontier zones.

•••

Hence transport and communication routes, due to their prominent role in travel writings and mission reports, their central position on sketches and maps, and the mobilisation they generated, to measure them and invent corresponding map symbols, became a key geographical object. River and land transport routes, and telegraph lines, took up most of the space on maps, illustrating how the territory as a whole was read in terms of its communications. These produced an organised, linear representation of the imperial territories. After centuries of development in Europe and in the Indochinese peninsula, the mapping of itineraries continued, along with detailed descriptions of the – segmented and measured – roads, with particular attention paid to crossing-points such as bridges (water) or tunnels (mountains). The British and the French adopted different approaches, such that even when the type of information gathered was the same, the format often differed, with the British preferring text, and the French, maps.

"Route Books" and maps described the route taken and the reconnoitred roads, but they also showed the path of planned roads. These gave rise to long feasibility studies and reports in which French and British policies were sometimes compared. When we analyse preferred itineraries, and planned and actually realised communication routes, it becomes clear that roads actually *produced* and defined imperial territories. A key feature on maps, roads shaped the territories, enabling their control and access for military and economic purposes, which is why they were important regardless of who built them (the Europeans, the Chinese, the Siamese, or local chiefs). In the competition for new territories, the province of Yunnan was particularly coveted by both the British and the French, and brought to the fore the issue of the access to territories opened up by roads, or their actual integration into the empire: the British were interested in the supposed wealth of the southern Chinese provinces, whereas Doumer's policy suggests that he actually wanted to integrate these same areas into French Indochina. The communication routes actually built reveal something of this polarity. Only the French railway was built, but it alone was not capable of achieving the inversion of space that would have drawn Yunnan into the French empire. This failure was due to insufficient flows along the line, and also to the fact that there was no real political will behind the project. So, despite concentrating all the efforts of the Europeans, the province of Yunnan did not, in the end, fulfil the hopes it had held out initially. Other frontier regions were also coveted, and roads were driven through to them in an attempt to orientate these marginal lands inwards. On a much smaller scale, roads gave substance to borders because they directed and restricted the local populations' movements. We shall examine, in the next chapter, to what extent population movements were really constrained by borders, and whether the latter – as marked on maps and also on the ground – produced significant territorial changes.

CHAPTER 8

Locating, Demarcating, and Crossing the Border

> We had consulted atlases – published in Germany, naturally – and noted that French Laos, Cambodia and Siam – that is the name of their barbarous country – actually meet at a number of points. Thereupon, some diplomats drew on paper what they declared to be the frontier of these three States, aided by their imagination. There were blue, red, and bistre lines, which apparently represented rivers, roads and mountains, and strange names were placed on them, in order to give a semblance of plausibility to these imaginings, the strangeness of the names being the sign of their veracity. [...] But some poor sports had qualms, whereupon our diplomats, with their customary candour, asked themselves whether what they had imagined really existed. I was instructed to go and take a look. Before I left, I was told not to attach undue importance to geographical details, for with a little adroitness I would easily find on the ground everything I had wanted to put there, and, should any difficulties arise, the spirit of diplomacy would light my path, and the Holy Trinity would infallibly come to my aid. These words caused me some apprehension. I might say that I am something of a heretic, and have no access to certain forms of assistance. [...] Everything was very simple at first. The Government of Siam had dispatched some officials with whom I entertained pleasant relations, and we carved up the country with great skill in order to disoblige no one.[1]

∴

Colonel Bernard's irony in this quotation, his aversion to diplomats and his praise of work in the field, should be read in the immediate context of the 1904 border convention between Siam and France. But his statements also usefully

1 AIP, Fonds Fernand Bernard, BEF2, Dossier 2 sur la Commission française de délimitation entre l'Indochine et le Siam, Avant-propos du rapport de mission, p. 8–9.

 | DOI:10.1163/9789004469853_010

highlight the multifaceted and uneven construction of borders, as the site of multiple and often divergent interests. For borders have to pass from diplomatic treaties signed in capital cities to the frontier zones where the decisions taken are concretely implemented.

Boundary marking became increasingly important in the northern Indochinese Peninsula after 1885. Whether this involved reactivating an ancient route or creating a whole new political space, such as Laos (founded in 1893), new territories were produced. Long considered artificial and external, demarcation in fact required the active involvement of local populations, and although the British and the French may have been defending divergent regional interests, their main, and shared, concern was to preserve peace and security along their borders. Good terms had to be established with Chinese and Siamese diplomats, which was no easy task. This raises the question of the border commissions' results, not only regarding a frontier's route, but also regarding the maps produced. Did Europeans and local populations contribute equally to mapping these regions? The Europeans attempted to compensate for their ignorance by undertaking detailed inquiries among the local populations at the level of the village or district, in order to establish the correct boundary line. So to what extent were these communities listened to? What role did they play in finalising the path of the border on their own lands? Did they play an essential role, did they simply rubber-stamp the result, or was it simply imposed on them? And once the border had been defined, and marked out, how was it guarded in the frontier zones and over a larger area? For without border controls, the newly created sovereign territories were nothing but words on paper. By examining the policies and practice of colonial administrators, we get a sense of the complex workings of these colonial states, the individuals involved, their scope of action and, lastly, their utter inability to enforce the demarcation lines they had agreed. The failures encountered reveal the power games played out in the frontier zones, which thwarted the territorial arrangements decreed from the centre. The only populations for whom the frontier seems to have had a concrete reality were those who crossed it illegally, on an occasional or daily basis.

1 Theoretical Variations

1.1 *The Concept of Frontier*

Two main conceptions of the frontier coexisted in the Indochinese Peninsula. They both originated in foreign logics of territory, and show that the Europeans were by no means the first to project their own representations onto these spaces.[2] The first model, which originated in India and was dominant in Burma, Siam and Laos, was the Mandala-State, in which multiple centres were structured by horizontal power relations. The State was grounded in a central power and in its relations with other surrounding centres. This often short-lived form produced unstable and ill-defined marginal zones over which the central power had little control. The second model, which cohabited with the first, originated in China, and could be found mainly in Vietnam, and so concerned the French exclusively. It consisted of an historically stable, unified central State which sought to bring under its control, and gradually integrate, the territories it conquered. It was a hierarchical, vertically structured State, which imposed tributary relations between the centre of power and the vassal states over which it ruled.

Hence, when the Europeans attempted to define linear borders in spaces where they had never had much importance, they were greeted with incomprehension and ignorance. It is again Colonel Bernard who makes this point, in the context of the French border demarcation with Cambodia in 1904:

> I inquired about the exact position of this frontier. The high-ranking official replied that he did not know, and he asked me to go and have a look myself, and to refrain from informing him in any detail. Until that time he had considered it humiliating to know the limits of his power, and he preferred to remain in doubt. I left, and along the way I consulted other people, who found my questions utterly surprising.[3]

In the Mandala-State system, the border was not conceived as an object in itself, and so presented no particular problem. It passed through the least densely populated areas, which were often mountainous, and were called "marches" – the outermost areas of empires, which largely escaped central control. When the linear frontier was introduced, these spaces became internal peripheries, which the central power had to monitor and integrate.[4] In the second, and contrasting, conception of the frontier, one can detect some continuity between how the Vietnamese State and the French Colonial State were organised, in relation to China.

---

2 Michel Bruneau, "La notion de frontière et sa signification dans la Péninsule indochinoise", Moussons, 3 June 2001, p. 33–55; Id., "Les États-nations de l'espace eurasiatique issus d'empires, de cités-États, d'États-mandala. Que doivent-ils au modèle européen occidental?", *L'Espace géographique*, 43/3, 2014, p. 251–264.

3 AIP, Fonds Fernand Bernard, BEF2, Dossier 2 sur la Commission française de délimitation entre l'Indochine et le Siam, Avant-propos du rapport de mission, p. 9–10.

4 J. Michaud, *Historical Dictionary of the Peoples of the Southeast Asian Massif*, *op. cit.*, p. 44–46.

1.2 *Relations with China*

After gaining independence from China in the 10th century, Vietnam sought to secure the limit between the two empires in the course its ensuing dynasties, in order to prevent Chinese territorial expansion southwards. Previously an administrative limit, this frontier thus became a real State border.[5] Its first Vietnamese depictions date from the fifteenth century, but no actual line is visible. Instead, the blank space on the map is simply filled in with the name of the neighbouring country, and the lists of villages on either side remain the surest way of dividing up the territory and determining which population is attached to which power. On the ground, the frontier was sometimes materialised by boundary markers, but more generally by prominent military forts, and frontier guards.[6] Hence the concept of boundary line existed in Chinese and Vietnamese representations, and had been applied in practice to the limit separating the two from the 10th century onwards. Thus the French did not so much bring about a theoretical revolution as prolong a centuries-old tradition, adopting all the familiar prerogatives of a strong and centralised State. The China-Vietnam border was quite the opposite of a boundary-line imposed from the outside, and in the Treaty of Tientsin, signed on 9 June 1885, the French simply acknowledged the existing line, and did not seek to change it. After ten years of delimitation and border-marking, the Franco-Chinese border was finally agreed in Peking, on 20 June 1895, under the aegis of Auguste Gérard, and except for minor details, it simply endorsed the route of the 10th-century frontier along its whole length, from the Red River to the Mekong.

However, there was no linear frontier to the west, between China and the newly-created Laos (1893). The China-Vietnam line could not simply be extended because "on the Laos side, to the west and south-west, there was absolutely no certainty, and it had not been possible to draw a boundary line."[7] This notion of linearity contrasted with a traditional terminology characterised by the vagueness of "confines", "periphery" or "outlying regions".

A similar complexity pertained between China and Burma. An Anglo-Chinese Convention was signed on 24 July 1886 in Peking, relating to Upper

5 Pierre-Bernard Lafont (ed.), *Les frontières du Vietnam, op. cit.*, p. 12.

6 On the dynasties of frontier guards and their relations with the colonial powers, see the recent thesis by Nguyen Thị Hải, "Monarchie et pouvoirs locaux au Vietnam. Le cas de la marche frontière de Cao Bằng (1820–1925)", supervised by Alain Forest and Emmanuel Poisson, University of Paris VII Denis-Diderot, viva 19 November 2015. See also, concerning another space, Sabine Dullin, *La frontière épaisse. Aux origines des politiques soviétiques, 1920–1940*, Paris, Éditions de l'EHESS, 2014.

7 Nguyễn Thế Anh, "Établissement par le Vietnam de sa frontière dans les confins occidentaux", in P.-B. Lafont (ed.), *Les frontières du Vietnam, op. cit.*, p. 185.

Burma and Tibet, whereby the British were to take control of the former, and China of the latter. A series of reports were sent in from the Burma-China frontier, in an attempt to determine its route. One of the obstacles were the Kachin lands to the north-east of Burma, claimed by both the British and the Chinese.[8] In a reconnaissance expedition of 1891–1892, Major Hobday mapped the border zones, while Lieutenant Eliott studied the boundary line's feasibility in the light of the distribution of Kachin and Chinese populations.[9] Their report maintains that not only is there no sign of Chinese sovereignty over the Kachin populations between Yunnan and Burma, but also that the local people described "a recognised border line with China all the way from Nmaikha to Shweli."[10] This is further evidence that the linear frontier was not a product of colonial rule. The report concluded that the British should quickly demarcate the border, while leaving the northern part vague, because the Chinese were increasing their pressure on the Upper Irrawaddy.

Another Anglo-Chinese Convention, concerning this time the Eastern frontiers of Burma, was signed in London on 1 March 1894, giving effect to Article III of the 1886 Convention. It was followed, on 4 February 1897, by another agreement, concerning the frontier, and trade, between China and Burma, which modified the agreement of 1894. In this 1897 document, the British introduced a qualitative distinction: they wanted the boundary line to be drawn at approximately latitude 25° N, while leaving a zone further north undefined until further surveys could determine the boundary's most accurate route.[11] The treaty itself clearly distinguished between the line, represented on the map appended to the Treaty, and the zone where no frontier could yet be determined. So the idea that a linear boundary wholly recast relations between China and Burma is not tenable, given that certain drawn sections already existed, at least on maps, before the arrival of the British, and the British themselves decided to leave certain sections unmarked, whether voluntarily or for lack of an alternative.

---

8 M. Sadan, *Being and becoming Kachin. Histories beyond the State in the Borderworlds of Burma, op. cit.*

9 Arthur Blewitt, *General report and diary of the expedition under Lieutenant L.E. Eliott, Assistant Commissioner, to the Upper reaches of Irrawaddy and subsequently to the Chinese frontier, 1890–91, with notes concerning the Kachins inhabiting that region*, Rangoon, 1891.

10 *Ibid.*

11 This 1897 agreement was in force right up to 1941. The whole length of the border between Burma and China south of 25°35'N (Manang Pum) was fixed in 1897. The northern part, known as the N'Mai Kha section, was delimited after the First World War.

1.3 *Frontier Zones and Pioneer Fronts*

In the rest of the Indochinese Peninsula, there were no preexisting limits between the sovereign territories, but instead, ill-defined peripheral frontier zones, marginal spaces, or "marches", with shifting contours depending on the balance of power.[12] The boundary markers that were sometimes placed on a limit between two territories were not conceived as part of a linear frontier, but only as signalling strategic points. Although these liminal territories were officially independent, they were in fact tributary to one or more larger power, making the task of the Europeans absolutely impossible as they tried to unpick the numerous political bonds between entities, and draw boundaries which were fair, as well as serving their interests.

The kingdom of Lanna (which had Chiang Mai as its capital), for example, had to make concessions to both Burma and Siam as its own power waned, and it was later incorporated into Siam and reduced to the rank of a province (1892).[13] In the western part of this kingdom, along the Salween River, the territory of the Karen people was a mobile frontier zone between Burma and Siam.[14] To the north-east, the group of tributary states (of which Kengtung was the largest) called the Sibsongbanna [Xishuangbanna] constituted a border zone between Siam, China and Burma. The allegiances were unstable, both to these three main powers, and also to local, weaker, but no less ambitious chiefs. Similarly for the Lao principalities along the Mekong River. The kingdoms of Luang Prabang and Vientiane, which emerged from the division of the Lan Sang kingdom at the end of the 17th century, were tributary to the Siamese, Burmese and Vietnamese courts, and sites of recurrent conflict.[15] Smaller principalities were also involved in these power struggles between multiple sovereignties. Laos, for instance, was not a homogeneous political entity before 1893, and was composed of kingdoms and principalities with tributary relations to neighbouring States. No precise name could be found for these ill-defined frontier zones, a fact clearly illustrated by a sketch drawn by Charles Lemire, a French administrator in Indochina from 1886 to 1894, entitled "*Régions entre le*

12 For a more detailed account of what follows, see T. Winichakul, *Siam Mapped: A History of the Geo-Body of a Nation, op. cit.*, p. 97–101.

13 On the history of Lanna, see the two articles by Pierre Lefèvre-Pontalis, a graduate of the French School of Oriental Languages ("Langues O'"), a diplomat and a member of the Pavie Mission and of several Boundary Commissions: Pierre Lefèvre-Pontalis, "Les Younes du royaume de Lan Na ou de Pape", *T'oung pao*, 11, 1910, p. 105–124, and 12, 1911, p. 177–196.

14 Ronald D. Renard, "The Delineation of the Kayah States Frontiers with Thailand: 1809–1894", *Journal of Southeast Asian Studies*, 18, 1987, p. 81–92.

15 Søren Ivarsson, *Creating Laos: The Making of a Lao Space Between Indochina and Siam, 1860–1945*, Copenhagen, NIAS, 2008, p. 29.

Tonkin, l'Annam et le Mékong, dépendant de l'Annam et reconnues en 1888–1889 par la mission Pavie" ["Regions between Tonkin, Annam and the Mekong, subordinated to Annam and surveyed in 1888–1889 by the Pavie mission"].[16] These Lao spaces, which remained nameless because they contained a multiplicity of fragmentary entities, were defined only by their neighbouring territories. Moreover, since these spaces were still independent, the thorny issue of boundary delimitation was overlaid by that of imperial expansionism which, once the Europeans arrived in the region, effectively transformed these zones into "pioneer fronts" between the British, the French and the Siamese, who each tried to extend their influence.

As such, the maps made in the late 1880s represented a fantasised and coveted territory rather than a real basis for negotiation; all parties were aware that it was a case of projection rather than certainty. However, no diplomatic confrontation concerning tributary relations occurred on the basis of maps or historical documents. In 1888, in the Thaeng confrontation between the Siamese and the French, the parties presented their maps as the basis for a compromise solution, but in actual fact the status quo was assured by a military presence on either side. Likewise, the French use of force in Bangkok, in 1893, was what really put a stop to the race for territory: political negotiation, and the use of maps, were no match for gunboat diplomacy.

The Franco-Siamese Treaty of 3 October 1893 made Laos into a French Protectorate, the left bank of the Mekong became French up to Luang Prabang, and 25 km of the right bank were declared neutral. This new organisation of Indochinese territories had a number of consequences. As the Mekong was henceforth an international border, the Lao principalities spanning the two banks became separated. This brought about a complete transformation in how Siam, which adopted the colonial powers' models, represented space. In its new perception of space, no fluctuating peripheral zones could be tolerated. So, in this case, the linear frontier was indeed conceptually alien, and the frontiers created were imposed from the outside, even if one must also examine how willingly the local elites accepted and adopted these changes.

Elsewhere in the peninsula final boundary agreements were also being negotiated at the time. At the date of the London Conference, in 1894, the French presence went as far as the Mekong, and the Chinese as far as the Irrawaddy (Burma). The last unresolved issue was the fate of the territories between the French and the British Empires in the Upper Mekong, particularly the principalities situated on both banks of the river, in the confines of Burma, China, Laos and Siam. From 1894 to 1895, an Anglo-French Commission worked on

16 ANOM, GGI 14460, Croquis de Charles Lemire, Vinh, 15 mai 1889, au 1:1,000,000.

delimiting the boundary between Burma and Laos, at the only point where the British and the French were in direct contact. It aimed to create an intermediary zone on the Upper Mekong between the two empires.[17] However, the term "intermediary zone" could refer to several different solutions: creating a buffer zone, incorporating the zone into one side or the other, or dividing it in two along the whole length of the Mekong, making the river into an international boundary.[18] The two powers decided that, given the lack of geographical information on the region, it should be mapped before delimitation could take place. The outcome of the mapping showed unambiguously that creating an intermediary zone was unworkable, so the ball was again in the court of the governments, who had to take the final decision, while the Anglo-French Commission simply submitted the same map. The issue was finally settled by the Anglo-French agreement of 15 January 1896: it fixed the Upper Mekong as the frontier between the two powers, redistributed the unassigned territories, guaranteed Siam's rights over the Menam Valley, and detemined the European spheres of influence in Siam.[19] The idea of preserving a frontier zone instead of drawing a line was thus no match for the realities on the ground, but at the same time, the European approach was less rigid than has sometimes been said, insofar as the linear frontier was not systematically tabled from the outset, even if it ended up winning the day.

Lastly, the Franco-Siamese Treaty of 13 February 1904 modified the boundaries of still contested provinces, and put an end to the demilitarised zone of 25 km, which France did not annex, since it had decided finally for the Mekong as its frontier. This Treaty was altered by a further one of 23 March 1907, by which the territories of Battambang and Angkor were restituted by Siam to Cambodia. By then, the frontiers of Burma and of French Indochina with Siam had been marked on maps, and the idea of linear frontiers had won out over the model applied up until then, that of indeterminate peripheral zones. In summary, European representations of space were imposed, and the need to delimit sovereign territories proved more powerful than the imperative

17 A. Pavie, *Mission Pavie en Indo-Chine, 1879–1895, op. cit.*, p. 335 *et suiv.*; BL, IOR, L/PS/20/83, Report on the Anglo-French Buffer State Commission, 1894–1895, by Captain Walker, Rangoon, November 1895.

18 A "buffer State" is a state situated between two larger political entities to avoid them coming into direct contact. Siam was the best example of a buffer State in the area, between the French and British empires.

19 On the diplomatic aspects, and with a very precise chronology, see D. Woodman, *The making of Burma, op. cit.*, p. 298–332; Nigel J. Brailey, *Imperial Amnesia: Britain, France and "the question of Siam"*, Dordrecht, Republic of Letters, 2009.

of preserving preexisting balances. However, this linear conception was only applied if the colonisers considered it essential in the particular case.

As we have seen, various types of frontier cohabited in the northern Indochinese Peninsula, including border lines, ill-defined zones and pioneer fronts. The linear frontier imposed by the Europeans was not simply an alien and artificial import, but sometimes had a long history locally, as the China-Vietnam border illustrates. The notions of exclusion and foreignness were familiar in these spaces before the Europeans drew borders to separate different sovereign territories. Delimitation was often the task of diplomatic treaties, but sometimes border sections were left undefined, suggesting once more that the European powers did not always rigidly apply their linear conception of the border. When circumstances obliged them to defer fixing a precise route on paper, they tolerated the persistence of imprecise areas until they could arrive at a border agreed consensually by all parties. However, in all cases, only concrete implementation could really breathe life into the lines marked out on diplomatic documents and maps. So how did demarcation work in the field?

2 Boundary Marking in Practice

Once the treaties had been signed, Commissions were generally instructed to confirm the route on the ground. This was the case for the Franco-Chinese Commission of 1885–1887, the Anglo-Siamese Commission of 1892–1893, the Anglo-French Commission of 1894–1895, the Anglo-Chinese Commission of 1897–1899, and the Franco-Siamese Treaty of 1904–1906.

2.1 *Who Draws the Boundary?*

These Boundary Commissions were composed of civilians and military personnel, explorers, colonial administrators of the spaces to be mapped, Chinese or Siamese officials, provincial court civil servants and provincial governors, and surveyors. Auguste Pavie, who headed the French side in the Anglo-French Commission when he was Minister Plenipotentiary in Bangkok, and who had been surveying the whole of the peninsula since the 1870s, is a particularly good example of a commissioner. His standing, and his vast geographical knowledge, made him an ideal representative of French interests. He had never met his British counterpart, James George Scott, before 1894, although they could have met in 1891 in the Sibsongbanna [Xishuangbanna]. Scott (1851–1935), the Superintendent of the Northern Shan States, was a member of the Anglo-Siamese Commission of 1892–1893, the Anglo-French Commission of

1894–1895, in the company of his wife, and of the Anglo-Chinese Commission of 1897–1899. The son of a Scottish Presbyterian pastor, his experience as a journalist gave him extensive knowledge of South-East Asian countries.[20] He was at Perak for the *Evening Standard* in 1875, in the aftermath of the British annexation of Malaya. He covered events in Burma for the *Daily News* until 1882, while teaching at St John's College in Rangoon. In his first book, *The Burman: His Life and Notions*, which he published under the Burmese pseudonym of Shway Yoe, he summarised his insights into the country.[21] The *Evening Standard* sent him to Tonkin in 1884 to cover the French campaign and to gather information, which was later used by French explorers, including Pavie.[22] On his return to Upper Burma in 1886, he joined the new civil administration, first as Deputy Commissioner in Mandalay, then in Meiktila. On becoming Superintendent of the Shan States, he travelled extensively throughout the region, publishing his *Gazette* in 1901.[23] He was thus an experienced Commission member, who had accrued detailed knowledge of the Indochinese peninsula, and played an important political role in Upper Burma.

Another much appreciated quality of members, in addition to their political influence and excellent understanding of the territory, was knowledge of languages. Failings, or absence, in this area created numerous problems for the work of the Commissions.

One of these problems concerned how the Commissions communicated internally. In the Franco-Chinese Commission of 1885, Sherzer and Haïtce knew Chinese, but as full Commission members, they could not also play the role of interpreters. Consequently, a certain Li, an engineer in the Chinese navy who had studied in France, became the official interpreter, and also James Hart, the advisor to the Chinese delegation (and brother of Robert Hart), especially for private conversations.[24]

For the Anglo-French Commission, Walker reported that Pavie, Lefèvre and Thomassin all spoke English; Thomassin spoke it exceptionally well, despite

---

20 See the biography of Scott by his last wife, based on his own writings: James George Scott, Geraldine Mitton, *Scott of the Shan Hills. Orders and Impressions*, London, John Murray, 1936; see also Andrew Marshall, *The Trouser People: A Story of Burma, in the Shadow of the Empire*, New York, Counterpoint, 2002. Scott's personal papers are held at the Cambridge University Library; his brother was the Head of St John's College there.

21 Shway Yoe, *The Burman: His Life and Notions*, London, Macmillan, 1882.

22 James George Scott, *France and Tongking: A Narrative of the Campaign of 1884 and the Occupation of Further India*, London, T. F. Unwin, 1885.

23 James George Scott, *Gazetteer of Upper Burma and the Shan States, op. cit.*

24 Sir Robert Hart was the Inspector-General of China's Maritime Customs Service from 1863 to 1911.

FIGURE 39 Photo: Auguste Pavie surrounded by members of the Franco-British Commission of the Upper Mekong, 1895 (*Front row*: Gurkhas; Dr Lloyd; Mr Warry; Dr Lefèvre; Captain Walker; Mr Caillat; Vietnamese militiamen. *Second row*: Gurkhas; Mr Stirling; Lieutenant Thomassin; Pandits; Mr Scott; Mr Pavie; Mrs Scott; Colonel Woodthorpe; Mr Lefèvre-Pontalis; Tchioum; Ngin)
SOURCE: Auguste Pavie, Géographie et voyages, II, *Exposé des travaux de la mission (troisième et quatrième périodes, 1889 à 1895)*, Paris, E. Leroux, 1906, p. 333

the fact that he had never visited England.[25] In 1904, Fernand Bernard complained about communication difficulties due to lack of linguistic competence, and welcomed the changed appointments decided by Siam in the following year: "General Chatidy Won understands French very well, and Mr Visuts Kosa knows our language perfectly."[26] Both had studied in France, while a third nominee, Lieutenant Narpat, had studied in Belgium and then joined a regiment of Belgian grenadiers. In Fernand Bernard's words, "These changes to the composition of the Commission show that the government of Siam is determined to have its interests defended."[27] Knowing the other party's languages is thus vitally important for the diplomatic strategies of boundary demarcation.

Apart from knowing the languages of the other party, there was also the issue of local languages. The assistant surveyor Huzarat Ali was described as

25 BL, IOR, L/PS/20/83, Report on the Anglo-French Buffer State Commission, 1894–1895, by Captain Walker, Rangoon, November 1895, p. 13.

26 AIP, Fonds Fernand Bernard, BEF1, Lettre du commandant Bernard au gouverneur général de l'Indochine, 8 February 1905.

27 AIP, Fonds Fernand Bernard, BEF1, Lettre du commandant Bernard au gouverneur général de l'Indochine, 25 November 1905.

having "an excellent colloquial knowledge of Yünnanese,"[28] a fact considered extremely useful in the area. When the delimitation work continued in 1905, Lieutenant Lakin was involved in negotiating the northern segment between China and Burma, for which his good knowledge of Hindi and Kachin was judged extremely helpful. In the Anglo-French Commission, the French were accompanied by Chinese, Vietnamese, and also, at the recommendation of the local chief Đèo Văn Trì, of Thai servants because of their knowledge of the language and customs of the territories to be crossed. This shows that it was necessary to communicate with the frontier-zone populations not only for information-gathering, but also more generally to help the commissioners' movements in the field. However, knowing the languages of the other party could not guarantee a Commission's success, which was always hard-won, and on which depended the shape of future countries [Figure 39].

2.2 *Negotiation and Disagreement*

Commission reports show that disagreements between European and Asian countries on the exact path of a boundary were legion. Some frontier segments were decided without much difficulty, but others generated endless discussions, and an outcome which revealed the power relations between the parties involved. These debates throw into relief the arbitrary nature of the boundary, insofar as some segments were decided without considering the whole line, which was determined by many factors, not all of which were grasped by the Europeans. What are we to make of the different strategies adopted, and what do they tell us about the scope of action of the Siamese and the Chinese?

In addition to the disagreements over the path of the boundary, or, more frequently, over areas of territory, arguments over place names could ruin a Commission's work. In 1893, when marking out the China-Vietnam border in the field, "after Lieutenant Jacob, assigned to survey the land between Lai Chau and Phong To, failed to come to an agreement with a mandarin, his opposite number, concerning the name of the watercourse, he returned to Mongtze".[29] In 1898, a similar conflict emerged between the Chinese and the British around the name and exact location of a place, with the result that this first mission ground to a halt.[30]

---

28 BL, IOR, L/PS/20/D103, Report of the Intelligence Officer on the Burma-Chinese Boundary Commission, Northern Party, 1898–1899, and on a reconnaissance in western Yunnan, by Captain Walker, p. 3.

29 ANOM, GGI 19601/62, Lettre du consul Rocher à M. Lemaire, ministre de la République française en Chine, Mongtze, 17 September 1893.

30 L/PS/20/D101, Report of the Intelligence Officer attached to the Southern Party, Burma-Chinese Boundary Commission, 1897–1898, by Captain Norie, Simla, 1898. Due to a disagreement at the start between the English and Chinese members on a particular

The thick red line to the south represents the border already agreed; the yellow line, the limit of the British "administered territories" (to the west); the red dotted line, the "ideal route" drawn by the British diplomat Ernest Mason Satow, which continues as far as the "confines of Tibet"; in green, a solid and a dotted line, two alternative routes proposed by the British; in blue, red and brown, three routes suggested by the Chinese [Figure 40].

Apart from the incomprehension frequently caused by names and their spelling, some border segments had to be left vacant until later, as occurred for the north of the Burma-China border. The reasons for this could be strategic and economic, but also ethnic. For example, in the dispute over Bạch Long, an "Annamite enclave," the Chinese held that this territory was inhabited mostly by Chinese, and should therefore be theirs. The French referred to official Chinese geographical records, and administrative and fiscal documents, to prove that the territory was Vietnamese. The demarcation of the boundary continued, in the field, but this part was left vacant until June 1887, when it was settled in Peking, with the Bạch Long enclave being returned to China as rectifying a detail of the already established frontier line. Disagreements of this sort were visible on the maps: they were either left blank along the problematic segments, in order not to create confusion, or else several different frontier routes were represented.[31] In this case, the Chinese justified their claims in terms of the supposed ethnic and linguistic complexities of certain areas, when in fact their reasons were economic, a fact which escaped the Europeans at the time.[32] There were also ongoing discussions concerning certain border villages claimed by the Siamese on ethnic grounds. They argued against a purely artificial line that would separate them from their own territory and split up the population. Finally, further reconnaissance had to be undertaken to settle the issues raised.

The difficulty of achieving a single borderline accepted by all parties underscores the relatively arbitrary and non-linear process of boundary delimitation. Sometimes the Europeans themselves did not know where their own interests lay, and so were unable to impose them on others. Disagreements

---

segment, the Commission was at a stalemate for a whole three months. See the map annexed to the Anglo-French Agreement of 1897, where the territories coveted are visible, between the line of the 1894 border and the proposed 1897 border, which Liu refused when *in situ* (Southern Party, 1897).

31 On outline or blank maps, see Isabelle Laboulais (ed.), *Combler les blancs de la carte. Modalités et enjeux de la construction des savoirs géographiques (XVI*[e]*–XX*[e] *siècle), op. cit.*

32 The jagged outline of the China-Vietnam border was due to the presence of mines, which the French were not aware of at the time of the delimitation, and which the Chinese thus managed to preserve on their territory. C. Fourniau, *Vietnam. Domination coloniale et résistance nationale, 1858–1914, op. cit.*

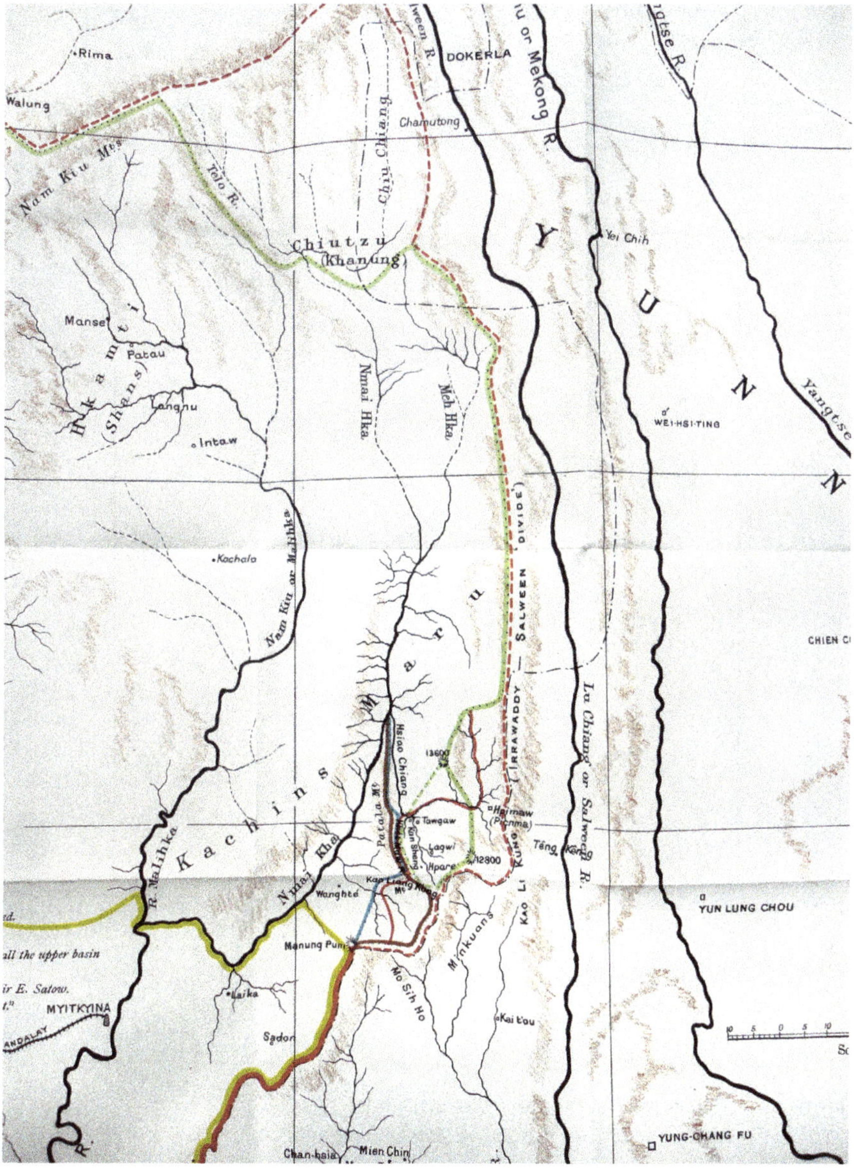

FIGURE 40 Sketch map showing suggested lines of demarcation of the Burma-China border, 8 March 1911 (detail)
SOURCE: British National Archives, Kew, FO/881/9885

FIGURE 41 Chinese map of the China-Vietnam border
SOURCE: "Carte de la frontière sino-annamite", EFEO, CAR. 2703

between the different protagonists thus complicated the task of large-scale mapping of boundaries.

The maps were often made by European mapmakers. Only a few were bilingual, for example those which have come down to us from the China-Vietnam [Annamite] Border Commission of 1896–1897 [Figure 41 and Figure 42]. On one side, the map is in Chinese and romanised Vietnamese, on the other, it

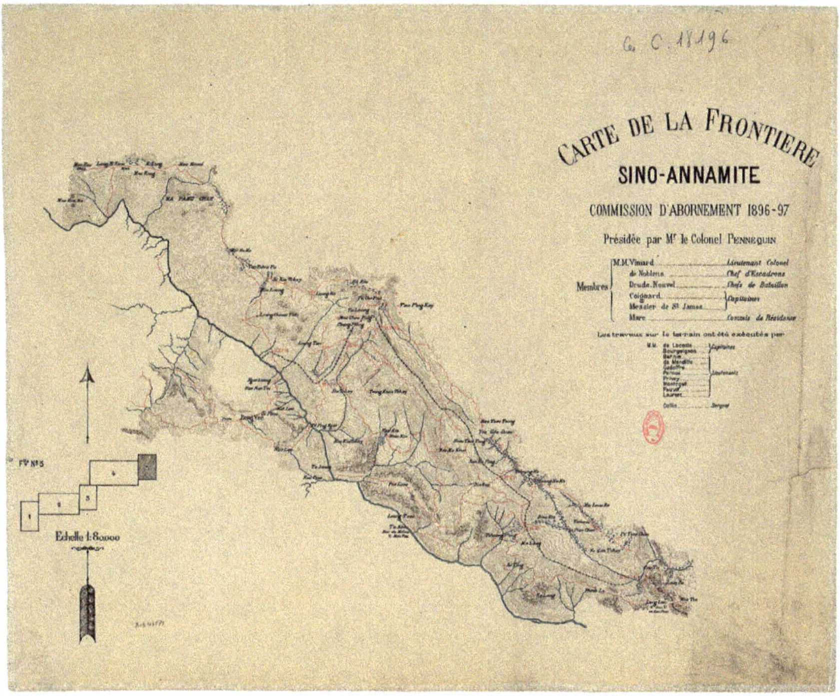

FIGURE 42 French map of the China-Vietnam border
SOURCE: "Carte de la frontière sino-annamite, commission d'abornement, 1896–97", BNF, Cartes et plans, GE C-18196, feuille 5

is in French and romanised Vietnamese.[33] A similar system was used in 1898 on the maps used to settle contested borders at the limits of, and within, the neutral 25-kilometre zone. One side was in French, and the other in Thai.[34] A French sketch map of the region of Pak Lay, drawn by the government commissioner Paul Macey, dated 19 August 1898, depicts the historic frontier between the kingdom of Luang Prabang and that of Nam through crosses which follow the path of the markers on the ground placed there 150 years

33 These two maps, numbered III and IV, are held at the EFEO. They belong to a larger set of five maps produced by the Chinese, of which sheets II and V are held at the ANOM, classified under shelf mark GGI 65354, along with some copies and preparatory maps. The Chinese characters in the top right-hand corner of map IV, Figure 41 above, can be translated as: "Map no. 4 of the frontier. Each square (5cm) corresponds to 10 *lis* (4km)", EFEO, CAR. 2703.

34 ANOM, GGI 14491, Délimitation de la frontière entre l'Indochine et le Siam. Royaume de Luang Prabang. Territoires contestés sur la rive droite du Mékong, Bo Sao et Ken Tao. ANOM, GGI 24917, Territoires contestés entre l'Indochine et le Siam, Luang Prabang.

FIGURE 43 French sketch map of the boundary between Luang Prabang and Nam
SOURCE: ANOM, GGI 24917, "Croquis de la région contestée", par le Commissaire du Gouvernement Paul Macey, Pak Lay, 19 August 1898

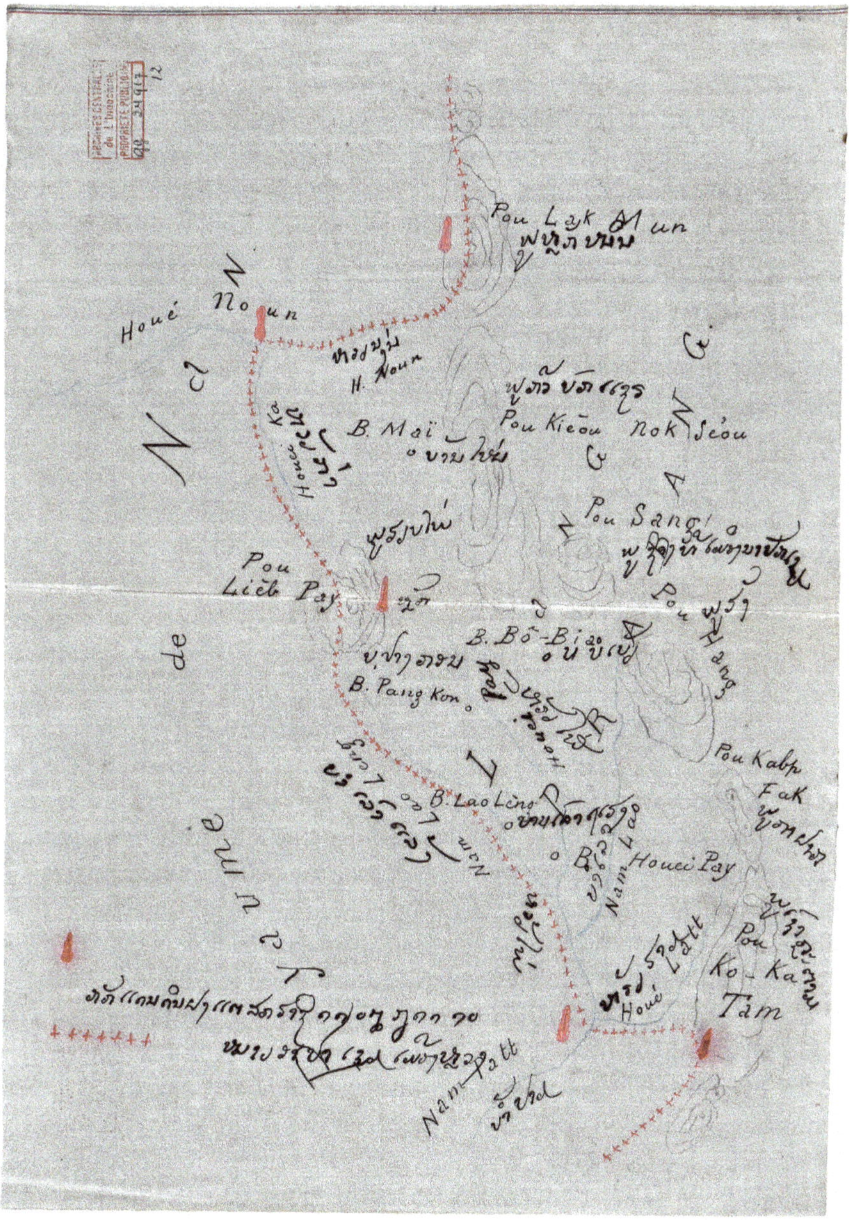

FIGURE 44 Siamese sketch map of the boundary between Luang Prabang and Nam
SOURCE: ANOM, GGI 24917

previously [Figure 43]. The Thai version retained the same line and the main place names, but translated the terms into Thai, with some exceptions, such as "Kingdom of Nam", which remained in French [Figure 44]. Only the Border Commissions produced these kinds of bilingual maps, which were otherwise unknown. This bilingualism reveals the specificity of the territory-building process, and the underlying diplomatic issues at stake. A related question is the role really reserved for local populations in this process, alongside the discussions between Commission members.

2.3 *The Role of Frontier-Zone Populations*

Analysing the relations between Boundary Commission members and frontier-zone populations highlights the process of boundary-building at a very local level. It can tell us about the population's scope of action and room for manoeuvre. Often, frontier-zone populations did not belong to the majority Han, Viet or Bamar groups, respectively, and were represented negatively by the central administration and the elites.

One of the first tasks which the Europeans undertook was to try to settle preexisting local border disputes. At the time of the Anglo-Siamese Boundary Commission of 1892–1893, the border near Möng Hang, a town of only fifty to sixty houses, was a conflict zone.[35] Captain Walker, for the British, described a territorial dispute of 25 years' standing between the Siamese and the Burmese, who each positioned their border markers to correspond to their claimed line. The local chiefs, meanwhile, simply tried to extend their own territory, and treated the border as a pioneer front. The town's inhabitants destroyed the Siamese markers, and the Burmese retaliated. Finally, the watershed was accepted as the limit by both parties. So, prior to the arrival of British territorial interests, certain borders, and border dynamics, were already in place. The task of the British on their arrival was thus to get a single borderline acknowledged by all the parties.

For the primary imperative was to preserve peace along the border. Pavie was confident that he had achieved precisely that when he stated that the border demarcation took place at record speed and that "the populations agreed so unanimously that no difficulty can arise in the future on this account."[36] The reality was a little more complex, but it is true that the local populations' demands were taken into account whenever these were not at odds with European or Chinese interests, and would not compromise peace and safety on the frontier. For the Anglo-French Commission, Walker explained that

35 BL, L/PS/20/92, Anglo-Siamese Boundary Commission, Southern Party, 1892–1893, by Captain Walker, Rangoon, September 1893, p. 5.

36 A. Pavie, *Mission Pavie Indo-Chine, 1879–1895 ..., op. cit.*, p. 387.

the line of the watershed between the Irrawaddy and the Salween had been chosen as the one best able to ensure justice and peace for the inhabitants.[37] Should this be interpreted as a genuine concern for the interests of the local people or as a wish to arrive at a compromise that would preserve British goals intact? The two were not necessarily incompatible, and sometimes European and local interests coincided.

When the conflicts between local chiefs were too heated, the Europeans decided not to listen to them at all, giving rise to absurd solutions on the ground. Thus, when determining the frontier between Longzhou and Guangxi, the Border Commission requested that three roads previously used by tributary envoys and missions be used as reference points, but since the roads split up into paths as from Lạng Sơn, it was impossible to draw a single line. In May 1886, therefore, the villages previously under Vietnamese jurisdiction became Chinese.[38] Similarly for the 25 villages at Kokang, where the officials there did not manage to come to an agreement with the Mengting authorities, with the result that these mostly Chinese populations were incorporated into Burmese territory.[39] Both the Chinese and the Europeans tried to instrumentalise local disputes and the unrest of the previous decades, which had often already muddied the waters. Lack of historical evidence to support local chiefs' claims was also taken as a good enough reason for changing the route of a border, regardless of the local chief's views.[40]

Local populations and their leaders were sometimes manipulated, to provide support for the Europeans or the Chinese in opposition to the other party. The British often complained that the Chinese and Siamese, who were closer to the local populations simply because of their shared language, used these populations to mislead them. They claimed they had been unable to establish relations with the local inhabitants because the Chinese or Siamese had prepared the ground in advance. Norie observed, at the beginning of March 1898, that the Commission was making progress, despite the fact that the Chinese commissioner Chen was not being cooperative, and that the *Sawbwa* of Nantien was

37 BL, IOR, L/PS/20/83, Report on the Anglo-French Buffer State Commission, 1894–95, by Captain Walker, p. 21–22.

38 Davis Bradley Camp, *States of Banditry: The Nguyen Government, Bandit Rule, and the Culture of Power in the Post-Taiping China-Vietnam Borderlands*, Seattle, University of Washington, 2008, p. 325.

39 BL, IOR, L/PS/20/D103, Report of the Intelligence Officer on the Burma-China Boundary Commission, Northern Party, 1898–99, and on a reconnaissance in western Yunnan, by Captain Walker, p.22.

40 BL, IOR, L/PS/20/95, Report of the Intelligence Officer accompanying the superintendant of the Northern Shan States, on his tour in 1896–1897.

"perverting local evidence to declare that names are different from what they are known to be and thus putting the Chinese boundary further to the West than intended."[41] So there were often underhand dealings going on between the Chinese, the local chiefs and the British, without it being clear whether the local chief was bribed in this instance by the Chinese, or whether he was acting on his own initiative.

The Europeans' ignorance of the exact path of the borders they were attempting to establish obliged them to turn to the local elites. This was also the reason why they consulted the existing documentation, and inquired about the power relations existing historically between different territories, in an attempt to clarify the different claims. So, insofar as the central authorities did not know where the frontier ran, and since they wanted peace along their borders, there was a real vested interest in listening to the local population and its demands. These two factors explain why the Europeans were attentive to the local inhabitants, and why the work of delimitation could be carried out on good terms with them, and even be assisted by the local elites, when they understood the benefits they could reap from it. The boundary line was thus not simply an alien feature imposed from the outside. Consultations with local populations preceded or accompanied the topographical surveys, and show that the Europeans attempted to work from very local knowledge. Their attention to the spatial arrangement of villages also belies the idea of the purely artificial line, since they adapted the latter to the realities of the terrain and the daily life of the inhabitants. Although drawing a line profoundly alters the conception of space compared to previous models, the border did not in itself abolish the interactions and trade in these peripheral territories; the stroke of a pen on a map is too static an object to express the lived reality of a frontier.

3 Border Controls and Infringements

3.1 *Marking the Border: Boundary Markers and Their Efficacy*

Placing boundary markers makes the border visible in the landscape. However, this did not settle all the associated problems, and very often the markers were moved or disappeared, especially when the limit drawn was contested. For instance, when the limits between French and Siamese territory were marked

41 BL, IOR, L/PS/20/D101 Report of Intelligence Officer attached to the Southern Party, Burma-Chinese Boundary Commission, 1897–1898, by Captain Evelyn William Medows Norie, p. 18.

out, between 1893 and 1902, conflicts crystallised between Luang Prabang, on the French side, and Nan, on the Siamese side:

> During the month of August, men sent by the king of Nan entered the territory of Luang Prabang, in the region of Muong Loung, and moved a frontier post they then carried to Phu Sang, with a view to extending the limits of the kingdom of Nan to the detriment of our populations … As if moving a post were not enough, they also planted a new one at Phou Lah Munn on the main road from Nan to Mekong, a few metres behind the borderline of trees that separates the two principalities at this point.[42]

The international border between Siam and French Indochina was thus based on a much older line between two principalities. The Siamese tried to shift it to their advantage in a disputed area for which many maps were produced, as mentioned above. This game of cat and mouse with their foreign counterparts in attempts to move the markers became common practice in the ensuing years. The French police held the Siamese police to be more responsible for these actions than the local villagers, who were grouped together as auxiliaries under the command of their local chief, for border surveillance:

> The auxiliaries have been present for a number of months now, yet mandarins sent by the Siamese have been able to come and visit all the frontier points, to position markers and remove placards of protest, without ever being seen by these frontier guards, who were only informed long after the fact by the rumours circulating.[43]

The British were faced with a similar problem in the contested territory of Kokang situated on the Burmese border to the north-east of the Shan States. In 1898, the Chinese removed border markers that a diplomatic agreement had ratified one year previously. They were then beaten off by the chief of Somu, south of Kokang, which left thirty of them dead. The reaction of the British authorities was somewhat moralistic:

> Moving back the boundary of a neighbour's field is generally a game one does not play with impunity on a distant border, but luckily, in the

42 ANOM, GGI 20878, Déplacement de poteaux frontière par des habitants du royaume de Nan, en vue d'agrandir leur territoire au détriment de Luang Prabang.

43 ANOM, GGI 65353, Report on the Government Commissioner's visit to Paklay and Xieng Khan, to the Higher Committee of Upper Laos, 1899.

> present case, it can have escaped no one's notice that it is dangerous, and we cannot but point to the case of the chief of Somu in this regard – who, incidentally, is no more delighted to be incorporated into British possessions than he was to come under Chinese rule.[44]

So in this case it is not the colonial powers who ensure that borders are respected, but a third party, here a local chief of one of the Wa States, which the British had difficulty conquering and incorporating into their empire. The British maintained order on this distant, and reputedly dangerous, boundary, not by imposing their will or their troops, but due to local power politics and antagonisms which preceded the arrival of the Europeans, and which they could exploit. Securing the border and keeping the peace were not achieved here by the Colonial State through the mediation of the local elites it governed, but rather by a local chief who lived within colonial territory, but who acted on his own initiative to preserve a certain balance of power favourable to him.

3.2 *Enforcing the Border*

The gap between what was represented on the maps and the reality of the border on the ground caused the police countless problems, even using the most precise cartographic methods. In the early years, one could blame the bad quality of the maps, as occurred after the Anglo-Siamese Boundary Commission of 1892–1893. Certain triangulation points were inaccessible because of the mist, and this had compromised the continuity of the geodesic reference network:

> This was undoubtedly very prejudicial, because the actual boundary line remained inaccurate on the survey sheets, although the line had been decided on the ground, and the actual local topography obtained was correct.[45]

Hence the boundary was marked and defined on the ground, but it could not be transferred with precision onto the maps. This problem was only resolved when the geodesic reference networks were completed and the large-scale maps of the boundary could be inserted within them. Overall, colonial officials considered maps to be useless. This necessarily alters how we should understand domination and control through cartography.

---

44 ANOM, GGI 54143, Information on the Burma-China boundary delimitation by the Chinese and British governments. Extract from the *Rangoon Gazette* of 8 March 1898.

45 BL, IOR, L/PS/20/92 Anglo-Siamese Boundary Commission, Southern Party, 1892–93, by Captain Walker, p.18.

Both the British and the French had to deal with illegal frontier crossings and the phenomenon of gangs. Although boundary demarcation managed to reduce the official Chinese and Siamese incursions, it did not put a stop to the various trafficking operations and infractions, which the police had trouble controlling.

In the early 1890s, when the frontier regions of Tonkin were being pacified by the French, the activities of rebel gangs increased to levels equivalent to those of the 1860s. Colonial reports are full of accounts of pillage, of villages burnt down, of cattle and human beings stolen – all of which reflected the extent of these gangs' power in the frontier zones.[46] The number of rebel gangs in Tonkin was estimated at 132 in Hải Phòng province, and 80 in Sơn Tâ province.[47] The creation of armed groups, and the tactic of dissemination – whereby small units operated from nodal points, rather then sending impressive, but inefficient, columns of troops – led to a more effective response to the insurgent gangs.[48] The armed units, constituted as an auxiliary police force in 1892–1893 by arming the villagers of Upper Tonkin, were deployed along the border between China and Tonkin.[49] They were recruited among the local population, and chosen by the local and district authorities. Their existence typified the way the colonial powers delegated peace-keeping operations. During the Pacification, some Chinese gangs and other groups deemed by the French authorities to be rebels had found support among the local population, sometimes simply to be able to obtain provisions at the markets without trouble. The armed auxiliaries, for their part, leaned not only on the local population, but also on the local elites, in order to put a stop to the gangs' activities. An emblematic case of this role of the local elites in maintaining law and order is Đèo Văn Trì's switch from opponent to supporter of the French.[50] The expression "transactions of collusion" ["*transactions collusives*"][51] can thus be applied to relations in the northern Indochinese Peninsula, where the French did not

46 B. C. Davis, *States of banditry ..., op. cit.*, p. 351-355.

47 ANVI, RST 76319, Rapport général sur la piraterie dans les provinces du Tonkin, 1st May 1891.

48 On the pacification of Tonkin, and the leading role played in this b y Pavie and Pennequin, whose method inspired Gallieni's later strategies in Madagascar, see J.-F. Klein, *Pennequin, une "éthique coloniale française?" Penser la pacification, dépasser la colonisation (1849–1916) ..., op. cit.*

49 ANOM, GGI 65374, 121, Rapport du colonel Hocquart, chef d'état-major, président de la commission chargée de la révision du règlement de police frontière, sur l'organisation des partisans.

50 P. Le Failler, *La rivière noire. L'intégration d'une marche frontière au Vietnam ..., op. cit.*

51 E. Blanchard, Q. Deluermoz, J. Glasman, "La professionnalisation policière en situation coloniale" ..., *op. cit.*

really care what went on as long as security reigned in the borderlands and throughout the Upper Region.

On the British side, border control was also an important issue, for which a dedicated police force was created in 1887, when Upper Burma was endowed with a force of 12,386 Indian and 5,000 Burmese policemen, headed by 80 European civilian and military officials.[52] These figures show not only the very small numbers of Europeans, but also the use of Indians as well as Burmese, and a sort of delegated Indian imperialism in Burma.[53] The idea behind this was to reduce the role of the army in favour of the police, while keeping the two roles distinct, and to balance the numbers. The Indian police was generally given responsibility for law and order, prison guarding, and for holding posts where the army would be unwelcome. But since the Indians did not know the local languages or customs of Upper Burma, they were not given civil policing tasks, for which special local police units were created, composed of Burmese. Another reason also lay behind the creation of this force:

> There are certain sub-montane posts in Upper Burma which it is necessary for us to occupy and hold either with troops or police, but the climate of which is unhealthy for any but natives of the country. The Chief Commissioner is of the opinion, and has represented it to us, that the force of indigenous Burman Police (including Shans, Karens, and men of other well-disposed tribes) should be strong enough to hold some, at all events, of these posts, as well as to perform the ordinary duties of a civil police force.[54]

The populations of the region were thus considered vital for maintaining law and order along the border, not for strategic reasons, but due to the difficulties of the terrain and above all the climate. In reality, however, things were a little more complex, and using Burmese in the police force proved tricky for the British, who were more familiar with working with Indians. Local elites and inhabitants, who actually lived in the borderlands, and were integrated into local power relations, seemed the best choice for imposing order in the frontier zone. The British and the French thus ended up with this system of delegated

52 BL, IOR, L/PJ/6/195 file 321, Letter from Dufferin, Gvnt of India, Dpt of Finance and Commerce to Viscount Cross, Secretary of State for India, Calcutta, 1st February 1887.

53 Thomas R. Metcalf, *Ideologies of the Raj*, Cambridge, Cambridge University Press, 1994.

54 BL, IOR, L/PJ/6/195 file 321, Letter from Dufferin, Gvnt of India, Dpt of Finance and Commerce to Viscount Cross, Secretary of State for India, Calcutta, 1st February 1887.

powers, while also maintaining more formal relations of cooperation with the police forces of the neighbouring powers.

3.3 *Cross-Border Cooperation: Hot Pursuit and Extradition*

When boundary delimitation started, the police forces on either side did not work together, and indeed had no particular contact with each other. The situation changed gradually. The extradition of criminals to or from China was defined in article XV of the Anglo-Chinese treaty, which was signed in London in 1894. The article provided that:

> Should criminals, subjects of either country, take refuge in the territory of the other, they shall, on due requisition being made, be searched for, and, on reasonable presumption of their guilt being established, they shall be surrendered to the authorities demanding their extradition.[55]

The requisition could be triggered by a request made by any official of either government to the nearest border police unit in the country in which the criminal had found refuge.[56] In Upper Burma and Tonkin, there were recurrent raids of armed bandits into British territory, including an incursion of 300 men in March 1901.[57] Despite the many complaints, the British did not take these threats very seriously until 1902, when they signed an agreement with the Chinese to increase collaboration on border surveillance.[58] Its second clause provided for Chinese frontier posts to be set up, and cooperation between the authorities, with a view to cracking down on raids, particularly the attacks by Chinese Kachins against Burmese Shans. The British had wanted to establish a right of hot pursuit, that is, authorise the police to pursue marauders across the border into the neighbouring territory for a certain distance.[59] This was already permitted in the areas where the frontier had not yet been drawn, but it had never before been agreed elsewhere, because international law forbade

---

55 BL, IOR, V/27/160/15, Note on the rules concerning the extradition of criminals to and from Siam and China and the issue of passports to Siam and China, Rangoon, 1910.

56 This was the official rule. We have been unable to assess whether it was applied, and how many extraditions took place during this period.

57 BL, IOR, L/PS/7/131/402, Entrance of Chinese soldiers into British territory, 7 March 1901; BL, IOR, L/PS/7/131/430, Entrance of Chinese in arms into British territory, March 1901.

58 BL, IOR, L/PS/7/144/616–617, Yeh, Sub-prefect of Tengyueh, with Prefectural Bevet Rank, Superintendant of customs, addresses the following official dispatch to the Deputy Commissioner, Bhamo, the Superintendant of the Northern Shan States, and the Tengyueh Consul Litton, 3 January 1902.

59 BL, IOR, L/PS/7/144/616–617, Letter from the Chief Secretary to the Government of Burma to the Secretary to the Government of India, Foreign Office, Rangoon, 10 February 1902.

the pursuit of criminals across a recognised border. The British Government of India was nonetheless extremely cautious, stating that British police officers should cross the border only in exceptional circumstances.[60] As such, the border represented a divide rather than an interface between two countries, and its defence remained essentially the business of each sovereign territory separately.

The British and French Residents and Commissioners wanted to settle their problems without have to go through their representative in Peking. The Resident of Lang Son complained in 1906 that the regulations concerning the border police were unclear:

> That is why I did not mention extradition, which the authorities on the Chinese frontier request daily when Chinese subjects who have fallen foul of the justice system of their own country flee to Tonkin. I believe I am not mistaken in saying that the law on extradition is not applied in the colony despite the fact that in many circumstances Chinese individuals have been extradited on the request of the authorities concerned. Be that as it may, there is, in my opinion, a shortcoming here which should be remedied.[61]

In the case of the China-Vietnam border, although extradition could be justified by invoking Article 1 of the Treaty of Tientsin of 9 June 1885, and Articles 16 and 17 of the Trade Agreement of 25 April 1886, there was, strictly speaking, no extradition agreement specific to that particular border. Police reports, however, contain some instances of cooperation. For example, in 1912, bandits entered a house in Yunnan, killed the daughter and grandson, wounded three men, and stole an ox, fabrics and two dresses.[62] Two of the assailants were identified as living in Tonkin, one of them being a local chief. The Chinese police commissioner made a request to the French border police to arrest the two bandits, one of whom was apprehended. This shows that, in the absence of fixed legal rules concerning hot pursuit or extradition, there was nevertheless

60 BL, IOR, L/PS/7/144/616–617, Letter from the Secretary to the Government of India in the Foreign Office, Sir Barnes, to the Chief Secretary to the Government of Burma, Fort William, 26 March 1902.

61 ANV1, RST 76382, Lettre du lieutenant-colonel Hocquart, résident, au résident supérieur du Tonkin, Lang Son, 2 February 1911.

62 ANOM, GGI 19670, traduction d'une lettre du commissaire des Affaires étrangères chinois au délégué français, Yunnanfou, 21 June 1912.

some cooperation between police forces on the French and Chinese sides, and that border control depended largely on the individuals involved.[63]

An extradition agreement existed between Siam and Burma, however, since the British had moved faster in this domain than the French. Extradition of criminals to or from Siam figured in the Treaty of 1883, the supplementary article of 1885, and the Indian Extradition Act of 1903:

> A Political Agent has power to issue a warrant for the arrest of any person other than a European British subject accused of committing any offence specified in the first Schedule attached to the Act in the State for which he is a Political Agent. Persons other than European British subjects may be: (a) Subjects of Siam or of a third Power; or (b) Native British subjects.[64]

In the first case, according to the Treaty of 1883, and the Supplementary Article of 1885, fleeing criminals should be handed over to the Siamese authorities. In the second case, the Political Agent may hand them over to the British or the Siamese authorities, or else to the Consul-General in Bangkok. To extradite European British subjects, however, one should refer to the British Government of India. The Burmese courts thus had no jurisdiction over European or Siamese or other British subjects, if they committed a crime in Siam: they must be extradited or released. The fact that European British subjects were not covered by these Articles and were directly under central control reflects the hierarchies introduced by colonial rule. The different categories mentioned here are defined in law, even if they are not entirely respected in practice. It is clearly assumed that foreign nationals or local populations, and not European British subjects, are the principal trouble-makers on the borders, who thus risk being extradited. The French did not introduce a right to hot pursuit across the Franco-Siamese frontier until 1920, and then only for the specific case of theft of pack animals or canoes.[65] Siamese incursions into Karen and Shan territories, and into Tonkin and Annam, to steal elephants, recurrently gave rise to complaints from the British and the French. Elephant theft was endemic

63 The ability of provincial governors to set their own political agenda, far from Peking, has been demonstrated in the case of another province in south-west China, Sichuan. See Samuel A. M. Adshead, *Province and Politics in Late Imperial China. Viceregal Government in Szechwan, 1898–1911*, London, Curzon, 1984.

64 BL, IOR, V/27/160/15, Note on the rules concerning the extradition of criminals to and from Siam and China and the issue of passports to Siam and China, Rangoon, 1910.

65 Ministère des Affaires étrangères du Laos, Recueil des traités franco-siamois délimitant la frontière de l'Indochine et du Siam (lao-thai), 1886–1946, Vientiane, Publication du Département des traités et du droit, 1996, p. 296.

in the area, and required the intervention of the police. In 1911, the Siamese Under-Secretary of State for Foreign Affairs expressed his satisfaction at the decrease in the number of complaints received on this subject.[66] He reported that in 1909–1910, of the 44 elephants stolen, 33 had been retrieved and 11 lost, whereas in the following year, 15 elephants were stolen, 13 of them retrieved, and only two lost, a success he attributed to the national and local police on both sides of the border, which shows that collaboration between police forces was both possible and could be beneficial. The right to hot pursuit introduced ten years later supplemented these measures with wholly new tools for maintaining order on the frontier and fighting illegal trafficking. Yet it was not so much the border markers, forts, and passport checks that determined successful (or unsuccessful) peace-keeping, but rather the Europeans' understanding of the local strategies and practices in play.

3.4 *The Fiction of a Frontier?*

In 1891, a French missionary in Siam criticised the map of the country drawn by Father Launay, particularly the borders with the Lao principalities.[67] He explained the complexity of the situation and all the questions it raised:

> For Father Launay, there are Lao populations only on the left bank of the Mé Không above M. Nakhon, precisely in a region where the people are not at all Lao and do not call themselves Lao. If we inquire of their nationality, they all reply, with the exception of some recently settled isolated families here and there, that they are Sos or Kalongs, or Jos, or Khas, or else, further north, Karens, Phu thang, etc. So where are the Lao? Some are on the left bank of the Mé Khong, above and below Vien Chan [...]. On the Burmese side, Xiang mai [Chiang Mai] is Lao, although their language is slightly different from ordinary Lao; their writing is also slightly different from Siamese, and from the Lao of Vien Chan. I can read the Lao of Vien Chan and Siamese quite easily, but I cannot understand the writing used in Xiang mai.[68]

This missionary's comments are revealing on several accounts. He depicts the different populations as he himself understands them, from an external

66 BL, IOR, L/PS/10/98 2622/1906 Pt 4, Letter from Phya Phipat Kosa, Under-Secretary of State for Foreign Affairs, to Mr. Beckett, Foreign Office, Bangkok, 13 June 1911.

67 Adrien Launay, *Atlas des missions de la Société des missions étrangères*, Lille, Desclée de Brouwer, 1890.

68 MEP, vol. 896, Siam, 1889–1913, Lettre de M. Prodhomme à M. Péan, 30 August 1891, 73 (5) à (7).

viewpoint, but also in terms of how they view and describe themselves. By comparing the two versions – the European and the local discourse – he concludes that no Lao people live in the areas in question. When he mentions the isolated families which had settled in the territory recently, he suggests that boundary-marking has produced population movements and migration across frontiers. Lastly, he frames his question in terms of nationality, which brings back the issue of where the populations feel they belong, and of building a centralised State in relation to its multi-ethnic peripheries. The newly created countries of Siam, Burma and Laos seem, despite clear international borders, to have populations that exist only on paper. On the ground, the only thing that counts is ethnic, not national, identity, and this is how the populations define themselves.

Ethnic groups that spanned national borders often defied the authority of the centralised State, and could spark territorial conflicts. Indeed, how draw the "approximate boundary between Ava Shans and Siam Shans,"[69] which figures on Yule's map of 1857, when the written names of the ethnic groups in question extend over the borderline? The French encountered the same problem in attempting to establish the China-Vietnam border, which initially destroyed the territorial integrity of the local chief Đèo Văn Trì's lands, a decision later reversed.[70] Frontier demarcation did not settle these issues, which remained acute throughout this period. In 1913, the French Resident in Laos complained that the populations of Sibsongbanna [Xishuangbanna] did not respect the borders:

> It is undeniable that our domination in this region encounters very great difficulties due to the delimitation of 1894–1895. The Sib Song Pannas, peopled by the Lu race, were arbitrarily divided up between England, France and China. China has moved the corner of Pou Eul and I-Hou forwards into our territories, and so has completely cut off communication between Muong-Sing and Muong-Hou. We cannot really maintain control over the region until the border has been rectified and our authority can be imposed over the whole Lu region, with a territory which is less carved up.[71]

69 "Approximate boundary between Ava Shans and Siam Shans", *New Map of Burma and the Regions Adjacent*, by Captain Henry Yule, 1857; MEP, K MYA 3.

70 P. Le Failler, *La rivière Noire. L'intégration d'une marche frontière au Vietnam, op. cit.*

71 ANOM, GGI 19674, Extrait d'un rapport no 25 sur la situation politique du Laos, 4e trimestre 1913.

Two arguments are combined here: one concerns ethnic identity, which does not recognise national frontiers, with the result that the central government cannot impose its authority effectively over the Sibsongbanna [Xishuangbanna] region; the other concerns the geographical aberrations created by drawing territorial boundaries, and the resultant problems in governing these territories. From the coloniser's perspective, wherever ethnic identity is stronger than national identity, borders are weakened, and not respected at local level. From the local population's perspective, the newly established international frontier introduces major changes, especially in taxation, since every State has its own system.[72] The issue of tribute was raised by Captain Walker, during the Anglo-French Commission of 1894–1895:

> Here they told me with great glee that they had been gravitating backwards and forwards from Möng Long Pukkha into Hsib Hsawng Panna for many years to avoid paying tribute, remaining on one side of the boundary until their tribute was demanded from them and then crossing the boundary.[73]

The inhabitants' actions to avoid paying tribute were thus carried out long before the arrival of the Europeans, and border delimitation. However, this phenomenon of fleeing taxation did not come to an end when the Europeans began governing the region. In 1906, the French commander Bernard indignantly described the exodus of Lao populations towards Siam, in the opposite direction from a similar movement in the 1890s. He blamed Siam's integration policy, but also the over-strict French government, which laid too many constraints on the population:

> In Middle Laos, [the Siamese] make greater efforts not only to win over the population, but also to attract the inhabitants of French Laos. The taxation is low, not collected, and requisitions are rare. The government waives all taxation for immigrants, gives them property rights, and also provides advances of money or rice:[74]

---

72 On the Han efforts to control the south-western borders of China and set up a viable taxation system in the modern period, see C. Patterson Giersch, *Asian Borderlands. The Transformation of Qing China's Yunnan Frontier, op. cit.*

73 BL, L/PS/20/83, Report on the Anglo-French Buffer State Commission, 1894–1895, by Captain Walker, November 1895, p. 21.

74 AIP, Fonds Fernand Bernard, BEF1, Lettre du commandant Bernard au gouverneur général de l'Indochine, Nam, 3 January 1906.

Taxation is again presented as the prime cause of population movements, and Bernard mentions hundreds of families who left for the Siamese side, suggesting just how ineffectual the frontier is. In order to counter the attraction of Siamese Laos, Bernard suggests, French Laos should move towards greater economic freedom, an end to requisitions, and a very simple administrative system, employing only a handful of civil servants. Population movements are thus considered to result from French governmental policy. The attempt to limit and manage these was a real test of the country's sovereignty.

•••

When the British and the French arrived in the Indochinese Peninsula, they were faced with such a variety of border situations that they sought to impose linear borders, precisely in order to determine the limits of the different sovereign powers. They thus created discrete States, marked on the land by posts, forts and checkpoints, which put an end to pioneer fronts and spaces bound to a number of different sovereign entities. But to what extent were these dividing lines imposed? The analyses undertaken show the different actors involved in producing the territories of the Indochinese Peninsula at the time, and suggest that borders were not simply imposed, but rather occasioned a large number of interactions and confrontations. In 1885, the French had just ended two years of war against China, and they wished to emphasise their new ascendancy over Vietnam as well as protect the newly conquered territories from incursion from the north. Delimitation, and a clear boundary accepted by all, was vital to them. Conceiving the frontier as a line certainly confounded local representations and relations to the land, when the border cut through complementary entities such as, for example, passing between a village and its rice paddies, but one should not essentialise these moments. What transpires above all is the pragmatism of the members of the mixed commissions, insofar as their primary goal, once they had secured their positions and marked their borders, was to maintain law and order. Colonial and local governors had to work together to preserve the peace. The main difficulty was the limited amount of communication between the centres of power and the border posts, especially the most distant ones, and also between border posts on either side of the line. To enforce the boundary required international legal instruments and police cooperation, which sometimes went as far as the right to hot pursuit and extradition, but not without some failures. The variety of arrangements throws a more nuanced light both on the colonial government's impotence in the face of the huge number of incursions (ubiquitous in the archives), and on its omnipotence, when it in fact had few representatives in these marginal lands. Encroachments, illegal border crossings and large-scale cross-border migrations constantly defied the lines on the maps, whose geographical paths had been officially recognised, and which had sometimes also marked the landscape physically, yet which, in certain cases, still did not correspond to the concrete realities on the ground.

CHAPTER 9

Logics of Rule and Territorial Anomalies

Although the imperial territories of the northern Indochinese Peninsula gradually acquired roads, agreed boundaries, and more precise limits on the maps, they nonetheless remained largely beyond the control of the Europeans. They were frontier zones, which did not obey the same types of territorial dynamics as the more central lands. Immediately after the annexations, in 1885, an administrative system had to be set up, with its internal structures and overall framework. No prior plan, containing precise instructions, settled the issue of whether the government would be civil or military: this was, rather, the fruit of circumstance and of the arguments that won out. Throughout the colonial period, the two systems coexisted, under British and French sovereignty. Despite a certain standardisation achieved over time, the disparities between different territories persisted, and even increased, because they were integral to the form of political domination involved.

Far from unifying the annexed territories, the administrative choices made often reinforced their heterogeneity and deepened spatial distinctions already present, with significant political and social effects. In Upper Burma, the directly administered British territories bordered on those not in this category, which were left in the hands of local leaders, for pragmatic reasons, while the British kept an eye on their activities. Likewise in northern Laos and in Tonkin, administered by the military, but where in practice power was vested in local representatives, unlike in the provinces under civil government, that is, in the rest of the French colonial territories.[1] These different situations corresponded geographically to the Delta region and the mountain regions. They originated in disagreements about the form colonisation should take, with some arguing for a restricted territorial hold, and others advocating the extension of pacification operations into the frontier zones. Although the borderlands were technically under British and French sovereignty, they were largely ruled through delegated powers. The modes of territorial government, rather than corresponding to a pre-defined plan, were the outcome of negotiation, with

1 The history of the territories administered by the army in Indochina has yet to be written. For an introduction to the question, see the entry by Philippe Le Failler in Christophe Bertrand, Caroline Herbelin, Jean-François Klein, *Indochine. Des territoires et des hommes, 1856–1956*, Paris, Gallimard, 2013, p. 212; J.-F. Klein, *Pennequin, une "éthique coloniale française?", op. cit.*; for monographs on a particular military territory, see P. Le Failler, *La rivière Noire. L'intégration d'une marche frontière au Vietnam, op. cit.*; and the thesis already cited, by Nguyen Thị Hải.

 | DOI:10.1163/9789004469853_011

on-going tensions between different protagonists – civil and military governors, European and local populations, and so forth. These disagreements also resulted in constant changes to internal administrative boundaries, which were contested by those who had to live with them daily.

In addition to this already complex spatial organisation, micro-territories emerged from border demarcation, when, even after a line had been agreed upon in a treaty, certain territories remained superimposed or overlapping due to the many sovereign entities involved. Sometimes the treaty itself gave rise to enclaves or mixed-status territories. Hence the situation in the colonial lands had more in common with a legal and administrative patchwork, a set of pieces tacked together without much regard for harmonisation, than with a homogeneous system.[2] The lack of uniformity in the borderlands was blatantly visible on European maps in the form of enclaves and in the mountain areas, where the spatial distribution of ethnic groups, in the northern part of the Peninsula, was a major challenge for the Europeans. Indeed, how represent the human factor, notably the irreducibly diverse ethnicities of the region? How define the territories of an ethnic minority when it pays little attention to instituted frontiers and itself combines several identities? The situation became even more inextricable if the local populations' sense of belonging to a particular territory was taken into account, that is, their feeling, and not the presence or absence of a boundary marker. Moreover, many of these groups contested the sovereignty imposed upon them to a greater or lesser extent. We shall examine below how the Europeans attempted to govern this conflictual patchwork of territories in the border regions, and the extent to which British and French methods varied.

1 The Specific Status of the Frontier Provinces

Two distinct forms of government were set up, in both the British and the French territories. A direct, civil administration governed the flat lands inhabited by the majority ethnic group; and a military administration controlled the mountain regions where the minority groups lived. The form of administration chosen was not planned in advance, and it was invariably part of wider debates, in Britain and France, on the extent and limits of colonisation.

---

2 Lauren Benton, *A Search for Sovereignty: Law and Geography in European Empires, 1400–1900*, Cambridge, Cambridge University Press, 2009.

1.1 *The Argument of Geography*

The division between plain and mountain corresponded to other binary oppositions, such as civilised and barbarian, full and empty, rich and poor. These were harnessed to justifying different administrative statuses, although the primary reasons were in fact practical. There was nothing new about the debate between those who advocated stabilising a smaller set of annexed territories, and those who advocated expansion; the issues had already been raised for India and Algeria, and are germane to the imperial project itself.[3] Territorial expansion, and with it, the introduction of a hierarchy between different territories and people, is a basic feature of empires.[4] It was thus perfectly natural for the debate to emerge in French Indochina, insofar as the issue of continuing or halting territorial expansion, and divergent views on the subject, are essential moments of empire-building.

The British justified the Anglo-Burmese and Anglo-Afghan wars by the perceived need to form a glacis of territories to protect the Raj in the face of Russian ambitions in Central Asia, and French, Siamese and Chinese interests in the Indochinese Peninsula. Successful territorial conflict was also considered to be a way of opening up a hinterland, a vision which the French and the British shared, for example regarding Yunnan. There was nothing self-evident about annexing the Shan States, but their proximity to China was used to justify their annexation into the Raj. French expansionism likewise brandished the argument of geography, putting forward predominantly strategic reasons, as in the conquests up until then. This policy was first contested when Annam-Tonkin was annexed, a move rejected in 1873, despite Dupuis and Garnier's capture of Hanoi Citadel. This episode brought about the fall of Jules Ferry's government on 30 March 1885, and marked a pause in official French support of expansionist policies, even though annexation continued on the ground.[5] The debate resurfaced concerning the Upper Region – the mountainous lands bordering on China, comprising two thirds of Tonkin – an area that remained under military control, following a decree of 24 May 1886, due to the many problems still encountered there. The general de Courcy recommended confining French efforts to the Delta region, whereas others maintained that security in the Delta depended on further pacification in the borderlands, and hence that quashing the seeds of revolt in the Upper Region could alone

3 On the discussions in India, see Christopher Bayly, *Imperial Meridian: The British Empire and the World, 1780–1830*, London, Longman, 1989; for Algeria, see H. Blais, *Mirages de la carte. L'invention de l'Algérie coloniale, op. cit.*

4 J. Burbank, F. Cooper, *Empires in World History. Power and the Politics of Difference, op. cit.*

5 Charles-Robert Ageron, *France coloniale ou parti colonial?*, Paris, PUF, 1978.

guarantee peaceful relations in the Delta zone. They regarded the Upper Region as a defensive glacis for the Tonkin delta, a buffer zone which could guarantee security throughout the Protectorate. In the case of India, and likewise for Algeria and French Indochina, it was the advocates of expansion who had the last word.

In Burma, as in Tonkin, the British and French quickly discovered that the central administration really only had control over the majority ethnic group, and had little purchase on the ethnic minorities of the mountain regions. They based their policies on this inherited situation. Aware that in mountain areas, the social and political organisation shaped by power relations between local chiefs was so complex that they would have to invest men and money well beyond their original budget in order to gain influence there, they decided to consider these inhabitants as buffer populations between the central territories and the neighbouring powers, insofar as these regions were also frontier zones. This was already accepted practice in British India, where the princely states were treated as a protective shield, and the "martial races" were recognised for their particular qualities.[6] The idea for this came from the Indian Rebellion of 1857, and was later used to justify the preferential recruitment of the Nepalese Gurkhas or the Punjab Sikhs into the police and army. In French Indochina, the Mường and Thai generated a similar enthusiasm on the part of the colonial officials, for their hardiness and their skills at warfare. Rather than being deployed all over the empire, they were used to create groups of partisans to maintain order in their territories.

In an organisational plan drawn up for these borderlands, probably in 1890, the French officer Pennequin proposed a special regime for the "outlying regions of Tonkin".[7] He explained that the "Thai race" populating the area detested both the Chinese and the Vietnamese, and that its feudal functioning was so strong "that the democratic regimes of China and Annam have never been able to destroy [it]."[8] The way forward, he concluded, was to strengthen the authority of the chiefs in order to win their allegiance.[9] This was one of Pennequin's key ideas, and it was associated with the conviction that a special regime was justified because the pacification of the borderlands could alone guarantee the security of the rest:

6 Gavin Rand, "'Martial Races' and 'Imperial Subjects': Violence and Governance in Colonial India, 1857–1914", *European Review of History/Revue européenne d'histoire*, 13/1, 2006, p. 1–20.

7 ANOM, GGI 14463, Projet d'organisation des régions frontière du Tonkin et du Haut-Laos (no date, probably 1890, by Pennequin).

8 *Ibid.*

9 On Pennequin's theories and his visionary role in pacification, see J.-F. Klein, *Pennequin, une "éthique coloniale française?", op. cit.*

> We will have a zone around the Delta where the Annamite will find neither refuge nor support. Towards China, there will be a protective zone, but at the same time it gives us access, because the populations on the other side of the border have the same customs. The same race (the Thai) lives along the banks of the Mekong and the territories that we claim in Siam.[10]

Hence, by 1890, the military territories appeared as both hermetically sealed, and porous, liable to expand beyond the limits of the empire. The representation of the ethnic minorities as a buffer population, some of whom would become partisans and fight alongside the French, casts them as "intermediaries", not only because they inhabited the borderlands administered by the army, but also because they were neither Chinese nor Vietnamese. The creation of military territories with a special status thus separated off these populations: "They are equally distant from the Chinese and the Annamites, and can be easily used by us, as a barricade against incursions from China, and also as a watertight seal by which to limit the effect on the Delta area of any disorders in the interior".[11] The idea of a protective glacis is very clear here, applying both to the territory and to its inhabitants.

1.2 *On-Going Military Rule*

Different types of rule were in force in the central provinces and in the marches of Burma and French Indochina. Since, in both cases, only the centre was income-generating and could thus justify investment by the metropole (in principle the colonies had to be self-supporting), it required a different mode of administration from the frontier zones, which not only brought no wealth to the empire, but also presented an endless string of problems. The large and heterogeneous borderlands were thus ruled differently from the other spaces.

Burma, a possession within the British Raj, was first governed by a Chief Commissioner, and then, in 1897, by a Lieutenant Governor, when it was promoted to become a major province in its own right. The area came under the authority of the Government of India, in the same way that Tonkin, Vietnam and Laos were ruled by a Governor General within the Indochinese Union. The administrative organisation of Burma, and the resulting distinctions between different territories within it, were borrowed from the system prevailing in the

---

10 ANOM, GGI 14463, Projet d'organisation des régions frontière du Tonkin et du Haut-Laos (no date, probably 1890, by Pennequin).

11 ANOM, GGI 65374, Dossier 1, 19, Lettre du gouverneur général de l'Indochine au ministre des Colonies, January 1912.

British Raj.[12] Thus, after the 1885 conquest, Upper Burma was divided into four civil divisions, composed of districts. Districts were the pivotal units between local and regional governing bodies, and were the responsibility of a Deputy Commissioner. The sub-divisions, while remaining under civil administration, changed many times over the period.

The frontier zones, by contrast, were classified as "excluded territories", "partially excluded territories" and "scheduled areas". The general rule for these spaces was that of non-interference, unless raids into the administered areas occurred. This system for classifying, distinguishing and hierarchising territories had been conceived for India, where it was in force, and it was then applied to Burma. The Shan States, the Karen and Chin territories, and the Kachin territories north of Myitkyina, at the frontier with Yunnan and Tibet, known as "the Triangle", as well as a few other smaller territories peopled by ethnic minorities, were accordingly set apart.[13] These territories had the same status as the princely states in India, and thus remained under their customary laws, as long as these did not contradict the British ones. As such, they were excluded from the measures applied in directly administered colonial spaces, namely policing, security guarantees, the construction of law courts and schools, tax collection and obligatory labour.[14] The local princes – the Shan *Sawbwas* and Kachin *Duwas* – thus remained largely independent, and preserved most of their prerogatives. The same was true in the French zones. The Upper Region was only partially under direct French administration, with powers delegated to local chiefs, who carried out most of the work, on condition that they remained faithful to the French.

The difference between the plains and the mountains, yet their contiguity, legitimated the two different types of management: "The mountainous region surrounding the Delta does not lend itself to being governed under the same regime as the Delta, due to its physical characteristics, which make it relatively impenetrable, and due to the living conditions of its populations."[15] The relief made access difficult, and economic activity unprofitable. Moreover, the area's widely spread and small population did not bode well for tax revenues, such

---

12 M. Sadan, *Being and becoming Kachin. Histories beyond the State in the Borderworlds of Burma, op. cit.*, p. 167.

13 Daniel George Hall, "British Burma, 1886–1942", in *A History of South-East Asia*, New York, Springer, 1981, p. 770–788.

14 The territories populated by ethnic minorities would later be collectively called "Frontier Areas". See Donald M. Seekins, *Historical dictionary of Burma (Myanmar)*, Lanham, Scarecrow Press, 2006, p. 18.

15 ANOM, GGI 14463, Projet d'organisation des régions frontière du Tonkin et du Haut-Laos (no date, probably 1890, by Pennequin).

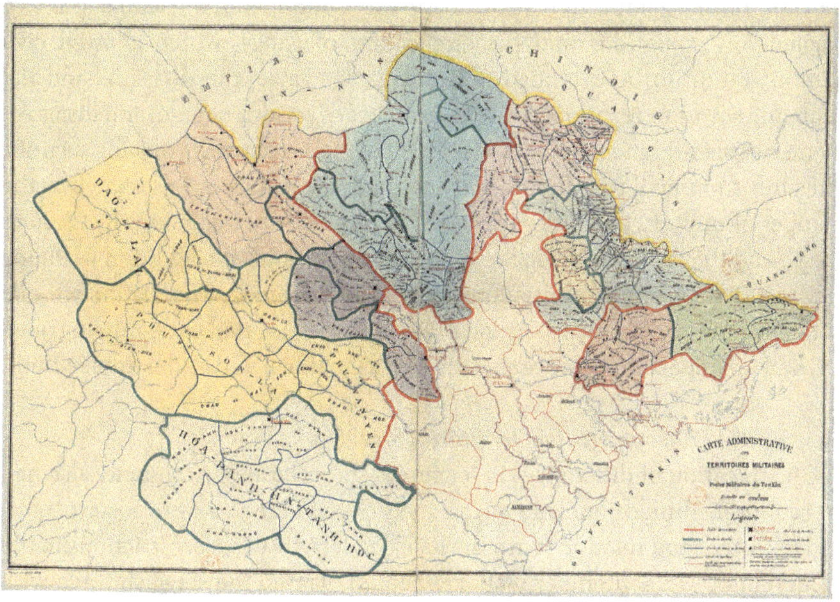

FIGURE 45 Administrative map of military territories and posts in Tonkin
SOURCE: BNF, Cartes et plans, GE C-1919

that, for essentially financial reasons, a similar type of colonial regime as in the Delta made little sense. In 1886, a French civil administration of the Tonkin Delta replaced the previously the military one, and four territories under military control were later created by the Governor General Jean-Louis de Lanessan (ordinances of 6 and 20 August 1891). This dual, civil and military administrative system, formalised in 1890, lasted in Tonkin until 1945.

The military territories in Tonkin underwent many changes. The initial four, from East to West, were Móng Cái, along Hạ Long Bay and the Chinese province of Guangdong; Cao Bằng, bordering on Guangxi and including Lạng Sơn; Hà Giang, next to Yunnan; and Lai Châu, a territory which included Điện Biên Phủ. The *Carte administrative des territoires militaires et postes militaires du Tonkin*, drawn on a scale of 1:500,000, and published by the *Bureau topographique des troupes de l'Indochine* in April 1894, illustrates the dual French administrative system, and the depth of the protective glacis created by it [Figure 45]. In 1905, the military territories were reduced to three, then to two in 1906, as unrest lessened. The later *Carte administrative des territoires militaires*, on a scale of 1: 500,000, published by the *Service géographique de l'Indochine* in September 1903, clearly shows the reduction in military-controlled territories along the whole length of the border: compared to the broad band of 1894, they are nothing but a thin strip.[16] This was doubtless due to the success of pacification operations and the eradication of gangs, which allowed civil administration to be extended northwards. However, the divisions and the changing status of territories were the subject of serious debate, and disagreement, between French colonial officials. The major concern was for security. The shift from civil to military rule occurred, for instance, in 1912, when the number of military territories increased to three due to incursions by Chinese gangs, and then to five during the First World War, after the addition of Phong Saly, a Laotian territory contiguous to China and Tonkin. The distinctive rule of the frontier provinces, where local chiefs were delegated wide-ranging powers, had an impact on tax revenues, as well as political and social implications.

1.3 *The Consequences of Territorial Status*

The justification of differences in territorial status based on ethnicity also had an economic dimension. One of the goals of the French was to restart trade, and to foster good relations with the local leaders, by easing restrictions in the borderlands.[17] In September 1896, Pennequin blamed the sluggish trade with Yunnan partly on the customs officers, and suggested that the military authorities be less strict on the sale of opium and alcohol.[18] This resulted, effectively, in shifting the customs inspections from the Chinese-Vietnamese border to the limit between the military and the civil territories, where, however, there were "too few border posts, too widely spaced, and without links between them".[19] In the military territories of Hà Giang and Lai Châu, the most northerly points at which the *Administration des Douanes et Régie* collected customs dues were Yên Bái on the Red River and Phủ Doãn on the Clear River. The military territory of Lạng Sơn was an exception in maintaining a tax office, and the post of Lào Cai was kept on, in order to inspect the goods passing in and out of Yunnan. Although this decision was not strictly about administrative divisions, it shows that the presence of customs units did not so much aim at preventing goods from the northern regions from infiltrating across the Chinese-Vietnamese border, as at separating off the territories under civil administration. The draft ordinance precisely stated that the military territories "form an external

16 BNF, Cartes et Plans, GE B-8235.

17 ANV1, RST 14405, Rapport à M. le secrétaire général de l'Indo-Chine, Hanoi, 19 December 1896.

18 The sale of these products was forbidden in British Burma: J. G. Scott, *Gazetteer of Upper Burma and the Shan States, op. cit.*, p. 515.

19 ANV1, RST 14405, Extraits du procès-verbal de la séance du Conseil de protectorat du 30 décembre 1896.

frontier zone for the Protectorate, and the limit of the civil territories."[20] This suggests that the frontier was no longer at the frontier, or rather that there were two paths for one border, corresponding to the distinction between the diplomatic limits and the customs barriers.

In British Burma, taxation arrangements depended on the territory. The Scheduled Districts Act of 8 December 1874 introduced a different system for spaces deemed "backward" (Backward Tracts), where the population, and the anticipated tax revenues, were lower than in other parts of the Raj. There were four different taxation methods in British India.[21] Two of these were indirect taxation: the levy paid by Princely States; and the *zamindari*, or "permanent settlement" system, the most extensive in "useful" India, which was established in Bengal as early as 1793.[22] There were also two systems of direct taxation. The *raiyatwari* system was used in the Deccan and Maratha territories, where the British found no equivalent for the *zamindars*. They assessed the lands of the village elites in a much more careful way, on a case by case basis. The other system was the *mahalwari*, applied in northern India between Benares and the Sikh Kingdom of the Punjab, where officials proceeded village by village, and respected the principle that land was held in common by the community. The British in Burma attempted to set up a *raiyatwari* system, whereby the local chiefs collected taxes on real estate, to which was added a tax on households inherited from the Burmese system, the *thathameda*.[23] The local Shan, Kachin, Chin and Karen chiefs, however, paid an annual tribute directly to the Government of India, like the Indian Princely States.[24] The balance of power between the local leaders, and their degree of integration into the colonial system, were partly linked to the administrative status of their territory.

The autonomy granted to local leaders could be used by them to win favour (or incur disfavour) with the colonial administration, with concrete consequences for them and their populations. In 1895, for instance, the Kachin territories passed from the status of non-administered territory to an indirect

20 *Ibid.*

21 Claude Markovits (ed.), *Histoire de l'Inde moderne, 1480–1950*, Paris, Fayard, 1994.

22 The *zamindars*, a landed elite, were recognised as owners of the land as of right. They levied taxes for the East India Company, and later for the government, which ensured that a fixed rate of taxation was applied. The British authorities hoped that this would transform them into a dynamic class, on the model of the English agricultural revolution, but they were disappointed in this because the Bengali *zamindars* took advantage of the system, and the majority of the peasantry continued to be exploited.

23 Thant Myint-U, *The Making of Modern Burma, op. cit.*, p. 229–231.

24 In 1886, the British estimated the population of the Shan States to be roughly 500,000 inhabitants. J. G. Scott, *Gazetteer of Upper Burma and the Shan States, op. cit.*, p. 520.

administration. This had the effect of taking tax collection out of the hands of Kachin *Duwas*.[25] The change took place in territories where the local leaders were well-disposed towards the British colonisers, and where the British thought the rural chiefs could provide them with useful information concerning the Chinese officials and military forces along the frontier. As a result of this separation of the Kachin territories into two, with one part under indirect administration, and the other without colonial administration, the most distant territories were increasingly marginalised, and their local elites neglected. By contrast, in the south and east of the Kachin territories, the indirect administration began to produce a network of local elites.[26] The choice of administrative arrangements was thus not simply aimed at rationalising territorial government, but it also had political implications, and played a crucial role in the relations between imperial powers and local elites. Local elites had greater freedom outside of a colonial administration, but they risked marginalisation; whereas signs of goodwill on their part could bring benefits to their lands through greater integration into the colonial system.

The anomalous administrative characteristics of borderlands could, however, create complications for the internal administrative boundaries between territories, and produce territorial exceptions.

2 Unstable Territorial Divisions

2.1 *Fluctuating Internal Administrative Boundaries*

The limits of administrative areas were very often criticised, and always from the bottom up, that is, either by the local elites or by the colonial administration closest to realities on the ground. The criticisms were passed up to the general administration, which tended to propose changes, either for security reasons or for pragmatic reasons of effective territorial management. For although the dividing lines were decided at the centre, only daily experience at the local level – by officials and the populations themselves – could show their failings or absurdities, for which the central administration had to find solutions.

The divisions gradually implemented by the colonisers kept as closely as possible to the preexisting lines, which were not always easy to establish. August Pavie, exploring Sibsongbanna [Xishuangbanna] in 1891, regretted that the limits were not better defined:

25 BL, IOR, L/PJ/6/392/1895/307, Kachin Hill Tribes Regulation, 1895.

26 M. Sadan, *Being and becoming Kachin. Histories beyond the State in the Borderworlds of Burma, op. cit.*, p. 168.

> As for the limits of each *panna*, and the *muongs* they contain, they cannot be less precise, since each of them has most often been altered by the death of a chief, or by civil wars. That is why the information from Phya Kao, Phya Proma Det and Phya Chai Yac does not entirely match.[27]

Subdivisions of military territories and of provinces under civil administration were also contested, not so much for their boundaries as for the authority to which the units were then attached. Thus some local populations who wanted greater protection were not keen to change to a civil administration, because they considered that administrators in military territory, reinforced by the presence of the army, were a safer option. Several cases of this kind arose, in the context of competition between military territories and civil provinces. The outcome often went against the dominant French policy of gradually phasing out military rule. On 12 June 1895, Gérard, the Commander of the *cercle* of Phố Bình Gia, passed on two requests to Gallieni, the commander of the second military territory. He had received these from the former inhabitants of Chợ Chu, in the province of Thái Nguyên, who had been moved to Chợ Mới after it had been ransacked in 1893 by the Lương Văn Sơn gang. They refused to return home for fear of the militia groups, and would only do so if their safety was guaranteed by the military. In the first request asked for their area to be assimilated into the *châu* of Bạch Thông, which had just been incorporated into the administrative *cercle* of Phố Bình Gia, in order to benefit from the protection of a military Commander.[28] The second request contained similar complaints from another head of canton, and four local dignitaries, who had been raided by the same gang in 1893, and who also asked to be attached to the *châu* of Bạch Thông.[29]

Gallieni, in line with Pennequin's approach, listened to the local population's demands, declaring that "All our efforts, to the west of the second territory, aim precisely at reducing the area where the gangs find shelter and enlarging accordingly the area where the inhabitants can take refuge whenever there are

---

27 ANOM, GGI 14388, Journal de séjour de MM. Pavie et Lefèvre-Pontalis à Xieng-Hung et Muong Hai. Exploration des Sip-Song Pan-Nas, 1er-5 avril 1891. Rédigé par Pierre Lefèvre-Pontalis, Xieng Hung, 7 April 1891.

28 ANOM, GGI 65373, Dossier B18, a translation of a letter from the authorities of the canton of Yen-do to the *quan-châu* of Bach Thong, on the subject of a request to have the canton of Yen-do placed under the authority of the *cercle* of Pho Binh Gia, signed by the names of the head of the canton (Ly truong) and two dignitaries: Luong dàng Duc, Nguyen dinh Chi, Luong dinh Dàng.

29 ANOM, GGI 65373, Dossier B18, translation of the letter from the head of the canton of Yên-Trach and the dignitaries of the *quan-châu* of Bach Thong, signed with the names of the head of the canton and four dignitaries.

incursions, protected by our stations."[30] This approach was not without territorial implications, and it was naturally criticised by the Thái Nguyên Resident, M. Sestier, who maintained that the two villages were adequately protected by the civil guard, and that they had been raided only once, moreover by partisans from the *cercle* of Chợ Mới, who had come marauding into the province of Thái Nguyên.[31] A sketch map was attached, which had no title, date, key, or orientation, showing a hatched yellow area, representing the second military territory, and its westwards continuation, the "Civil Territory", crossed by a red line that marked the limit with the third military territory [Figure 46]. The names of the two villages concerned do not appear, but Chợ Chu and Chợ Mới are marked. Thus the issue of security reveals the territorial competition between military and civilian areas, since it can lead to the expansion of the former to the detriment of the latter. It should be mentioned, however, that the administrative *cercle* of Chợ Mới was an exception, in that it was the only mountain region, in the early 1890s, to have a civil administration, as shown on Pierre-Paul Famin's *Carte générale du Tonkin*, drawn on a scale of 1:3,000,000.[32] The rule was otherwise that the Upper Region was controlled entirely by the military, which may explain why it condensed so many the problems.

On the British side, Upper Burma's internal boundaries were also changed frequently, in order to streamline its territorial organisation. In April 1895, Lord Elgin, the Governor of India, gave his agreement to the territorial changes requested by Burma's Chief Commissioner. The first change concerned the district of Bhamo, which bordered on Yunnan in the east, and was composed of four subdivisions: Bhamo, Shwegu, Myitkyina, and Mogaung. After the Kachin territories had gradually been brought under civil administration, other territories were added, reaching around 25,000 km2, which was too much for one administrator alone.[33] Moreover, the district capital was cut off from the rest of the territory for half of the year, which made the task impossible for the British officials on the ground. A solution was found by the Chief Commissioner, who proposed that Bhamo and Shwegu be made into a single district, in the south, and that Mogaung and Mytkyina together constitute the northern

30 ANOM, GGI 65373, Dossier B18, Lettre de Gallieni, commandant du iie Territoire militaire, Langson, 24 June 1895.

31 ANOM, GGI 65373, Dossier B18, Lettre du secrétaire général du gouverneur général de l'Indochine à propos de la cession de deux cantons de la province de Thái Nguyên au iie territoire militaire, 11 August 1895.

32 Pierre-Paul Famin, *Au Tonkin et sur la frontière du Kwang-si*, Paris, A. Challamel, 1895.

33 BL, IOR, L/PJ/6/397/817, Lettre de Thirkell White, Chief Secretary to the Chief Commissioner of Burma, to the Secretary to the Government of India, Rangoon, 4 August 1894, p. 5.

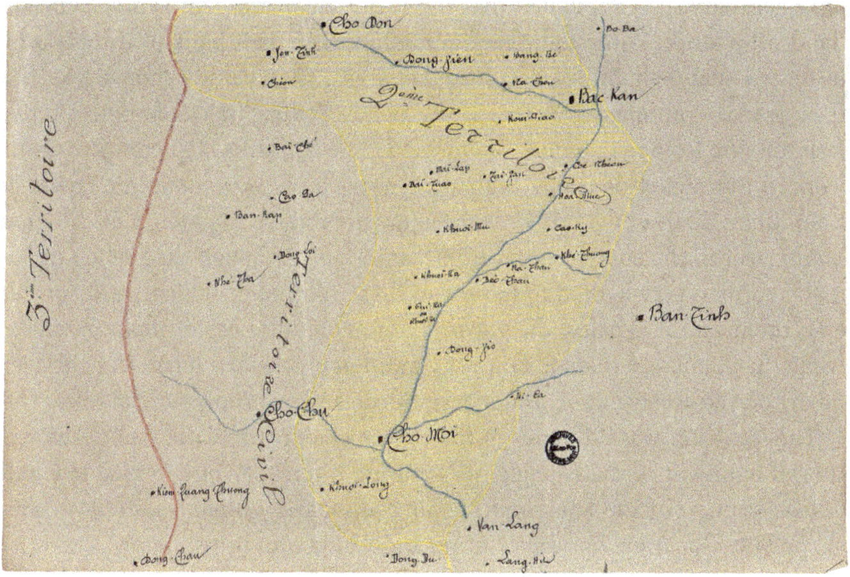

FIGURE 46 Request for an administrative border to be changed, 1895
SOURCE: ANOM, GGI 65373, Dossier B18, Lettre du secrétaire général du gouverneur général de l'Indochine à propos de la cession de deux cantons de la province de Thái Nguyên au IIe territoire militaire, 11 August 1895

district. These sorts of fusions and transfers of cantons, or equally the transformation of districts into subdivisions and subdivisions into cantons, were quite common.[34] They demonstrate how necessary it was for the administrators to fine-tune the bureaucratic arrangements in order to optimise government of these new territories. At the outset, for pragmatic reasons, internal boundaries followed previous lines, but they kept changing at the micro level, for strictly administrative and tax purposes.

2.2 *Enclaves and Spatial Discontinuities in the Frontier Zones*

In addition to the civil or military status of territories in Burma and French Indochina, some areas had a particular legal status for a certain period of time, which could be long or short. The French concessions on the right bank of the Mekong, for example, were enclaves within foreign territory, and as such had an exceptional status. Article 8 of the Franco-Siamese Agreement

34 This was also the case in Algeria and French West Africa. See H. Blais, *Mirages de la carte. L'invention de l'Algérie coloniale, op. cit.*; C. Lefebvre, *Frontières de sable, frontières de papier, op. cit.*

of 13 February 1904 stipulated that the Siamese government was to concede "lands of a surface area to be determined" on the right bank of the Mekong to set up inland shipping posts, and wood and charcoal warehouses, for the river trade. Commander Bernard, the head of the Franco-Siamese Border Commission from 1904 to 1907, delineated these enclaves after completing his work in the region of Luang Prabang. Seven points were noted in Article 8, from Xieng Khan in the north, near Luang Prabang, to the mouth of the Nam Moun river in the south, not far from Cambodia. Although the French diplomatic corps was delighted to have won these footholds in Siam, and considered them essential gains, the government made little use of them in reality, while the Siamese remained relatively indifferent, and the territorial enclaves finally played a very minor strategic, political, and even commercial role.

The situation was different on the British side, particularly due to the territorial anomaly of the triangle of Namwan, a territory of 220 km2 situated at the juncture of the Shan and Kachin States, and China. Due to its strategic position for trade between Burma and China, the British approached the Chinese government for an indefinite loan of the territory. An Agreement was signed in 1897, whereby the Chinese retained ownership of the area, but the British administered it, fearing that if they did not themselves control the area, trade and traffic could be interrupted.[35] They paid a rent of 1,000 rupees annually to China, and incorporated the territory administratively into the district of Bhamo. Unlike the French enclaves on the right bank of the Mekong, the Namwan territory was highly strategic, and remained a bone of contention between Burma and China.[36] Enclaves such as these, whatever their strategic importance, added to the complexity of territorial administrative regimes in Burma and French Indochina, adding further spatial disparities to what were already very different types of territory.

Other territorial discontinuities overlaid and confused the administrative categories. This was the case for certain territories on French maps beyond the control of the colonial government, grey areas that escaped colonial power and the law.[37] It is not that all the other territories were adequately controlled, but that representing territories on the colonial maps as being outside the law

35 35. BL, IOR, V/27/64/27, *Burma Gazetteer*; Bhamo District, vol. A, compiled by G. W. Dawson, Rangoon, 1912, p. 2.

36 This conflict has not yet been resolved, and has given rise to a huge number of publications, particularly by Chinese academics. See also Maung Aung Myoe, *In the Name of Pauk-Phaw: Myanmar's China Policy Since 1948*, Singapore, Institute of Southeast Asian Studies, 2011.

37 A "grey area" can be defined here as a "space of social deregulation, of a political or socio-economic nature, […] where the central institutions are unable to assert their

was quite an admission of the disorder reigning on the ground, and the consequent fragmentation of any imperial hold.

The end of the Franco-Chinese War in 1885 did not put a stop to the disturbances in the Upper Region, which continued for another 10 years. Yet the metropole simply insisted on an administrative system being set up, and would hear nothing of the difficulties encountered. On the ground, however, the activities of pacification were a daily effort. In 1889, General Bichot mentioned that to his astonishment the Governor General seemed to have no idea of the military situation in the northern regions of Tonkin, and he attributed this to the Residents, who only knew the Delta. He provided a map in support of his point:

> We are not dealing with *one gang*, as you seem to think, but with some forty gangs, and in order for you to judge the situation at a glance, and what we still have to achieve, I attach a map of the whole of Tonkin to this letter, showing the positions of these various Chinese, Annamite and mixed gangs, the names of their leaders, their areas of operation, etc.
>
> Apart from these gangs, which are known and classified, new ones are being formed every moment.[38]

The map mentioned by the General was therefore explicitly meant to represent the lawless areas in as much detail as possible. The *Carte des emplacements occupés par les pirates* [The Map of the Positions Occupied by the Pirates] made by French officers in 1889 for the purposes of "pacification" distinguished the Chinese gangs, in yellow, from the Vietnamese gangs, in green, while the mixed gangs, comprising both Chinese and Vietnamese, were represented by circles of both colours [Figure 47]. Whereas in the Delta region the circle is generally medium-sized, in the Upper Region, and particularly to the west of Cao Bằng, broad bands of yellow show the extent of the Chinese gangs in this frontier province. In addition to these zones, the whole of north-east Tonkin, and a broad corridor further west, orientated north-south, represented by black hatching, indicates "the zone the gangs pass through". It is a much larger area than that of the gangs' bases themselves, and it exceeds the limits of the brigades' action (in green). These zones of transit are crossed by

---

domination, which is assured by alternative micro-authorities." See Gaïdz Minassian, *Zones grises. Quand les États perdent le contrôle…*, Paris, Autrement, 2011.

38 ANOM, GGI 22335, Lettre du général Bichot, commandant en chef les troupes de l'Indochine, au gouverneur général de l'Indochine, sur les bandes de pirates au Tonkin, accompagné d'une carte, Hanoi, 18 September 1889.

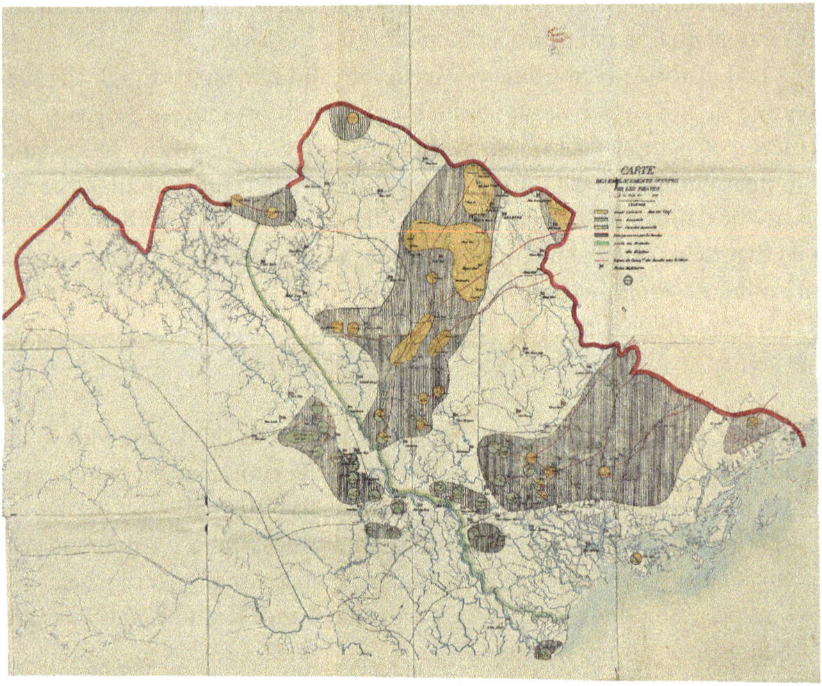

FIGURE 47 Map of the positions occupied by the "pirates", 1889
SOURCE: ANOM, GGI 22335, Lettre du général Bichot, commandant en chef les troupes de l'Indochine, au gouverneur général de l'Indochine, sur les bandes de pirates au Tonkin, accompagné d'une carte, Hanoi, 18 September 1889

communication routes with China, represented in red, which link the gangs to each other and to their support bases. The gangs are mostly present in these two zones, but also at the frontier with China, particularly in its unmarked sections, west of Hà Giang: the thick line representing the China-Vietnam border stops, and the gangs take over. Thus, while the French sought to represent the territory as organised, categorised and differentiated, the maps also reveal the large swathes of territory rendered inaccessible due to lawlessness, and the resulting enclaves, which broke up the colonial territory.

The attempt by the French to distinguish Chinese from Vietnamese gangs was a mark of progress, since initially they had been amalgamated under the name of "pirates", "bandits" or "rebels". Yet the Chinese armed groups had, in fact, little in common with those resisting colonial rule in Cần Vương.[39] Three

39 B. C. Davis, *States of Banditry: The Nguyen Government, Bandit Rule, and the Culture of Power in the Post-Taiping China-Vietnam Borderlands, op. cit.*

categories were finally created: Vietnamese groups hostile to colonisation, disorganised gangs who attacked the peasants, and organised gangs who came from China, were involved in cross-border traffic, and posed a real threat to trade.[40] To these categories corresponded territorial zones. Tonkin was divided into three areas: the Delta, where Vietnamese gangs operated; on its edge, a strip of land where there were both Chinese and Vietnamese; and the margins, inhabited by a Chinese population that either formed gangs, or came together occasionally for raids. Although the 1889 map did not show these three distinct areas clearly, it still conveyed the variety of the gangs, and the different territories on which they operated.

Not all protagonists had such a subtle approach to these categories, however. The Resident Superior Bonnal, declared in 1890 that "If you draw a large circle through Hong Hoa, Bac Ninh, Phu Lang Thuong, and Langson, you can consider that all the territory north of this circle towards the border with China is entirely in the hands of the Chinese gangs."[41] The whole of the Upper Region was thus defined as a grey area, which escaped the control of the central authorities. The zones where gangs operated continued to be mapped, however, for pacification purposes. The *Carte d'ensemble des bandes de pirates et des postes militaires français au Tonkin, à la date du 15 avril 1891* [General Map of the gangs of pirates and French military stations in Tonkin, as of 15 April 1891] represents in yellow the extensive territories of the gangs, even if they seem to have been eradicated in north-east Tonkin, and even if the transit zones are not represented, as though the gangs had more difficulty communicating among themselves – which was not necessarily the case.[42] The following year, the *Carte de l'emplacement des bandes pirates à la date du 1er mai 1892* [Map of the positions of gangs of pirates as of 1 May 1892] showed that several regions were still in the hands of Chinese gangs, but in clusters across the territory, and over a much smaller area. The gradual reduction in the space occupied by the gangs, in Tonkin's Upper Region, as pacification operations quelled pockets of resistance and brought territories under military control, was thus visible on the maps.

---

40 Nguyễn Thế Anh, "L'image de la piraterie tonkinoise dans la littérature coloniale", in Denys Lombard (ed.), *Rêver l'Asie. Exotisme et littérature coloniale aux Indes, en Indochine et en Insulinde*, Paris, Éditions de l'EHESS, 1993, p. 179–193.

41 ANV1, RST 24374, Rapport du résident supérieur Bonnal, 16 October 1890.

42 ANOM, GGI 66185, *Carte d'ensemble des bandes de pirates et des postes militaires français au Tonkin, à la date du 15 avril 1891*, sur papier, 50 × 75 cm.

2.3 *Governing an Ethnic Patchwork in the Borderlands*

The Europeans had a long tradition of interest in ethnic minorities, and travel writings abound with accounts of local tales collected in the field, customs, language, religion, clothing, systems of law and political organisation. Generally speaking, colonial explorers, missionaries, and scholars took the oral traditions seriously, noting them down in order to preserve them. Likewise for the colonial authorities, who invested a lot of effort getting to know these populations, by collecting and preserving their oral traditions. They could thus assess what they knew about these ethnic groups, and how they could best be approached.

This interest was not exceptional under colonial rule, as can be seen, for example, in East Africa in the early 20th century, where the French produced anthologies of local customs based on the oral traditions they had heard and noted down.[43] As the military operations of pacification decreased, they began to reflect on the importance of understanding the people they were now ruling. Anthropology can be seen as a "colonial science" here,[44] insofar as it objectified and categorised the borderland populations in order to administer them more effectively. Yet it was more out of curiosity than anything else that the French first carried out ethnographic surveys among the mountain dwellers who had remained beyond the reach of the central powers. The idea that these geographical enterprises were essentially colonial in character is not borne out by the choice of symbols, of scale, and of the information represented on the maps, nor does it take into account that mapping these populations was in itself a major feat.[45]

The British were much more advanced than the French in their methods of ethnographic inquiry. By 1874, the British Association for the Advancement of Science had developed notes and questionnaires, which were regularly revised, and which formed a much more substantial textual corpus than the single page of instructions provided by the Governor General of French Indochina.[46] Information concerning the populations was incorporated into the different reports and did not generally give rise to surveys requested independently by the authorities. The British also attempted to count – if not represent cartographically – those under their rule in British India, in two censuses, of 1891

43 M.-A. de Suremain, "Coutumes, coutumiers", *art. cit.*, p. 37–38.

44 For the case of Africa, see Emmanuelle Sibeud, *Une science impériale pour l'Afrique? La construction des savoirs africanistes en France, 1878–1930*, Paris, Éditions de l'EHESS, 2002.

45 These issues have been raised for later ethnographic maps from French West Africa. See Marie-Albane de Suremain, "Les cartes ethno-démographiques de l'Afrique occidentale: enjeux d'une construction 1952–1963", *Mappemonde*, 92/4, 2008.

46 British Association for the Advancement of Science, *Notes and Queries on Anthropology, for the Use of Travellers and Residents in uncivilized Lands*, London, Edward Stanford, 1874.

and 1901.[47] These were impressively precise, and showed an in-depth knowledge of the different ethnic groups which easily matched the French studies of the time.

In the censuses, which were conducted throughout British India, and not solely in the frontier zones, the questions were always the same. They concerned the main areas of life, such as the family, religion, language, caste, profession, marriage, and illness. When local princes were approached, the census also addressed the political history of the locality and of the reigning dynasties, with further questions on how land taxes were collected. The descriptions of the different ethnic groups and castes tended to be combined with summary judgements on their characteristics, as the Gazettes illustrate.[48] The British classified the results of their census inquiries by category, to make them useful for administration and tax collection. The *Burma Gazette* of 1912 mentioned, however, that one should not overstate the importance of the census figures because many areas were not covered in 1891, for example the district of Bhamo.[49]

Thus both maps and census reports aimed at providing more subtle classifications of territories and populations, to facilitate colonial rule. The results showed the considerable means employed, and the administration's efficiency. However, the limits of such operations also became evident: they were exercises devised by colonial officials rather than being defined by the realities on the ground.

As soon as Paul Doumer took office as Governor General of French Indochina in 1897 he instructed the commanders of the military territories to send back a report on each sector.[50] He appended a list of the items to be included: 1) a physical description, 2) a history of the sector's conquest, 3) its social, political and administrative organisation, 4) its economic situation, 5) its military organisation, 6) a conclusion, with remarks on the probable future of the region, along with any requests. This list is similar to the contents of monographs published at the time by both the British and the French.

47 C. C. Lowis, *Census of India, 1901. Burma. A.pt. 2. Imperial Tables*, India, Government Central Press, 1902.

48 J. G. Scott, *Gazetteer of Upper Burma and the Shan States, op. cit.*

49 49. BL, IOR, V/27/64/27, *Burma Gazetteer*; Bhamo District, volume A, compiled by G. W. Dawson, Rangoon, 1912, p. 27.

50 On the ethnographic surveys carried out during this period and presented here, see Jean Michaud, "French Military Ethnography in Colonial Upper Tonkin (Northern Vietnam), 1897–1904", *Journal of Vietnamese Studies*, 8/4, 2013, p. 1–46; Jean Michaud, Sarah Turner, Yann Roche, "Mapping Ethnic Diversity in Highland Northern Vietnam", *GeoJournal*, 57/4, 2002, p. 305–323.

These inquiries produced substantial results: thirty-five sector reports overall, of which ten were devoted to the first military territory, nine to the second and third territory, and seven to the last territory.

This survey was supplemented by a second one, commissioned in 1903 by the new Governor General, Paul Beau. It covered all the provinces of French Indochina, and concerned their inhabitants. The questionnaire, based on the one devised in 1900 by the *Ecole française d'Extrême Orient* (EFEO) for anthropologists, focused on the different populations, their customs, traditions, and cultural specificities.[51] A map and a report had to be produced in every sector:

> The map shall be on a scale of 1:100,000; this scale, which is the one used for some of the maps produced by the army's Geographical Department, seems to be the most suitable. It shall show the topographical distribution of the different ethnic groups in different colours.
>
> The report shall contain as comprehensive a list as possible of the different races or groups settled in the administrative area. It will supplement the information about their situation given on the map.[52]

Jean Michaud considers these surveys, and in fact particularly the second ones, to be genuinely ethnographic, different from those conducted by explorers and missionaries.[53] The reports were all written by military staff over a set period, using similar methods, and as instructed by their superiors. The results are therefore much more convincing than the findings of isolated missionaries and explorers, if only because of the number of participating heads of sector. However, realities on the ground and material obstacles thwarted the project's ambition of covering the whole territory, and in fact it was principally the military territories that handed in reports at the end.[54]

The final report, compiled by Lunet de Lajonquière from all the documents sent in, was published in 1904 in the form of a monograph of almost

51 Marcel Mauss, *Carnet. Instruction pour les collaborateurs de l'École française d'Extrême-Orient*, Saigon, Imprimerie Ménard et Legros, 1900; see Jean Michaud, *Incidental Ethnographers: French Catholic Missions on the Tonkin-Yunnan Frontier, 1880–1930*, Leiden, Brill, 2007, p. 226.

52 ANOM, GGI 23832, Lettre du gouverneur général Beau au général commandant supérieur des troupes, notes statistiques et ethniques demandées à tous, 7 June 1903.

53 J. Michaud, "French Military Ethnography in Colonial Upper Tonkin (Northern Vietnam), 1897–1904", *art. cit.*, p. 34.

54 ANOM, GGI 23832, Minute du gouverneur général Beau au résident supérieur du Laos à Vientiane, Hanoi, 29 April 1905.

1,000 pages.[55] Attached to it was a general map of Tonkin drawn on a scale of 1:500,000 by Friquegnon. Only the military territories showed the ethnic distribution of the population, using colours for the major groups, and hatching or points for sub-groups – "race" and "tribe" were the terms used at the time. The work was published two years later by the publishers Leroux. It contained a map on the scale of the military territories of Tonkin, which showed clearly that it was impossible to map the ethnographic complexity of the frontier zones – let alone of the whole of the Indochinese Peninsula:

> The ethnographic problem of the Indo-Chinese peninsula is very complex; it is far from having been solved. We have hardly begun to understand the dominant traits of this human patchwork, which exhibits such astonishing variety between Yun-Nan and the Island of Singapore, the South China Sea and the Bay of Bengal.[56]

Étienne Lunet de Lajonquière's 1:100,000 scale *Carte ethnographique du Tonkin septentrional* [Ethnographic Map of Northern Tonkin] has a note explaining that "Different shades represent the zones inhabited by the different ethnic varieties. For this map to give a more precise picture, it should be borne in mind that in some of the regions indicated, the population's density varies considerably" [Figure 48]. As is often the case, the memo supplements the visual representation to make it easier to understand, as well as pointing to its limitations. A careful colour code showed the different ethnic groups – a total of 27 – on the map, as well as their main language groups. The map focused on the military territories, and hence solely on the French possessions, leaving the frontier provinces out of account. Only the rivers flowing into China were marked as extending across the border, but not the cross-border ethnic groups, suggesting that maps of this sort were designed purely for administrative purposes, and showed nothing beyond the colonial territory.

Lunet de Lajonquière was assisted in his work by other officials, who also published the results of their inquiries.[57] One of them, Édouard Diguet, had an important point to make on mapmaking:

55 Étienne Lunet de Lajonquière, *Ethnographie des territoires militaires, rédigée sur l'ordre du Général Coronnat, Commandant supérieur des troupes du groupe de l'Indo-Chine*, Hanoi, Schneider, 1904.

56 Étienne Lunet de Lajonquière, *Ethnographie du Tonkin septentrional*, Paris, E. Leroux, 1906, p. 1.

57 Auguste Bonifacy's contribution deserves a special mention, due to his insight and expertise. See particularly Auguste Bonifacy, *Les groupes ethniques du bassin de la Rivière claire* (*Haut-Tonkin et Chine méridíonale*), Paris, Société d'anthropologie, 1906.

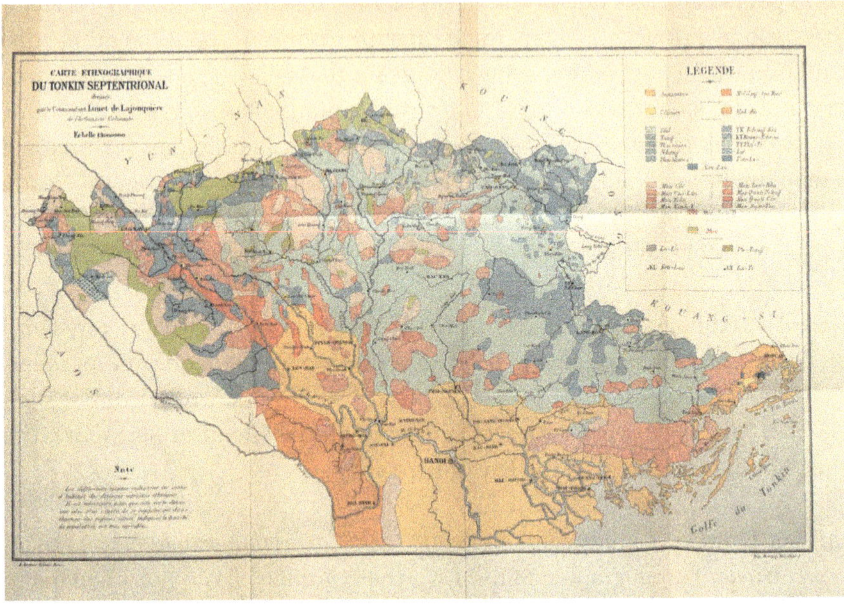

FIGURE 48 Ethnographic Map of Northern Tonkin on a scale of 1:100,000, by Étienne Lunet de Lajonquière
SOURCE: Étienne Lunet de Lajonquiere, *Ethnographie du Tonkin septentrional*, Paris, E. Leroux, 1906, *in fine*

> We thus become aware of the strange fact that the different ethnic groups are not juxtaposed like the squares on a draughtboard, as they are in other countries., but instead they are *superposed on one another* on the terraces formed by the country's relief. Such that, if I may overstate the case a little, were one to draw a map of the topographic contours one would find the Annamites massed between the sea shore and 100 metres, the Thai with the Meo above the 100 m mark, and the Lolos between 400 and 1500 metres.[58]

This is an interesting comment. None of the ethnographic maps reflected it, however, if only because they never showed contour lines.[59] This means that the human complexity of the Indochinese territories was represented only

58 Édouard Diguet, *Les montagnards du Tonkin*, Paris, Challamel, 1908, p. 9–10.

59 For a 3-D representation, see the maps made by Yann Roche, on the basis of data from 1989 and 1994, in Jean Michaud, Sarah Turner, Yann Roche, "Mapping Ethnic Diversity in Highland Northern Vietnam", *art. cit.*, p. 320–321.

partially, for many a long decade, due to the maps' representation of two dimensions only.

•••

Thus, while the imperial powers were demarcating international borders, they were also drawing limits internally within their new territorial possessions. In both Burma and French Indochina the requirements of territorial administration led to separate systems for the border provinces, which reinforced and institutionalised pre-existing territorial distinctions. In Burma, civil and military provinces were created in 1885, and in Tonkin, military territories were created in 1891. The reasons were the same in both cases, and mirrored the situation in India, Algeria and French West Africa: areas considered useful (that is, containing resources), often inhabited by the country's majority ethnicity, were placed under civil government, whereas the confines were left under military control, which was both more authoritarian, to ensure security, and less intrusive for the local populations. These were generally minority ethnic groups whose way of life and political organisation were less suited to direct rule.

The British and the French thus had very similar strategies of territorial administration. This belies the received wisdom that the French tended to establish direct rule, and the British, indirect rule.[60] However, it is useful to compare the situation in the borderlands of these empires. Although in both cases the military presence was sparse, the French military territories were often pioneers in administrative arrangements. When the *Carte des communications du Tonkin*, [Map of Communications in Tonkin] was made, the territories under civil administration took their cue first from the military territories, and the many surveys carried out to ascertain the distribution of ethnic groups in French Indochina gave rise to only one map, of the military territories, while Annam and Laos were hardly treated at all. The reasons for this could be the efficiency of the military coupled with their obedience to orders, and, for Laos, the lack of resources. However, it would be misleading to think that the plethora of documents on the military territories expressed an over-investment by the colonial powers in these areas. This is not borne out by the facts, and in the case of Burma, the non-administered territories cannot be interpreted in these

60 The theorisation of indirect rule was a later phenomenon, namely through Frederick Lugard's *The Dual Mandate in British Tropical Africa*, London, W. Blackwood and Sons, 1922. For an overview of this issue, see V. Dimier, *Le gouvernement des colonies. Regards croisés franco-britanniques, op. cit.*

terms, and were clearly neglected: the lack of materials concerning these areas are a sure sign of the lack of colonial administrators. That said, the Burmese territories under civilian rule were not much better off. Overall, the salient difference to emerge between French and British administrative methods was not the investment made in different types of territory, but the role played by maps. The French produced them in large numbers, and used them as work tools, while the British, although just as obsessional about territorial administration as the French, used them much less. The received idea that the French system was overblown compared to the pragmatic and less costly British one should also be qualified. The British census reports, which were used to optimise territorial administration by showing the population's distribution, mobilised bureaucratic efforts as much, if not more so, than the ethnographic map of Tonkin, commissioned by the French for similar purposes.

All these administrative measures were an attempt to impose order, as though delimiting ill-defined territories and mapping a mobile population could somehow bring these under control. Border demarcation was not enough, and the European powers sought additionally to mark administrative divisions and sub-divisions, carried away by their obsession with lines, circumscriptions, and boundary markings. The initiatives obeyed no defined policy nor overall plan, and finally it was circumstances alone that determined the forms taken by the frontier territories at different times. Government officials had a very free hand, which added to the improvised and local character of some of the decisions taken. The limits of the military territories kept changing, as did their number, their names, and the size of their sub-divisions (*cercles, secteurs*), depending on the security situation, the geography, population movements, and the availability of the officials. Some areas went from a military to a civil administration, and vice versa, some were joined together, and then separated, military posts were created and then abandoned when they ceased to be useful. The on-going military administration of territories, temporarily down-scaled, then reinforced when the gangs sprang back into action, showed that it was impossible to integrate the borderlands into the Empire, and to produce a homogeneous territorial administration. The internal boundaries drawn were considered useless by the local populations, who anyway paid little attention to them, or else contested them, and always found them wanting. They were the most visible sign that the frontier zones were largely ungovernable.

Conclusion

So we have seen that over a span of 30 years, set within a longer history, European and Asian protagonists represented and gave shape to the northern borderlands of the Indochinese Peninsula through intense map-making activities. The interactions of the British and French empires, Siam and China, at this time of demarcation of sovereign territories and intense map production come into sharpest focus precisely in these frontier zones. Generally speaking, as we have seen, the British and the French had similar map-making goals and practices, employed similar methods, and were interested in similar spatial phenomena. Thus it was possible to delineate a connected history of imperial geographical policies, and, as we hope to have shown, to describe the unique play of forces at work in this scenario of territory-building.

1 A Military Map-Making Venture rather than a Colonial Programme

The British and French exploratory expeditions, little studied to date, were, as we have seen, more often the result of individual initiatives whenever the circumstances allowed, than of a solid political agenda. A genuine scientific curiosity, supported by the Geographical Societies of London and Paris, by economic interests and by the presence of established colonies in India and Cochinchina, encouraged the race for territorial possessions in these areas of imperial rivalries. With each new expedition, itineraries were added, and the Europeans' knowledge expanded. However, nothing compares with the French Mekong Expedition (1866–1868), and the Pavie Mission (1879–1895) (for which there were no equivalents on the British side): their scale, the quality of their final results, and the scope of their reception, in the home country and beyond, remain unparalleled. Although the Pavie Mission was not as large as the scientific campaigns in Egypt and Algeria, or the Morea expedition, its role in the spatial production of these northern territories was absolutely central, not only due to the maps it generated, but also due to the territorial annexations into the French empire that it accompanied.

The relative indifference of the British to explorers and their maps was doubtless due to the fact that the Raj already boasted a dedicated Geographical Department, the Survey of India, founded to triangulate and map the whole of India. Immediately after the annexation of Upper Burma, in 1885, British topographical brigades from India were deployed for systematic surveys there.

 | DOI:10.1163/9789004469853_012

The French *Bureau topographique des troupes de l'Indochine* was operative from 1886, but it was not until 1899 that the *Service géographique de l'Indochine* became autonomous – an exception within the French Empire. It was in charge of mapping the Upper Region of Tonkin, which began in 1901. Apart from this staggered chronology, there were many similarities between the French and the British in this field: in both empires, cartographers were military personnel, with some local labour used to different degrees, as we have seen, and in both, the units were under-staffed and under-resourced. The presence of Geographical Departments suggests that systematically mapping the territories was backed by political will, but on the ground the map-making programmes were beset by failings and disagreements.

The history of Geographical Departments obliges us to shift our perspective and examine geography's role on the imperial and international stage. The colonial Geographical Departments inherited the training given and the standards laid down in the home country, especially, for France, the *Service géographique de l'armée*, and, to a lesser extent, the *Service des colonies*. The Ordnance Survey maps made for the metropole or for other colonies provided models for the work in the Indochinese Peninsula, but due to the latter's specificities, new techniques for fieldwork and for printing maps were also tried out. The way people and models circulated between the colonies, the efforts made to standardise maps, and the exchanges between different departments on technical issues all suggest that the most productive scale of analysis is the empire as a whole. The Geographical Departments were sometimes in conflict, but they also fostered Franco-British scientific exchange and collaboration with their Chinese and Siamese counterparts, which likewise were the fruit of circumstance, and did not obey a real plan or pre-defined strategy.

2 Between European and Indochinese Approaches: Colonial Encounters on Unfamiliar Terrain

The difficulties of the map-making enterprise, which showed up in the inadequacies of the finished maps, and compromised their real power, were evident in the topographers' daily work. All the processes of the scientific work, from measuring angles and carrying out surveys to getting the documents to their destination were hampered by the relief, the climate, and the insecurity. The topographers' principal, and unanticipated, obstacle was the unfamiliarity of the territory, which they had difficulty interpreting, and for which they had to constantly adapt their acquired knowledge. The French and the British imported methods learnt in the home country, and tried, in principle, to fit Indochinese geography into preconceived cartographic categories. However, given the absolute foreignness of the land for the topographers, they did not, in practice, simply go about applying a colonial science, but drew on the local population to improve their powers of observation and their calculations.

True to a practice that was not essentially colonial, the information sought was above all place names, the state of the roads, and the location of crossing points. Reports of the inquiries carried out reveal the different sorts of encounter involved, which were often requested by the French and the British with a particular end in view. They involved local interlocutors, or else travellers and regional elites, whose information was considered more useful by the Europeans. The complexity of the situation on the ground and the multiplicity of languages in use were two of the reasons why the French and the British had difficulty understanding, and therefore representing, these territories. The large number of toponyms for the same place was particularly bewildering, for which the only solution was to read the map in tandem with an accompanying text.[1]

Another specificity of the Indochinese Peninsula was the abundance of written geographical sources, found or commissioned by the Europeans. This documentation sheds new light on how indigenous knowledge was understood and used in a colonial situation. It was, in fact, not used very differently from oral data, but it supplemented it, added complexity, and so gave the oral records greater weight. Locally-made maps, for example, served to find out the same information as oral inquiries: the paths of roads and borders, and the names of places. As with the oral sources of information, the written ones tended to disappear in the final product, which only retained the elements useful for territorial organisation.

3 The Problematic Relation between Map and Territory

Topographical maps were thus produced by and for the Europeans principally to help them administer their territories. The borderlands, represented through lines, ill-defined zones, and demarcated areas, appeared to be organised around main roads and urban centres that had clear outlines and were structured in various ways around internal boundaries. Although the maps

1 This lack of correspondence between the name on the map and the name really used is still the case today, particularly in Vietnam.

visibly improved over the thirty-year period, their real impact on the territories and the border populations is harder to assess.

It is tempting to apply Chad Haines's conclusions to the northern Indochinese Peninsula: it is the roads built rather than the maps drawn that bring the border into existence, and integrate outlying areas into the empire by linking them to the centres of power.[2] Itineraries – measured, divided into sections, and abundantly explained – were the main object of descriptions as from the conquest, and in 1914 British road maps were still being updated and republished. Yet despite the road-building that successfully brought certain territories out of their isolation, this did not always have the effect of integrating the territories into the empire. A case in point is Yunnan, which was finally a colonial myth more than a territorial reality; the real territorial purchase of this powerful colonial fantasy was negligible. This should remind us not to assume that abundant representation necessarily implies concrete spatial effects.

Although the connection between maps and territory was particularly strong over the thirty years of border demarcation in the northern Indochinese Peninsula, here again we should be prudent in making links between maps of frontiers and their reality on the ground. Before the conclusion of treaties, maps represented the territories desired, and thus an image corresponding to the fantasies of each party. Moreover, British and French boundary commissioners had to contend with the local population's ignorance of the borders of sovereign territories, which often had no real existence for them. The maps produced by the Europeans, based on topographical surveys and on information from inhabitants or local elites, showed the final, unnegotiable, borderline. Yet neither the border police nor other forms of surveillance managed to enforce these lines.

Lawlessness, combined with the specific social organisation of the ethnic minorities that lived in these territories, not to mention the mountainous terrain, explain the choice by the imperial powers of a special administrative status for the border provinces, military rather than civil. This implied, in fact, a reduced administration, with, as the maps show, different prerogatives and means of control from civil-rule territories. However, the maps are silent on how power was really exercised in these spaces. Only a few corridors, like fine lines, linked these borderlands to each other, and most of these areas remained in reality beyond the control of the Europeans.

2 C. Haines, "Colonial Routes", *art. cit.*

4 Mirror Images: The Need for a Connected Imperial History

Our research into imperial histories has adopted comparison not only as a method, but also as the subject-matter of our study. Historiographical works and available sources demonstrate that the British and French approaches in cartography for this period are so similar, in their representations and practices, that it is possible to speak of a "European cartography". Yet rather than simply make a point by point comparison between the French and the British on these topics, it seemed more fruitful to attempt a connected history, which would include Asian protagonists as well.

The main difference between the European powers was the moment when they began mapping the territories after their annexation in 1885. Although map-making was an older and more deeply rooted tradition for the French than for the British, the latter had a head start in the Indochinese Peninsula, given the presence of the neighbouring Survey of India. That made mapping Burma a simpler task, not only for practical reasons, such as the availability of trained staff, but also because issues of principle such as the virtues of triangulation or the attention paid to the borderlands in the full-coverage mapping project had already been addressed. The *Service géographique de l'Indochine*, by contrast, was a brand new institution, and an exception within the French empire, so it took longer getting off the ground. Despite doing the same work, therefore, the two institutions did not have the same range nor the same history, and they should not be put on exactly the same plane. This marks a first limit in the comparison between the two. Another distinction was the fact that the British used many more indigenous staff than the French, and the links between the British and the Siamese Royal Survey Department were considerably stronger than on the French side. Another difference concerns the favoured medium of representation: in both cases, the priority was to facilitate the administration of the new territories, but the French tended to use maps for this, whereas the British relied on texts for describing the road system, administrative boundaries, and the local populations, when the census results were not transferred onto maps.

Another dimension of our research, beyond similarities and differences between imperial practices, is analysing the reliance of all empires on, precisely, the practice of comparison. This is comparison no longer as a method, but as a subject of study in itself. The French and the British kept an eye on each other, and the Siamese observed both. The borrowings and exchanges which this gave rise to had not previously been addressed in a connected history. Contemporary comparative work on empires concentrates, rather, on the circulation of administrative models and management techniques within them, and on the links between different anti-imperial resistance movements.[3] Imperial policies may be compared in the different colonies of a single empire, or else between empires. Comparisons *within* an empire tend to produce a more unified set of representations and practices, which circulate without necessarily involving the metropole directly, as for example when the Government General of Indochina observes how things were done in Algeria. The principal idea in this case is to apply methods that had proved their worth in the home country, and also in other colonial territories. The comparisons I draw *between* empires, on the other hand, focus on the different colonial possessions in Asia, because these territories raise specific problems that had no equivalent elsewhere, as I have shown.

By using a comparative method, but also taking comparison as an object of study, I hope to have demonstrated that the concept and scale of 'the empire' are adequate for examining relations between the home country and the colonies. This approach also allows a different history of international relations to be sketched, one that does not destroy the national level – in this respect, the inclusion of Siam is essential – yet where the notion of empire is accepted as a pertinent category.[4] Moreover, it also endorses the possibility of analogies between empires that do not abandon the requirement of precise temporal and geographical frameworks.

5 Cartography in a Colonial Situation: The Difficulty of a History "of Equal Shares"

The many Asian maps used by the Europeans lent support to the idea of a connected approach to cartography in the Indochinese Peninsula, comparing European and Asian sources, which had previously been studied separately. Histories of cartography separate the two traditions, partly for simplicity and partly because the interactions between different cartographic practices or representations are sometimes unclear. It is also easy to confuse objects that are in reality very different, not only when comparing Asian and European

3 See particularly the conference on this theme: "Imperial Comparison", Oxford, All Souls College, 8–9 July 2016.

4 We agree entirely with P. Grosser on this point, in his programmatic article "Comment écrire l'histoire des relations internationales aujourd'hui? Quelques réflexions à partir de l'Empire britannique", *art. cit.*

cartographic production, but also by simply referring to 'Asian cartography', which is in fact very diverse.

Some of the objects present on European topographical maps are the same as those represented on Asian maps, such as the description of itineraries or the marks of borders and provincial boundaries. In their use of maps, the French and the British, along with the Chinese and Vietnamese, are particularly interested in communication routes, in measuring rivers and roads, in frontiers, and in how the territories are divided up administratively. Coupling the map with a complementary text is also a feature of Asian practice, and shows that the different spatial representations overlap. The differences between the European topographical maps and the maps produced by the various Asian courts are transcended by their shared objective of territorial administration.[5]

It is precisely with this objective of territorial administration in mind that hybrid maps were made, as seen in the preparatory sketches for the topographical maps, which mix languages and symbols, and which most probably had many authors, for example, local elites, interpreters, and colonial civil servants. It was important to locate these documents in the archives, and place them back in context. They constitute a rare testimony to colonial encounters through cartography, alongside the many other cartographic sources that cannot be drawn into comparison. This is why a history comprising "equal shares", using these cartographic sources, is impossible. Although many indigenous spatial representations such as cosmographic and mythological maps exist, they lie outside the colonial encounter, beyond any fruitful hybridisation, leading Romain Bertrand to conclude, in another context, that an encounter did not take place.[6] The sources we have discovered show that in the northern Indochinese Peninsula encounters certainly did take place, although a bias remains insofar as the hybrid maps are scholarly creations made by or for the French or British colonial administrators. We know nothing, however, about the reception and use of European maps by the populations of the borderlands, probably because it was infrequent or inexistent.

5 The Europeans translated and annotated Vietnamese Royal Geographies and Burmese and Siamese Annals and Chronicles in their attempts to identify "historic rights", and evoked these sources in their diplomatic discussions. It is striking that South-East Asian countries today refer to the French maps of that period (1885–1914) to justify their current territorial claims. Maps thus serve the same purposes today as yesterday, yet the idea that the map is a scholarly discourse produced at a particular moment in time about a particular situation is not accepted.

6 R. Bertrand, *L'histoire à parts égales, op. cit.* The author makes it clear that "a history with equal shares" does not mean that the sources are equally substantial, but that both visions are taken into account. He concludes that an East-West encounter never really took place, because it did not exist as such for the Javanese people he studied.

6 The Empire's Margins: Complex Objects Irreducible to Cartographic Representations

Although historical approaches generally neglect marginal spaces in favour of examining the policies conducted from the capital, and anthropological or political approaches are interested mainly in how the populations of these areas defy the centre, we have shown that these zones concentrate essential issues for the historian.[7]

The history of borders, for example, already has a substantial bibliography for Africa, but has no equivalent for the northern Indochinese Peninsula, from the perspective of connected histories. The received idea that the border process takes place in three stages – from no frontier line, to demarcation during the colonial period, to frontiers reopened and the resumption of flows of trade after decolonisation – is, as we have shown, too simplistic a model to account for what occurred in the northern Indochinese Peninsula between 1885 and 1914. Not only were certain borders already in place when the French and the British arrived, but the frontiers imposed were not entirely artificial and purely external. They were consequently maintained, and did not require the kind of ratification that took place in 1964 for the African Continent.[8]

The complexity of marginal space is not only apparent in border demarcation, but in the differentiation of these spaces from the central zones and also in the internal divisions of the borderlands themselves. Cartography is the first to show this because, with the exception of the exact path of the frontier, which is the object of very precise surveys, the marginal lands are rarely included in systematic, large-scale, mapping projects.[9] When decolonisation took place, the maps showed patchy coverage, with large blank areas in Burma and Laos, and in the Upper Region of Tonkin. An intense cartographic activity

7 In his book on the links between map-making and nation-building, Tongchai Winichakul considers the frontier zones only in relation to border demarcation. The book generated complementary case studies, mainly by anthropologists, on trans-border ethnic groups that ignore the boundaries of nation states. Andrew Walker, "Seditious State-Making in the Mekong Borderlands: The Shan Rebellion of 1902–1904", *Sojourn: Journal of Social Issues in Southeast Asia*, 29/3, 2014, p. 554–590.

8 This is not to say that the frontiers did not change at all. After Burma separated from the British Raj and become an independent colony in 1937, its borders with India were altered, in 1941. Border disputes broke out between several States, and still continue today, but no substantial changes have been made to the lines drawn during the colonial period.

9 The frontier zones sometimes develop in similar ways to the centre, in contrast to the zone between the two. We can see this today in Vietnam, but for different reasons: the development of tourism at Bảo Lạc and Sa Pa has given rise to many more infrastructures than in the provinces between Hanoi and the Chinese border.

occurred around borders when they were being demarcated, but alongside these lay vast neglected expanses, which largely escaped central control, and can thus constitute a relatively self-contained subject of inquiry. These sparsely populated territories, which the colonisers considered to be unprofitable, were hardly mapped at all, and there were fewer censuses there. The tax regime was also different, and the presence of a marked borderline, together with the handful of policemen, was far from sufficient, despite the appearances, to prevent border-crossings. The Shan and Kachin territories, for example, which the central powers had consigned to marginality, played a major role in cross-border flows.

These territories were indeed a law unto themselves, resistant to colonial administration and surveillance, which always involve counting and classifying. Benedict Anderson mentions the map, the census and the museum as three markers of this desire to impose an order, and Appadurai shows how the first two were instrumental in bringing populations to heel.[10] Cohn focuses on the activities of the British in India, namely their production of linguistic knowledge to compensate for their ignorance of local languages; their emphasis on law, as a way of fixing previously non-formalised codes; the historical reconstruction of India's past; and the issue of clothing. However, this approach ends up attributing more importance to the Europeans than they actually had, and it reinforces the myth of the domination of the military officers, when the facts on the ground, expressed in tensions of empire, show something quite different, even more so in the confines, which remained largely outside colonial control. Moreover, as we have shown, the hierarchies established under colonial rule, which were particularly visible in the different administrative statuses within the empire, drew on earlier hierarchies between minority and majority ethnicities, which persisted after colonisation, and are even visible today. Our research has aimed at restituting the complex, interconnected processes shaping these far-flung regions of empires, over a pivotal yet underrepresented period, and at resituating these processes in the long history of the northern margins of the Indochinese Peninsula. We hope thereby to have rescued these confines from the firm grip of a geography of ignorance.

10 Benedict Anderson, *Imagined Communities: Reflections on the Origin and Spread of Nationalism, op. cit.*; Arjun Appadurai, *Modernity at Large. Cultural Dimensions of Globalization* (1996).

Bibliography

1 Primary Sources

1.1 *Archival Material*

Paris. Bibliothèque nationale de France, Département des cartes et plans
Service hydrographique de la marine
Portofolios 179 and 180.
Archives de la Société de géographie de Paris
Maps from the "Bonaparte fonds"; Manuscript articles from the *Bulletin* and other writings from the Society's correspondents.

Paris. Archives of the Institut Pasteur
Fernand Bernard collection

Paris. Missions Étrangères de Paris
Correspondence with missionaries (Burma, Siam, Laos, Tonkin)

Paris. École française d'Extrême-Orient
Maps: Asia. China. Indochina. Laos. Siam. Vietnam. Louis Finot.

Vincennes. Service historique de la Défense
Marine. CC7 series: personal information on Louis Delaporte, Ernest Doudart de Lagrée, Francis Garnier, Auguste Gouin and Paul Néis.
Armée de Terre. 10H series: Indochina

Aix-en-Provence. Archives Nationales d'Outre-Mer
Ministerial Archives.
- Administration centrale: Service géographique des Colonies
- Série géographique, Indochine
- Série géographique, Birmanie. Siam

Fonds territoriaux. Indochine. Fonds du Gouvernement Général de l'Indochine (GGI)
Série F: Affaires politiques
Série P: Marine de guerre
Série Q: Bureau des Affaires militaires
Série R7: Géographie

Hanoi. Trung Tâm Lưu Trữ Quốc Gia I [National Archives of Vietnam, centre 1]
Fonds du Gouvernement Général de l'Indochine
Série C: Personnel
Série F: Affaires politiques
Série Q: Bureau des Affaires militaires
Série R7: Géographie
Fonds du service géographique de l'Indochine (série R)
Fonds de la Résidence supérieure au Tonkin
Série C: Personnel
Série F: Affaires politiques
Série R7: Géographie

Ithaca (USA). Cornell University
Jules Harmand collection #3481

London. British Library
India Office Records
L/MIL: Military Affairs
L/P&J: Political and Judicial Affairs
L/P&S/7: Political and Secret correspondence with India, 1875–1911
L/P&S/10: Political and secret separate (subject) files 1902–1931
L/P&S/18: Political and Secret memoranda 1840–1947
L/P&S/20: Boundary Commissions
V: Official Publications, 1760–1957
W, X, Y: Map collection

London. Archives of the Royal Geographical Society
Coryton collection
Younghusband collection (ar GB402 GFY/2/1)
HR Davies collection (ar GB402 HRD)

Kew. The National Archives
Records created and inherited by the Foreign Office (FO), by the War Office, Armed Forces, Judge Advocate General, and related bodies (WO) of the Cabinet Office.

Cambridge. University of Cambridge
The Scott Collection
Manuscript maps

Yangon. National Archives of Myanmar
Chief's secretary's Office, 1835–1948
Manuscript maps

1.2 Printed Material

ANDERSON, John, *A report on the expedition to western Yunan via Bhamo*, Calcutta, Office of the Superintendent of Government Printing, 1871.

ANTONY, Georges, *Le Service géographique et des missions. Historique, organisation et publications*, Levallois-Perret, Impr. Crété de l'Arbre, 1901.

BEAU, Paul, *Situation de l'Indo-Chine, 1902–1907*, Hanoi, 1908.

BERNARD, Fernand, *À l'école des diplomates: la perte et le retour d'Angkor*, Paris, les œuvres représentatives, 1933.

BERNARD, Fernand, "L'exploration de l'Indo-Chine et la délimitation de la frontière franco-siamoise", *La Géographie*, huitième série, tome XIX, janvier–juin 1909, p. 177–190.

BIGREL, Théophile, "Note sur une carte de la Cochinchine française", *BSG de Paris*, sixième série, tome VI, juillet–décembre 1873, p. 436-439.

BLANCK, Père, "Le Trane-Nigne a l'ouest du Tong-King", *BSG*, septième série, tome V, 1884, p. 434–452.

BLEWITT, Arthur, *General report and diary of the expedition under Lieutenant L.E. Eliott, Assistant Commissioner, to the Upper reaches of Irrawaddy and subsequently to the Chinese frontier, 1890–91, with notes concerning the Kachins inhabiting that region*, Rangoon, 1891.

BONIFACY, Auguste, *Les groupes ethniques du Bassin de la Rivière Claire* (*Haut Tonkin et Chine Méridíonale*), Paris, Société d'Anthropologie, 1906.

BONVALOT, Gabriel, *De Paris au Tonkin à travers le Tibet inconnu*, Paris, Hachette, 1892.

BONVALOT, Gabriel, *Across Thibet: being a translation of "De Paris au Tonkin à travers le Tibet inconnu"*, traduit par C. B. PITMAN, London, Cassell, 1891.

BOWERS, Alexander, *Bhamo Expedition. Report on the practicability of re-opening the trade route, between Burma and Western China*, Rangoon, American Mission Press, 1869.

BOWRING, John, *The kingdom and people of Siam. With a narrative of the mission to that country in 1855*, London, J. W. Parker, 1857.

BRENIER, Henri, *Essai d'atlas statistique de l'Indochine française*, Hanoi, Imprimerie française d'Extrême-Orient, 1914.

BRITISH ASSOCIATION FOR THE ADVANCEMENT OF SCIENCE, *Notes and Notes and Queries on Anthropology, for the Use of Travellers and Residents in uncivilized Lands*, London, Edward Stanford, 1874.

Bửu, Cầm, Đỗ, Văn Anh, Phạm, Huy Thúy, Tạ, Quang Phát, Trương, Bửu Lâm, *Hồng Đức bản đồ*, Sài Gòn, Tủ sách Viện khảo cổ Số III, 1962.

Caspari, Édouard, "Les cartes marines de l'Indo-Chine", *Annales de Géographie*, 15–80, 1906, p. 172-173.

Chambre de Commerce de Lyon, *La mission lyonnaise d'exploration commerciale en Chine, 1895–1897*, Lyon, A. Rey, 1898.

Chasseloup-Laubat, Prosper de, "Exploration du Mé-Khong", *BSG*, cinquième série, tome XIV, juillet–décembre 1867, p. 92–101.

Chú, Phan Huy, *Hoàng Việt địa dư chí* [皇越地輿志], [s.n.], [s.l.], 1833.

Colquhoun, Archibald Ross, *La Chine méridionale*, traduit par Charles Simond, Paris, H. Oudin, 1884.

Colquhoun, Archibald Ross, *Across Chrysê, being the narrative of a journey of exploration through the South China border lands from Canton to Mandalay*, London, S. Low, 1883.

Colquhoun, Archibald Ross, Hallett, Holt Samuel, *Report on the Railway Connexion of Burmah and China*, Allen, Scott & Company, 1888.

Crosthwaite, Charles, *The Pacification of Burma*, London, E. Arnold, 1912.

Da-Qing huidian [大清會典圖] [Đại Thanh hội điển đồ] [Répertoire des cartes de l'empire des Qing]. Dernière encyclopédie de la dynastie des Qing (1644–1911). Première version terminée en 1684. Seconde édition, à la demande de l'empereur Yongzheng (1723–1735) en 1732. Troisième édition à la demande de l'empereur Qianlong (1736–1795) en 1764, qui constitue une seconde version de l'ouvrage. Troisième version en 1818, qui inclut des cartes et des illustrations. Quatrième version achevée en 1901.

Da Qing yi tong zhi [大清一統志] [Đại Thanh nhất thống chí] [Géographie unifiée de l'Empire des Qing]. Géographie impériale des Qing dont la première version date de 1686. Deuxième version en 1784 à la suite de changements frontaliers et troisième version en 1842.

Devéria, Gabriel, *La frontière sino-annamite. Description géographique et ethnographique d'après des documents officiels chinois traduits pour la première fois*, Paris, E. Leroux, 1886.

Devéria, Gabriel, *Histoire des relations de la Chine avec l'Annam-Việtnam du XVIe au XIXe siècle, d'après des documents chinois traduits pour la première fois et annotés*, Paris, E. Leroux, 1880.

Diguet, Édouard, *Les montagnards du Tonkin*, Paris, Challamel, 1908.

Diguet, Édouard, *Étude de la langue Thai, précédée d'une notice sur les races des hautes régions du Tonkin, comprenant grammaire, méthode d'écriture Thai et vocabulaire*. Hanoi, Schneider, 1895.

"Disparition du service géographique de Madagascar", *Annales de Géographie*, 21–116, 1912, p. 180.

Doudart de Lagrée, Ernest, Garnier, Francis, Delaporte, Louis, *Voyage d'exploration en Indo-Chine*, Paris, Hachette, 1873 (Première publication: "Voyage d'exploration en Indo-Chine", in *Le Tour du Monde*, vol. XXII–XXV, 1870–1873; Réédition: Garnier, Francis, *Voyage d'exploration de l'Indo-Chine*, Genève, Olizane, 2009).

Doumer, Paul, *L'Indochine française (souvenirs)*, Paris, Vuibert, 1905.

Doumer, Paul, *Situation de l'Indo-Chine (1897–1901). Rapport*, Hanoi, FH Schneider, 1902.

Dupuis, Jean, *L'ouverture du fleuve rouge au commerce et les événements du Tong-Kin, 1872–1873. Journal de voyage et d'expédition*, Paris, Challamel aîné, 1879.

Dutreuil de Rhins, Jules-Léon, *Avertissement géographique et orthographique sur la carte de l'Indo-Chine orientale*, Paris, Imprimerie nationale, 1881.

Dutreuil de Rhins, Jules-Léon, "Notice géographique sur la rivière de Hué (royaume d'Annam)", *BSG*, sixième série, tome XV, janvier–juin 1878, p. 98–115.

Famin, Pierre-Paul, *Au Tonkin et sur la frontière du Kwang-si*, Paris, A. Challamel, 1895.

Gait, E., *Report on the census of Bengal, 1901*, Calcutta, Bengal secretariat press, 1903.

Galliéni, Joseph-Simon, *Trois colonnes au Tonkin (1894–1895)*, Paris, R. Chapelot, 1899.

Gallois, Lucien, "Cartes nouvelles d'Indo-Chine et de Chine", *Annales de Géographie*, 8–39, 1899, p. 267–271.

Garnier, Francis, "Des nouvelles routes de commerce avec la Chine", *BSG*, sixième série, tome III, janvier–juin 1872, p. 147–160.

Giblin, R. W., "Royal Survey Work", in *Twentieth Century Impressions of Siam*, London, Lloyds Greater Britain Publishing Company, 1908, p. 121-127.

Gordon, Robert, *Report on the Irrawaddy River*, Rangoon, 1879.

Gouin, Auguste, "Le Tonkin, le haut Fleuve Rouge et ses affluents", *BSG*, septième série, tome VIII, 1887, p. 547–565.

Grabowsky, Volker, Wichāsin, Rēnū, *Chronicles of Chiang Khaeng: A Tai Lü Principality of the Upper Mekong*, Honolulu, University of Hawaii Press, 2008.

Hallett, Holt S., *A thousand Miles on an elephant in the Shan States*, Edinburgh, W. Blackwood and Sons, 1890.

Harmand, Jules, *L'homme du Mékong: un voyageur solitaire à travers l'Indochine inconnue*, Paris, Phébus, 1994.

Harmand, Jules, *L'Indo-Chine française, politique et administrative, conférence faite à l'Association républicaine du centenaire de 1789*, Paris, C. Pariset, 1887.

Hoàng, Hữu Xứng, *Đại-Nam quốc cương-giới vựng biên* [Traité des frontières du Đại Nam], 1886.

HOCQUARD, Charles-Édouard, *Une campagne au Tonkin*, Paris, Hachette, 1892. Réédition annotée par Philippe Papin, Paris, Arléa, 1999.

HOLDICH, T. H., "African Boundaries, and the Application of Indian Systems of Geographical Survey to Africa", *Proceedings of the Royal Geographical Society and Monthly Record of Geography*, 13–10, October 1891, p. 596.

KIPLING, Rudyard, *Kim*, London, Macmillan and Co, 1901.

LAUNAY, Adrien, *Atlas des missions de la Société des Missions-Etrangères*, Lille, Desclée de Brouwer, 1890.

LÊ, Quang Định, *Hoàng Việt Nhất Thống Dư Địa Chí* [皇越一統輿地志], Hải Phòng, 2005.

LEFÈVRE-PONTALIS, Pierre, "Notes sur quelques populations du nord de l'Indo-Chine (1ère série)", *Journal asiatique*, 1892.

LOCKHART, William Stephen, WOODTHORPE, Robert Gosset, *The Gilgit Mission, 1885–86*, London, Eyre and Spottiswoode, 1889.

LOWIS, C. C., *Census of India, 1901. Burma. A.pt. 2. Imperial tables*, Inde, Government Central Press, 1902.

LUBANSKI, Jules-Clément-Ladislas, "Note sur le Service géographique de l'Indo-Chine", in Paul DOUMER (ed.), *Situation de l'Indo-Chine (1897–1901). Rapport*, Hanoi, FH Schneider, 1902.

LUGARD, Frederick, *The Dual Mandate in British Tropical Africa*, London, W. Blackwood and Sons, 1922.

LUNET DE LAJONQUIÈRE, Étienne, *Ethnographie du Tonkin septentrional*, Paris, E. Leroux, 1906.

LUNET DE LAJONQUIÈRE, Étienne, *Ethnographie des territoires militaires, rédigée sur l'ordre du Général Coronnat, Commandant supérieur des troupes du groupe de l'Indo-Chine*, Hanoi, Schneider, 1904.

LYAUTEY, Hubert, *Lettres du Tonkin et de Madagascar (1894–1899)*, Paris, Armand Colin, 1920.

MADROLLE, Claudius, *De Marseille à Canton, guide du voyageur : Indo-Chine, canal de Suez, Djibouti et Harar, Indes, Ceylan, Siam, Chine méridionale*, Paris, Comité de l'Asie française, 1902.

MALTE-BRUN, Conrad, *Précis de la géographie universelle*, Paris, F. Buisson, 1812, vol. III, *Description de l'Asie, excepté l'Inde.*

MANGRĀI, Sao Sāimöng, *The Pāḍaeng Chronicle and the Jengtung State Chronicle Translated*, Ann Arbor, University of Michigan, 1981.

MARTONNE, Édouard de, MARTIN, Jean (eds), *La carte de l'empire colonial français*, Paris, G. Lang, 1931.

MAUSS, Marcel, *Carnet. Instruction pour les collaborateurs de l'École française d'Extrême-Orient*, Saigon, Imprimerie Ménard et Legros, 1900.

MCCARTHY, James, *Surveying and exploring in Siam*, London, J. Murray, 1900.

MILL, Hugh Robert, *The record of the Royal Geographical Society, 1830–1930*, London, The Royal Geographical Society, 1930.

MINISTERE DES AFFAIRES ÉTRANGÈRES DU LAOS, *Recueil des traités franco-siamois délimitant la frontière de l'Indochine et du Siam (lao-thai), 1886–1946*, Vientiane, Publication du Département des Traités et du Droit, 1996.

MINISTÈRE DES COLONIES, *Catalogue des cartes publiées par le service géographique des colonies*, Paris, impr. A. Gentil, 1907.

MOUHOT, Henri, *Voyage dans les royaumes de Siam, de Cambodge, de Laos et autres parties centrales de l'Indo-Chine. Relation extraite du journal et de la correspondance de l'auteur*, Paris, L. Hachette & cie, 1868 (Première édition: « Voyage dans les royaumes de Siam, de Cambodge, de Laos et autres parties centrales de l'Indo-Chine », in *Le Tour du Monde*, 1863).

NÉIS, Paul, "Sur les frontières du Tonkin", *Le Tour du Monde*, 55, 1888, p. 321–416.

NGUYEN, Trãi, *Nguyễn Trãi toàn tập* [Œuvres complètes de NT], vol. 6, *Địa dư chí* [géographie]. Traduction vietnamienne de Văn Tân, Đào Duy Anh, Trần Văn Giáp. Hanoi, Viện Sử Học, Nhà xuất bản KHXH 2e édition, 1976.

NOTTON, Camille, *Annales du Siam. 1, Chroniques de: Suvaṇṇa khamdëng, suvaṇṇa k'ôm khăm Siṅhanavati ; 2, Chronique de La:p'un ; 3, Chronique de Xieng Mãi*, Paris, Limoges, Nancy, Imprimerie Charles Lavauzelle, 1926–1932.

ORLÉANS, Henri d', *Du Tonkin aux Indes, janvier 1895–janvier 1896*, Paris, C. Lévy, 1898.

ORLÉANS, (Henri d'), "Du Tonkin au Yunnan", *BSG*, septième série, tome XVI, 1895, p. 389–404.

ORLÉANS, Henri d', *Autour du Tonkin*, Paris, C. Lévy, 1894.

PHASUK, Santanee, STOTT, Philip, *Royal Siamese Maps: War and Trade in Nineteenth Century Thailand*, Bangkok, River Books, 2004.

PAVIE, Auguste, *À la conquête des cœurs. Le pays des millions d'éléphants et du parasol blanc, les pavillons noirs, Déo-van-tri*, Paris, Bossard, 1921.

PAVIE, Auguste, *Mission Pavie en Indochine, 1879–1895*, Paris, E. Leroux, 1898–1919, 11 volumes (dont *Atlas de la mission Pavie en Indochine*, Paris, A. Challamel, 1903; *Géographie et voyages. II, Exposé des travaux de la mission (troisième et quatrième périodes, 1889 à 1895)*, Paris, E. Leroux, 1906).

PHAYRE, Arthur P., *History of Burma*, London, Trübner, 1883.

POTTIER, Edmond, "Notice sur la vie et les travaux de M. Gabriel Devéria", *Comptes rendus des séances de l'Académie des Inscriptions et Belles-Lettres*, 44–2, 1900, p. 127–143.

QUOC SU QUÁN (Bureau d'Histoire de l'État), *Đại Nam nhất thống chí* [大南一統志] [Encyclopédie géographique, économique, sociale et culturelle du Đại-Nam, par province], à la demande de Tự Đức, 1882.

QUOC SU QUÁN (Bureau d'Histoire de l'État), *Đồng Khánh địa dư chí* [Géographie descriptive de l'empereur Đồng Khánh], 1886–1888. NGÔ, Đức Thọ, PAPIN, Philippe, NGUYỄN, Văn Nguyên (eds), *Đồng Khánh địa dư chí* [*Géographie descriptive de l'empereur Đồng Khánh*], Hanoi, Institut Hán Nôm; EFEO, 2003.

RATCHASOMPHAN, Sænluang, WYATT, David K., *The Nan chronicle*, Ithaca, Cornell University Press, 1994.

RICHARDS, John, *The Gulf of Siam Pilot. From the Survey made in H.M.S. Saracen between the years 1856–58*, London, Hydrographic Office, 1863.

ROCHER, Émile, *La province chinoise du Yün-nan*, Paris, Ernest Leroux, 1879.

ROMANET DU CAILLAUD, Frédéric, "Ethnographie du Tong-King", *BSG*, septième série, tome II, juillet–décembre 1881, p. 252–261 and p. 267–271.

ROUGET, « Étude sur la cartographie de l'Indochine française », *Annales de Géographie*, 15–79, 1906, p. 26–42.

ROUX, Émile, *Aux sources de l'Irraouaddi, voyage de Hanoï à Calcutta par terre, illustré de cent dessins ou gravures directes d'après les photographies rapportées par l'auteur*, Paris, Hachette, 1897.

SCONCE, G. Colquhoun, *Journal of the Salween Surveying Expedition*, Calcutta, Military Orphan Press, 1865.

SCOTT, James George, *Gazetteer of Upper Burma and the Shan States*, Rangoon, Government printing, 1901.

SCOTT, James George, *France and Tongking: A Narrative of the Campaign of 1884 and the Occupation of Further India*, London, T. F. Unwin, 1885.

SCOTT, James George, MITTON, Geraldine, *Scott of the Shan Hills. Orders and Impressions*, London, J. Murray, 1936.

SLADEN, Edward Bosc, "Burma: Exploration via the Irrawaddy and Bhamo to South-Western China", *Proceedings of the Royal Geographical Society of London*, vol. 15, no. 5, 1870, p. 343–364.

SPEARMAN, Horace Ralph, *The British Burma gazetteer*, Rangoon, Government Press, 1879.

ST. JOHN, Wallace, *Josiah Nelson Cushing, Missionary and Scholar, Burma*, Rangoon, American Baptist Mission Press, 1912.

The French in Indo-China. With a Narrative of Garnier's Explorations in Cochin-China, Annam, and Tonquin, London, Nelson, 1884.

THINGYAN, Sithu Gamani, *Zinme Yazawin: Chronicle of Chiang Mai*, Yangon, Universities Historical Research Centre, 2003.

"Travaux du bureau topographique des troupes de l'Indo-Chine", *Annales de Géographie*, 7–36, 1898, p. 457–462.

VERNE, Jules, *Aventures de trois Russes et de trois Anglais*, Paris, J. Hetzel, 1872.

WATSON, C. E., FEDDEN, Francis, *Journal of the Salween Surveying Expedition During the Season 1864–65. With Route Map*, Calcutta, Military Orphan Press, 1865.

WHEELER, James Talboys, *Journal of a Voyage up the Irrawaddy to Mandalay and Bhamo*, Rangoon, J. W. Baynes, 1871.

WILLIAMS, Clement, *Through Burmah to western China. Being notes of a journey in 1863 to establish the practicability of a trade-route between the Irawaddi and the Yang-Tse-Kiang*, Edinburgh, London, W. Blackwood and sons, 1868.

WOODTHORPE, Robert Gosset, *The Lushai expedition, 1871–1872*, London, Hurst and Blackett, 1873.

WOODTHORPE, Robert Gosset, "Explorations on the Chindwin River, Upper Burma", *Proceedings of the RGS*, XI, 1889, p. 197–216.

WYATT, David K., 'ARUNRAT WICHĪANKHĪEO, *The Chiang Mai chronicle*, Chiang Mai, Silkworm Books, 1995.

YOE, Shway, *The Burman, his life and notions*, London, Macmillan, 1882.

1.2.1 *Journals*

Bulletin de la Société de géographie, Paris, 1858–1899.

La Géographie, Paris, 1900–1914.

Proceedings of the Royal Geographical Society, London, 1879–1892.

Bulletin de l'École française d'Extrême-Orient, 1885–1914.

The Geographical Journal, London, 1893–1914.

2 Secondary Sources

ADSHEAD (Samuel A. M.), *Province and Politics in Late Imperial China. Viceregal Government in Szechwan, 1898–1911*, London, Curzon, 1984.

AGERON, Charles-Robert, *France coloniale ou parti colonial?*, Paris, PUF, 1978.

AMSELLE, Jean-Loup, "Ethnies et espaces. Pour une anthropologie topologique", in Jean-Loup Amselle, Elikia M'Bokolo, *Au cœur de l'ethnie. Ethnie, tribalisme et État en Afrique*, Paris, La Découverte, 1999 [1985], p. 11–48.

ANDERSON, Benedict, *Imagined Communities: Reflections on the Origin and Spread of Nationalism*, London, Verso, 1983.

ANDERSON, David M., KILLINGRAY, David, *Policing the Empire: Government, Authority, and Control, 1830–1940*, Manchester, Manchester University Press, 1991.

Andrews, John Harwood, *A Paper landscape: the Ordnance Survey in Nineteenth-century Ireland*, Dublin, Four Courts Press, 2002.

Appadurai, Arjun, *Modernity at Large. Cultural Dimensions of Globalization*, Minneapolis, University of Minnesota Press, 1996.

Atwill, David G., *The Chinese Sultanate. Islam, Ethnicity, and the Panthay Rebellion in Southwest China, 1856–1873*, Stanford, Stanford University Press, 2005.

Aung-Thwin, Michael, Aung-Thwin, Maitrii, *A History of Myanmar since Ancient Times: Traditions and Transformations*, London, Reaktion Books, 2013.

Avila, Isabelle, "Les cartes de l'Empire dans les atlas britanniques, (1884–1914)", in Blais, Hélène, Deprest, Florence, Singaravélou, Pierre (eds), *Territoires impériaux. Une histoire spatiale du fait colonial*, Paris, Publications de la Sorbonne, 2011, p. 169–188.

Baigent, Elizabeth, "Clements Robert Markham (1830–1916)", in *Oxford Dictionary of National Biography*, Oxford, Oxford University Press, 2006.

Baird, Ian, "Different Views of History. Shades of Irredentism along the Laos-Cambodia Border", *Journal of Southeast Asian studies*, 41–2, 2010, p. 187–213.

Balandier, Georges, "La situation coloniale: approche théorique", *Cahiers internationaux de sociologie*, 11, 1951, p. 44–79.

Barkey, Karen, "Trajectoires impériales: histoires connectées ou études comparées?", *Revue d'histoire moderne et contemporaine*, 54–4bis, 2007, p. 90–103.

Barrau, Jacques, "Culture itinérante, culture sur brûlis, culture nomade, écobuage ou essartage?", *Études rurales*, 45, 1972.

Barrow, Ian J., *Making history, drawing territory. British mapping in India, c. 1756–1905*, New Delhi, Oxford University Press, 2003.

Barth, Volker, Cvetkovski, Roland, *Imperial Co-operation and Transfer, 1870–1930: Empires and Encounters*, London, Bloomsbury Academic, 2015.

Bat, Jean-Pierre, Courtin, Nicolas (eds), *Maintenir l'ordre colonial. Afrique et Madagascar, XIXe–XXe siècle*, Rennes, Presses Universitaires de Rennes, 2012.

Baud, Michel, Schendel, Willem van, "Toward a Comparative History of Borderlands", *Journal of World History*, 8–2, 1997, p. 211–242.

Bayart, Jean-François, "Hégémonie et coercition en Afrique subsaharienne", *Politique africaine*, 110, 2012, p. 123–152.

Bayart, Jean-François, "Les Études postcoloniales. Un carnaval académique", *Politique étrangère*, 4, 2010, p. 912–918.

Bayart, Jean-François, Bertrand, Romain, "De quel "legs colonial" parle-t-on?", *Esprit*, décembre 2006, p. 134–160.

Bayly, Christopher, *The Birth of the Modern World, 1780–1914. Global Connections and Comparisons*, Oxford, Blackwell Publishing, 2004.

BAYLY, Christopher, *Imperial meridian: the British Empire and the World, 1780–1830*, London, Longman, 1989.

BELHOSTE, Bruno, *La formation d'une technocratie. L'École polytechnique et ses élèves de la Révolution au Second Empire*, Paris, Belin, 2003.

BÉNAT TACHOT, Louise, GRUZINSKI, Serge (eds), *Passeurs culturels. Mécanismes de métissage*, Paris, Éd. de la Maison des sciences de l'homme, 2002.

BENTON, Lauren, *A Search for Sovereignty: Law and Geography in European Empires, 1400–1900*, Cambridge, Cambridge University Press, 2009.

BERDOULAY, Vincent, *La formation de l'École française de géographie, 1870–1914*, Paris, Bibliothèque nationale, 1981.

BERQUE, Jacques, "Qu'est-ce qu'une tribu nord-africaine?", *Éventail de l'histoire vivante, Hommage à Lucien Febvre*, Paris, Armand Colin, 1953, vol. 1, p. 261–271.

BERTIN, Jacques, *Sémiologie graphique. Les diagrammes, les réseaux, les cartes*, Paris, Gauthier-Villard, 1964.

BERTRAND, Christophe, HERBELIN, Caroline, KLEIN, Jean-François, *Indochine. Des territoires et des hommes, 1856–1956*, Paris, Gallimard, 2013.

BERTRAND, Romain, *L'histoire à parts égales. Récits d'une rencontre Orient-Occident (XVI*e*–XVII*e *siècle)*, Paris, Seuil, 2011.

BERTRAND, Romain, *Mémoires d'empire. La controverse autour du «fait colonial»*, Bellecombe-en-Bauges, Éd. du Croquant, 2006.

BERTRAND, Romain, *État colonial, noblesse et nationalisme à Java. La tradition parfaite*, Paris, Karthala, 2005.

BESSE, Jean-Marc, BLAIS, Hélène, SURUN, Isabelle (eds), *Naissances de la géographie moderne, 1760–1860: lieux, pratiques et formation des savoirs de l'espace*, Lyon, ENS éditions, 2010.

BLAIS, Hélène, *Mirages de la carte. L'invention de l'Algérie coloniale*, Paris, Fayard, 2014.

BLAIS, Hélène, "Frontière coloniale", in Sophie DULUCQ, Jean-François KLEIN, Benjamin STORA (eds), *Les mots de la colonisation*, Toulouse, Presses universitaires du Mirail, 2008.

BLAIS, Hélène, "Les représentations cartographiques du territoire algérien au moment de la conquête. Le cas de la carte des officiers d'état-major (1830–1870)", in Pierre SINGARAVÉLOU (ed.), *L'empire des géographes. Géographie, exploration et colonisation, XIX*e*–XX*e *siècle*, Paris, Belin, 2008, p. 124–134.

BLAIS, Hélène, "Les enquêtes des cartographes en Algérie ou les ambiguïtés de l'usage des savoirs vernaculaires en situation coloniale", *Revue d'histoire moderne et contemporaine*, 54–4, octobre-décembre 2007, p. 70–85.

BLAIS, Hélène, *Voyages au grand océan. Géographies du Pacifique et colonisation, 1815–1845*, Paris, CTHS, 2005.

Blais, Hélène, "Comment trouver le "meilleur nom géographique"? Les voyageurs français et la question de la dénomination des îles océaniennes au XIXe siècle", *L'Espace géographique*, tome 30–4, 2001, p. 348–357.

Blais, Hélène, Deprest, Florence, Singaravélou, Pierre (eds), *Territoires impériaux. Une histoire spatiale du fait colonial*, Paris, Publications de la Sorbonne, 2011.

Blais, Hélène, Laboulais, Isabelle (eds), *Géographies plurielles. Les sciences géographiques au moment de l'émergence des sciences humaines, 1750–1850*, Paris, L'Harmattan, 2006.

Blanchard, Emmanuel, "Introduction", *Genèses*, n° 86–1, avril 2012, p. 2–7.

Blanchard, Emmanuel, Deluermoz, Quentin, Glasman, Joël, "La professionnalisation policière en situation coloniale: détour conceptuel et explorations historiographiques", *Crime, Histoire & Sociétés / Crime, History & Societies*, 15–2, décembre 2011, p. 33–53.

Blanckaert, Claude, *Le terrain des sciences humaines, instructions et enquêtes, XVIIIe–XXe siècle*, Paris, L'Harmattan, 1996.

Boilley, Pierre, *Les Touaregs Kel Adagh. Dépendances et révoltes: du Soudan français au Mali contemporain*, Paris, Karthala, 1999.

Bonnemaison, Joël, *Gens de pirogue et gens de terre*, Paris, Éd. de l'ORSTOM, 1996.

Boucheron, Patrick, Delalande, Nicolas, *Pour une histoire-monde*, Paris, PUF, 2013.

Bourguet, Marie-Noëlle, Lepetit, Bernard, Nordman, Daniel, Sinarellis, Maroula (eds), *L'invention scientifique de la Méditerranée. Égypte, Morée, Algérie*, Paris, EHESS, 1998.

Brailey, Nigel J., *Imperial amnesia: Britain, France and « the question of Siam »*, Dordrecht, Republic of Letters Publishing, 2009.

Brailey, Nigel J., "The Railway-Oceanic Era, the India-China and India-Singapore Railway Schemes, and Siam", in Thomas G. Otte, Keith Neilson (eds), *Railways and International Politics. Paths of Empire, 1848–1945*, London, Routledge, 2006, p. 94–111.

Bravo, Michael T., "Ethnographic Navigation and the Geographical Gift", in David N. Livingstone, Charles W. J. Withers (eds), *Geography and Enlightenment*, Chicago, University of Chicago Press, 1999, p. 199–235.

Brebion, Antoine, *Bibliographie des voyages dans l'Indochine française du IXe au XIXe* siècle, New York, B. Franklin, 1970.

Brebion, Antoine, *Dictionnaire de bio-bibliographie générale, ancienne et moderne de l'Indochine française*, Paris, Société d'éditions géographiques, maritimes et coloniales, 1935.

BRET, Patrice, "Le Dépôt général de la guerre et la formation scientifique des ingénieurs-géographes militaires en France (1789–1830)", *Annals of Science*, 48–2, 1991, p. 113–157.

BROC, Numa, *Dictionnaire illustré des explorateurs français du XIXe siècle*, CTHS, 1988.

BROCHEUX, Pierre, HÉMERY, Daniel, *Indochina: An Ambiguous Colonization, 1858-1954*, Berkeley, University of California Press, 2009.

BRUNEAU, Michel, "Les États-nations de l'espace eurasiatique issus d'empires, de cités-États, d'États-mandala. Que doivent-ils au modèle européen occidental?", *L'Espace géographique*, Tome 43–3, 2014, p. 251–264.

BRUNEAU, Michel, *L'Asie d'entre Inde et Chine. Logiques territoriales des États*, Paris, Belin, 2006.

BRUNEAU, Michel, "La notion de frontière et sa signification dans la Péninsule indochinoise", *Moussons*, 3, juin 2001, p. 33–55.

BRUNEAU, Michel, "Pierre Gourou (1900–1999)", *L'Homme. Revue française d'anthropologie*, 153, 2000, p. 7–26.

BRUNEAU, Michel, DORY, Daniel, *Géographies des colonisations, XVe–XXe siècles*, Paris, L'Harmattan, 1994.

BRUNET, Roger (ed.), *Géographie Universelle*, Paris, Belin, 1994.

BRUNET, Roger, *Les mots de la géographie. Dictionnaire critique*, Montpellier, Reclus, 1992.

BUCKLAND, Charles Edward, *Dictionary of Indian Biography*, London, S. Sonnenschein, 1906.

BURBANK, Jane, COOPER, Frederick, *Empires in World History. Power and the Politics of Difference*, Princeton, Princeton University Press, 2010.

BUTLIN, Robin Alan, *Geographies of Empire. European Empires and Colonies c. 1880–1960*, Cambridge, Cambridge University Press, 2009.

CAIN, Peter J., HOPKINS, Antony G., *British Imperialism, 1688–1990*, London, Longman, 1993.

CAMPBELL, Alexander, "Education in Burma", *Journal of the Royal Society of Arts*, 49, 1946, p. 438–448.

CARTER, Paul, *The Road to Botany Bay: An Exploration of Landscape and History*, London, Faber & Faber, 1987.

CASAJUS, Dominique, "Las, le temps non, mais nous nous en allons …", *Textuel*, 56, 2008, p. 185–210.

CHALINE, Jean-Pierre, *Sociabilité et érudition. Les sociétés savantes en France. XIXe–XXe siècles*, Paris, CTHS, 1995.

CHANDLER, David Porter, *A History of Cambodia*, Boulder, Westview Press, 2008 [1983].

CHANG, Wen-Chin, *Beyond Borders. Stories of Yunnanese Chinese Migrants of Burma*, Ithaca, Cornell University Press, 2014.

CHAPUIS, Olivier, *À la mer comme au ciel. Beautemps-Beaupré & la naissance de l'hydrographie moderne, 1700–1850*, Paris, Presses de l'Université Paris-Sorbonne, 1999.

CHOUVY, Pierre-Arnaud, *Les territoires de l'opium. Conflits et trafics du Triangle d'Or et du Croissant d'Or (Birmanie, Laos, Thaïlande et Afghanistan, Iran, Pakistan)*, Genève, Olizane, 2002.

CLARENCE-SMITH, William Gervase, « Editorial note. Zomia and beyond », *Journal of Global History*, 5–2, 2010, p. 185–186.

CLAYTON, Daniel, "Le passé colonial/impérial et l'approche postcoloniale de la géographie anglophone", in Pierre SINGARAVÉLOU (ed.), *L'Empire des géographes. Géographie, exploration et colonisation. XIXe–XXe siècle*, Paris, Belin, 2008, p. 219–234.

CŒDÈS, George, *Les peuples de la péninsule indochinoise*, Dunod, Paris, 1962.

CŒDÈS, George, *Les États hindouisés d'Indochine et d'Indonésie*, Paris, E. de Boccard, 1948.

COHN, Bernard S., *An anthropologist among the historians and other essays*, Delhi, Oxford University Press, 1987.

COLLIER, Peter, "The Colonial Survey Committee and the Mapping of Africa", International Symposium on "Old Worlds-New Worlds", The History of Colonial Cartography, 1750–1950, Utrecht University, 21–23 August 2006.

COLLIER, Peter, INKPEN, Rob, "The Royal Geographical Society and the development of surveying 1870–1914", *Journal of Historical Geography*, 29–1, janvier 2003, p. 93–108.

COLLIGNON, Béatrice, *Les Inuit. Ce qu'ils savent du territoire*, Paris, L'Harmattan, 1996.

COOPER, Frederick, *Le colonialisme en question. Théorie, connaissance, histoire*, Paris, Payot, 2010.

COOPER, Frederick, STOLER, Ann Laura (ed.), *Tensions of Empire. Colonial Cultures in a bourgeois World*, Berkeley, University of California Press, 1997.

CORDIER, Henri, *Bibliotheca indosinica. Dictionnaire bibliographique des ouvrages relatifs à la péninsule indochinoise*, Paris, Imprimerie nationale, 1912.

CÔTE, Marc, *L'Algérie ou l'espace retourné*, Paris, Flammarion, 1988.

CUSHING, Josiah Nelson, *A Shan and English dictionary*, Rangoon, C. Bennett, American Mission press, 1881.

DAGENS, Bruno, "Angkor, instrument politique. Avant, avec et après le protectorat", in Hugues TERTRAIS (ed.), *Angkor VIIIe–XXIe siècle. Mémoire et identité khmères*, Paris, Autrement, 2008.

DAINVILLE, François de, *Le langage des géographes, Terres, signes et couleurs des cartes anciennes, 1500–1800*, Paris, Picard, 1964.

Darwin, John, *The Empire Project: the Rise and Fall of the British World-system, 1830–1970*, Cambridge, Cambridge University Press, 2009.

Darwin, John, "Imperialism and the Victorians: The Dynamics of Territorial Expansion", *The English Historical Review*, 112–447, 1997, p. 614–642.

Daughton, James Patrick, *An Empire Divided: Religion, Republicanism, and the Making of French Colonialism, 1880–1914*, Oxford, Oxford University Press, 2006.

Davis, Bradley Camp, *Imperial Bandits. Outlaws and Rebels in the China-Vietnam Borderlands*, Seattle, University of Washington Press, 2017.

Davis, Bradley Camp, *States of banditry: The Nguyen government, bandit rule, and the culture of power in the post-Taiping China-Vietnam borderlands*, University of Washington, 2008.

Debarbieux, Bernard, "Construits identitaires et imaginaires de la territorialité: variations autour de la figure du "montagnard"", *Annales de géographie*, n° 660–661, 2008, p. 90–115.

De Koninck, Rodolphe, *L'Asie du Sud-Est*, Paris, A. Colin, 2012.

Descours-Gatin, Chantal, Villiers, Hugues, *Guide de recherches sur le Vietnam. Bibliographies, archives et bibliothèques de France*, Paris, L'Harmattan, 1983.

Devillers, Philippe, *Français et annamites. Partenaires ou ennemis?*, Paris, Denoël, 1998.

Dimier, Véronique, *Le gouvernement des colonies. Regards croisés franco-britanniques*, Bruxelles, Université de Bruxelles, 2004.

Dion, Isabelle, *Auguste Pavie, l'explorateur aux pieds nus. Cambodge-Laos*, Aix-en-Provence, Archives nationales d'outre-mer, 2010.

Doherty, Gillian M., *The Irish Ordnance Survey: History, Culture, and Memory*, Dublin, Four Courts Press, 2006.

Donaldson, John W., "Politics and Scale in Boundary-making: the Work of Boundary Commissions", *Journal of Historical Geography*, vol 34, issue 3, July 2008, p. 393–396.

Driver, Felix, *Geography Militant: Cultures of Exploration and Empire*, Oxford, Blackwell Publishing, 2000.

Driver, Felix, Jones, Lowri, *Hidden Histories of Exploration: Researching the RGS-IBG Collections*, London, Royal Holloway / University of London, 2009.

Dubois, Jean, *Dictionnaire de linguistique*, Paris, Larousse, 1974.

Dullin, Sabine, *La frontière épaisse. Aux origines des politiques soviétiques, 1920–1940*, Paris, EHESS, 2014.

Dulucq, Sophie, Soubias, Pierre (eds), *L'espace et ses représentations en Afrique subsaharienne*, Paris, Khartala, 2004.

Dulucq, Sophie, Klein, Jean François, Stora, Benjamin (eds), *Les mots de la colonisation*, Toulouse, Presses universitaires du Mirail, 2008.

Dunlop, Catherine, *Cartophilia. Maps and the search for identity in the French-German borderland*, Chicago, University of Chicago Press, 2015.

Durand, Frédéric, "Atlas et ouvrages sur la cartographie de l'Asie du Sud-Est. Esquisse de perspective bibliographique", *Péninsule*, 54, 2007, p. 159–170.

Duroselle, Jean-Baptiste, *Tout empire périra. Une vision théorique des relations internationales*, Paris, Publications de la Sorbonne, 1981.

Edney, Matthew H., *Mapping an empire: the geographical construction of British India, 1765–1843*, Chicago, University of Chicago Press, 1997.

Erbig, Jeffrey Alan. *Where Caciques and Mapmakers Met. Border Making in Eighteenth-Century South America.* Chapel Hill, University of North Carolina Press, 2020

Espagne, Michel, Werner, Michael (eds), *Transferts. Les relations interculturelles dans l'espace franco-allemand, XVIIIe et XIXe siècle*, Paris, Éd. Recherche sur les civilisations, 1988.

Evard, Maurice, "Charles Jacot-Guillarmod", in M. Schlup (ed.), *Biographies neuchâteloises*, VG. Attinger, 2001, vol. 3.

Farrington, Anthony, *Delineating British Burma.*, Leiden, IDC Publishers, 2005.

Febvre, Lucien, *Le problème de l'incroyance au XVIe siècle. La religion de Rabelais*, Paris, A. Michel, 1942.

Feldmann, Hans-Uli, "Der Topograph Charles Jacot-Guillarmod. Einer der Pionniere der Schweizer Felsdarstellung", *Cartographica Helvetica*, 31, 2005.

Fierro, Alfred, *Inventaire des photographies sur papier de la Société de géographie*, Paris, Bibliothèque nationale, 1986.

Fierro, Alfred, *Inventaire des manuscrits de la Société de géographie*, Paris, Bibliothèque nationale, 1984.

Fierro, Alfred, *La Société de géographie, 1821–1946*, Genève, Droz, 1983.

Fisher, John, "Lord Curzon and British Strategic Railways in Central Asia before, during and after World War I", in Thomas G. Otte, Keith Neilson (éd.), *Railways and International Politics. Paths of Empire, 1848–1945*, London, Routledge, 2006, p. 137–155.

Forbes, Andrew D. W., Henley, David E. F., *The Haw. Traders of the Golden Triangle*, Chiang Mai, Asia Film House, 1997.

Forest, Alain, *Le Cambodge et la colonisation française. Histoire d'une colonisation sans heurts (1897–1920)*, Paris, l'Harmattan, 1979.

Fortunel, Frédéric, "Les plateaux méridionaux d'Asie du Sud-Est continentale, de la marginalité à l'interconnexion", *Péninsule*, n° 57, 2008, p. 19–41.

Foucher, Michel, *Frontières d'Afrique. Pour en finir avec un mythe*, Paris, CNRS Éditions, 2014.

FOUCHER, Michel, *Fronts et frontières. Un tour du monde géopolitique*, Paris, Fayard, 1988.

FOUCHER, Michel, *L'invention des frontières*, Paris, FEDN, 1987.

FOUCHER, Michel (ed.), *Asies nouvelles*, Paris, Belin, 2002.

FOURNIAU, Charles, *Vietnam. Domination coloniale et résistance nationale, 1858–1914*, Paris, Indes savantes, 2002.

FOURNIAU, Charles, *Le contact colonial franco-vietnamien. Le premier demi-siècle, 1858–1911*, Aix-en-Provence, Publications de l'Université de Provence, 1999.

FOURNIAU, Charles, "La frontière sino-vietnamienne et le face à face franco-chinois à l'époque de la conquête du Tonkin", in Pierre-Bernard LAFONT (ed.), *Les frontières du Vietnam*, Paris, L'Harmattan, 1989, p. 85–103.

FOX, Jefferson, "Siam Mapped and Mapping in Cambodia: Boundaries, Sovereignty and Indigenous Conceptions of Space", *Society and Natural Resources*, 15, 2002, p. 65–78.

FRANCHINI, Philippe (ed.), *Tonkin 1873–1954. Colonie et Nation. Le delta des mythes*, Paris, Autrement, 1994.

FRANK, Robert, *Pour l'histoire des relations internationales*, Paris, PUF, 2012.

GAZQUEZ, Denis, *Publications officielles de l'Indochine coloniale. Inventaire analytique (1859–1954)*, Paris, Bibliothèque nationale de France, 2004.

GHEERBRANDT, Jean-Laurent, *Pavie, le grand humain de l'Indochine*, Paris, Éditions de l'Empire Français, 1949.

GIERSCH, C. Patterson, "Coton, copper and caravans: trade and the transformation of Southwest China", in Eric TAGLIACOZZO, Wen-Chin CHANG (eds), *Chinese Circulations: Capital, Commodities, and Networks in Southeast Asia*, Durham, Duke University Press, 2011.

GIERSCH, C. Patterson, *Asian Borderlands. The Transformation of Qing China's Yunnan Frontier*, Cambridge, Harvard University Press, 2006.

GIRARDET, Raoul, *L'idée coloniale en France* de *1871 à 1962*, Paris, La Table ronde, 1972.

GODLEWSKA, Anne, SMITH, Neil (eds), *Geography and Empire*, Oxford, B. Blackwell, 1993.

GOH, Geok Yian, *The wheel-turner and his house: kingship in a Buddhist ecumene*, DeKalb, NIU Press, 2015.

GOMANE, Jean-Pierre, *L'exploration du Mékong. La mission Ernest Doudart de Lagrée-Francis Garnier, 1866–1868*, Paris, Harmattan, 1994.

GORSHENINA, Svetlana, *Asie centrale. L'invention des frontières et l'héritage russo-soviétique*, Paris, CNRS, 2012.

GOSCHA, Christopher E., *Vietnam. A new History*, New York, Basic Books, 2016.

GOSCHA, Christopher E., *Vietnam or Indochina? Contesting concepts of space in Vietnamese nationalism, 1887–1954*, Copenhagen, NIAS, 1995.

GOSCHA, Christopher E., IVARSSON, Søren (eds), *Contesting visions of the Lao past: Lao historiography at the crossroads*, Copenhagen, NIAS, 2003.

GOUDINEAU, Yves, LORRILLARD, Michel (eds), *Recherches nouvelles sur le Laos*, Paris, École française d'Extrême-Orient, 2008.

GOULD, Peter, BAILLY, Antoine (eds), *Le pouvoir des cartes. Brian Harley et la cartographie*, Paris, Anthropos, 1995.

GOUROU, Pierre, *Les pays tropicaux. Principes d'une géographie humaine et économique*, Paris, PUF, 1947.

GRABOWSKY, Volker, "Chiang Khaeng 1893–1896. A lue principality in the Upper Mekong Valley at the centre of Franco-British Rivalry" in Christopher E. GOSCHA, Søren IVARSSON (eds), *Contesting visions of the Lao past: Lao historiography at the crossroads*, Copenhagen, NIAS, 2003.

GRATALOUP, Christian, *L'invention des continents. Comment l'Europe a découpé le Monde*, Paris, Larousse, 2009.

GRÉMONT, Johann, *Maintenir l'ordre aux confins de l'Empire. Pirates, trafiquants et rebelles entre Chine et Viêt Nam 1895-1940*, Paris, Maisonneuve et Larose/ Hémisphères, 2018.

GROSSER, Pierre, "Comment écrire l'histoire des relations internationales aujourd'hui? Quelques réflexions à partir de l'Empire britannique", *Histoire@Politique*, 10, janvier 2010.

GUHA, Ranajit, *Dominance without hegemony. History and power in colonial India*, Cambridge, Massachussets, Harvard University Press, 1997.

GUILHOT, Nicolas, *Histoire d'une parenthèse cartographique. Les Alpes du nord dans la cartographie topographique française aux 19e et 20e siècles*, thèse de doctorat d'histoire, sous la direction de Girolamo Ramunni, Université Lyon II-Lumière, 2005.

GUILLEMOT, François, *Viêt-Nam, fractures d'une nation. Une histoire contemporaine de 1858 à nos jours*, Paris, La Découverte, 2018.

HAINES, Chad, *Nation, Territory, and Globalization in Pakistan:Traversing the Margins*, London, Routledge, 2012.

HAINES, Chad, "Colonial Routes: Reorienting the Northern Frontier of British India", *Ethnohistory*, 51–3, 2004, p. 535–565.

HALL, Daniel George, *A History of South-East Asia*, London, Macmillan, 1955.

HALL, Daniel George, "British Burma, 1886–1942", in *A History of South-East Asia*, New York, Springer, 1981, p. 770–788.

HARLEY, John Brian, *The New Nature of Maps. Essays in the History of Cartography*, Baltimore, Johns Hopkins University Press, 2001.

HARLEY, John Brian, WOODWARD, David (eds), *History of Cartography*, Chicago, Chicago University Press, 1987-.

HARLEY, John Brian, WOODWARD, David (eds), *History of Cartography*, Chicago, University of Chicago Press, 1992, vol. 2, book 2, *Cartography in the Traditional East and Southeast Asian Societies*.

HARVEY, Godfrey Eric, *History of Burma. From the Earliest Times to 10 March, 1824, The Beginning of the English Conquest*, London, Cass, 1925.

HEADRICK, Daniel R., *The Tools of Empire. Technology and European Imperialism in the Ninteenth Century*, Oxford, Oxford University Press, 1981.

HEFFERNAN, Michael, "Shifting Sands: The Trans-Saharan Railway", in Stanley BRUNN (ed.), *Engineering Earth: The Impacts of Megaengineering Projects*, New York, Springer, 2011, p. 617–626.

HÉMERY, Daniel, "Inconstante Indochine ... L'invention et les dérives d'une catégorie géographique", *Revue française d'histoire d'outre-mer*, 87–326, 2000, p. 137–158.

HILL, Ann, *Merchants and Migrants: Ethnicity and Trade among Yunnanese Chinese in Southeast Asia*, New Haven, Yale University Press, 1998.

HOPKIRK, Peter, *The Great Game: On Secret Service in High Asia*, Oxford, Oxford University Press, 1990.

HOPKIRK, Peter, *Trespassers on the Roof of the World: The Race for Lhasa*, London, J. Murray, 1982.

HOSTETLER, Laura, *Qing Colonial Enterprise. Ethnography and Cartography in Early Modern China*, Chicago, University of Chicago Press, 2001.

HOSTETLER, Laura, "Contending Cartographic Claims? The Qing Empire in Manchu, Chinese, and European Maps", in James R. Akerman (ed.), *The Imperial Map. Cartography and the Mastery of Empire*, Chicago, University of Chicago Press, 2009, p. 93–132.

HOWE, Stephen, (ed.), *The New Imperial Histories Reader*, London, Routledge, 2010.

HURD, John, KERR Ian J., *India's railway history: a research handbook*, Leiden, Brill, 2012.

IVARSSON, Søren, *Creating Laos: the making of a Lao space between Indochina and Siam, 1860–1945*, Copenhague, NIAS, 2008.

JACOB, Christian, *L'empire des cartes. Approche théorique de la cartographie à travers l'histoire*, Paris, Albin Michel, 1992.

JIENNINGS, Eric, *Imperial Heights. Dalat and the Making and Undoing of French Indochina*, Berkeley, University of California Press, 2011.

JENNINGS, Eric, *Curing the colonizers: hydrotherapy, climatology, and French colonial spas*, London, Duke University Press, 2006.

JONES, Max, "Measuring the World: Exploration, Empire and the Reform of the Royal Geographical Society, c. 1874–93", in Martin DAUNTON (ed.), *The Organisation of Knowledge in Victorian Britain*, Oxford, Oxford University Press, 2005.

JONES, Max, *The last great quest: Captain Scott's Antarctic sacrifice*, Oxford, Oxford University Press, 2003.

KATO, Kumiko, "Qing China's View of Its Border and Territory in Southernmost Yunnan in the 1830s: Analyses of Historical Sources Concerning Sipsongpanna", *Journal of the School of Letters*, 11, mars 2015, p. 1–18.

KEITH, Charles, *Catholic Vietnam: a Church from Empire to Nation*, Berkeley, University of California Press, 2012.

KERR, Ian J., *Engines of Change: The Railroads that Made India*, Greenwood Publishing Group, 2007.

KERR, Ian J., *Building the railways of the Raj, 1850–1900*, Delhi, Oxford University Press, 1995.

KHIN MAR MAR, *Guide to the archival sources of the British administration period, 1826–1948: Myanmar*, Yangon, Ministry of National Planning and Economic Development, National Archives Dept, Myanmar, 2003.

KIM, Nanny, *Mountain Rivers, Mountain Roads: Transport in the Southwest of Qing China, 1700–1850*, Leiden, Brill, 2012.

KIRK-GREENE, Anthony, *Britain's imperial administrators, 1858–1966*, Basingstoke, Macmillan, 2000.

KLEIN, Jean-François, *Pennequin, une "éthique coloniale française?" Penser la pacification, dépasser la colonisation (1849–1916)*, Université Paris-Sorbonne, 2014.

KLEIN, Jean-François, SUREMAIN, Marie-Albane de, SINGARAVÉLOU, Pierre, *Atlas des empires coloniaux, XIXe–XXe siècles*, Paris, Autrement, 2012.

KRATOSKA, Paul H., RABEN, Remco, SCHULTE NORDHOLT, Henk, *Locating Southeast Asia: Geographies of Knowledge and Politics of Space*, Singapore University Press, 2005.

KUMAR, Dharma (ed.), *Cambridge Economic History of India, Volume II (1757–1970)*, Cambridge, Cambridge University Press, 1983.

KYONG-MCCLAIN, Jeff, DU, Yongtao (eds), *Chinese history in geographical perspective*, Lanham, Lexington Books, 2013.

LABOULAIS, Isabelle (ed.), *Les usages des cartes. XVIIe–XIXe siècle*, Strasbourg, Presses universitaires de Strasbourg, 2008.

LABOULAIS, Isabelle, "Les cartes géographiques du fonds Montbret", *Annales historiques de la Révolution française*, 343, mars 2006, p. 175.

LABOULAIS, Isabelle (ed.), *Combler les blancs de la carte. Modalités et enjeux de la construction des savoirs géographiques (XVIe–XXe siècle)*, Strasbourg, Presses universitaires de Strasbourg, 2004.

LACOSTE, Yves, *La géographie, ça sert, d'abord, à faire la guerre*, Paris, Maspero, 1976.

LAFONT, Pierre-Bernard (ed.), *Les frontières du Vietnam*, Paris, L'Harmattan, 1989.

LAFONT, Pierre-Bernard (ed.), *Bibliographie du Laos*, tome I (1866–1961), Paris, EFEO, 1978.

LAGIRARDE, François, "Temps et lieux d'histoires bouddhiques: à propos de quelques "chroniques" inédites du Lanna", *BEFEO*, 94, 2010, p. 59–94.

LAKROUM, Monique, "Les projets français pour le transsaharien (XIXe–XXe siècle). Un challenge pour l'industrie", *Culture technique*, 19, 1989, p. 295–302.

LAMBERT, David, LESTER, Alan (eds), *Colonial Lives across the British Empire: Imperial Careering in the Long Nineteenth Century*, Cambridge, Cambridge University Press, 2006.

LATOUR, Bruno, *Science in Action: How to Follow Scientists and Engineers through Society*, Harvard University Press, 1987.

LATOUR, Bruno, "Les "vues" de l'esprit: une introduction à l'anthropologie des sciences et des techniques", *Culture technique*, 14, 1985, p. 4–29.

LATOUR, Bruno, "Comment redistribuer le Grand Partage?", *Revue de Synthèse*, 110, Avril-juin 1983, p. 203–236.

LAWRANCE, Benjamin, OSBORN, Emily, ROBERTS, Richard (eds), *Intermediaries, interpreters, and clerks: African employees in the making of colonial Africa*, Madison, The University of Wisconsin Press, 2006.

LAZARUS, Neil, *Penser le postcolonial: Une introduction critique*, Paris, Éditions Amsterdam, 2006.

LEACH, Edmund, *Les systèmes politiques des hautes terres de Birmanie. Analyse des structures sociales kachin*, traduit par Anne GUÉRIN, Paris, F. Maspero, 1972.

LE FAILLER, Philippe, *La rivière noire. L'intégration d'une marche frontière au Vietnam*, Paris, CNRS Éditions, 2014.

LEFEBVRE, Camille, *Frontières de sable, frontières de papier. Histoire de territoires et de frontières, du jihad de Sokoto à la colonisation française du Niger, XIXe–XXe siècles*, Paris, Publications de la Sorbonne, 2015.

LEFEBVRE, Camille, "Itinéraires de sable. Parole, geste et écrit au Soudan central au XIXe siècle", *Annales. Histoire, Sciences Sociales*, Vol. 4, 2009, p. 797–824.

LEFEBVRE, Camille, SURUN, Isabelle, « Exploration et transferts de savoir: deux cartes produites par des Africains au début du XIXe siècle », Mappemonde 92–4, 2008.

LEIDER, Jacques, "Taxation et groupes de service sous la royauté arakanaise. Un rapport d'enquête de 1803 dans un manuscrit de la Bibliothèque nationale de Paris", *Aséanie*, 1–1, 1998, p. 67–89.

LEIDER, Jacques, "La route de Am. Contribution à l'étude d'une route terrestre entre la Birmanie et le golfe du Bengale", *Journal Asiatique*, 282, 1994, p. 335–370.

LEJEUNE, Dominique, *Les sociétés de géographie en France et l'expansion coloniale au XIXe siècle*, Paris, Albin Michel, 1993.

Le Muséum au premier siècle de son histoire, Paris, Éditions du Muséum d'histoire naturelle, 1997.

Lenoir, Désirée, *Le consul qui en savait trop. Les ambitions secrètes de la France en Chine (1886–1904)*, Paris, Nouveau monde éditions, 2011.

Leroi-Gourhan, André, *Le Geste et la Parole*, Paris, Albin Michel, 1965, vol. II.

Lester, Alan, *Imperial Netwoks: Creating Identities in Nineteenth Century South Africa and Britain*, London, Routledge, 2001.

Lévy, Jacques, Lussault, Michel, *Dictionnaire de la géographie et de l'espace des sociétés*, Paris, Belin, 2003.

Lewis, G. Malcolm (ed.), *Cartographic Encounters: Perspective on Native American Mapmaking and Map Use*, Chicago, Chicago University Press, 1998.

Livingstone, David N., *Putting Science in its Place: Geographies of Scientific Knowledge*, Chicago, University of Chicago Press, 2003.

Livingstone, David N., *The Geographical Tradition: Episodes in the History of a Contested Enterprise*, Oxford, Blackwell, 1992.

Loiseaux, Olivier, "Jules Hansen (1849–1931): dessinateur-géographe", *Revue du Comité français de Cartographie*, 210, décembre 2011, p. 137–148.

Lombard, Denys (ed.), *Rêver l'Asie. Exotisme et littérature coloniale aux Indes, en Indochine et en Insulinde*, Paris, EHESS, 1993.

Lombard, Denys, *Le carrefour javanais. Essai d'histoire globale. Vol. 1: Les limites de l'occidentalisation; Vol. 2: Les réseaux asiatiques; Vol. 3: L'héritage des royaumes concentriques*, Paris, Éditions de l'EHESS, 1990.

Lombard, Denys, "Les concepts d'espace et de temps dans l'archipel insulindien", *Annales*, 41–6, 1986, p. 1385–1396.

Lorcin, Patricia, *Kabyles, Arabes, Français: identités coloniales*, Limoges, Pulim, 2005.

MacKenzie, John M., *Imperialism and Popular Culture*, Manchester, Manchester University Press, 1986.

Margolin, Jean-Louis, Markovits Claude François, *Les Indes et l'Europe. Histoires connectées, XVe–XXIe siècle*, Paris, Gallimard, 2015.

Markovits, Claude (ed.), *Histoire de l'Inde moderne, 1480–1950*, Paris, Fayard, 1994.

Marshall, Andrew, *The trouser people: a story of Burma, in the shadow of the empire*, New York, Counterpoint, 2002.

McDougall-Waters, Julie, "British School Atlases, 1880–1930: Questions of Relevance, Credibility and Authorship in the Production of Geographical Knowledge", *Imago Mundi*, 66–1, 2 janvier 2014, p. 82–94.

Mehaud, Catherine, *Manuscrits de la Société de géographie concernant l'Asie et l'Océanie*, Paris, Bibliothèque nationale, 1979.

Merle, Isabelle, ""La situation coloniale" chez Georges Balandier. Relecture historienne", *Monde(s). Histoire, Espaces, Relations*, 4–2, 2013, p. 211–232.

MESSAOUDI, Alain, *Les arabisants et la France coloniale. Savants, conseillers, médiateurs, 1780–1930*, Lyon, ENS, 2015.

METCALF, Thomas R., *Imperial Connections: India in the Indian Ocean Arena, 1860–1920*, Berkeley, University of California Press, 2007.

METCALF, Thomas R., *Ideologies of the Raj*, Cambridge, Cambridge University Press, 1994.

MICHAUD, Jean, "French Military Ethnography in Colonial Upper Tonkin (Northern Vietnam), 1897–1904", *Journal of Vietnamese Studies*, 8–4, 2013, p. 1–46.

MICHAUD, Jean, *Incidental ethnographers: French catholic missions on the Tonkin-Yunnan frontier, 1880–1930*, Leiden, Brill, 2007.

MICHAUD, Jean, *Historical Dictionary of the Peoples of the Southeast Asian Massif*, Lanham, Toronto et Oxford, Scarecrow Press, 2006.

MICHAUD, Jean, OVESEN, Jan, *Turbulent Times and Enduring Peoples: Mountain Minorities in the South-East Asian Massif*, Richmond, Routledge, 2000.

MICHAUD, Jean, TURNER, Sarah, ROCHE, Yann, "Mapping ethnic diversity in highland Northern Vietnam", *GeoJournal*, 57–4, 2002, p. 305–323.

MIDDLETON, Dorothy, "Guide to the Publications of the Royal Geographical Society 1830–1892", *The Geographical Journal*, 144–1, March 1978, p. 99–116.

MILLER, David Philip, REILL, Peter Hanns (eds), *Visions of empire: voyages, botany and representations of nature*, Cambridge, Cambridge University Press, 1996.

MINASSIAN, Gaïdz, *Zones grises. Quand les États perdent le contrôle...*, Paris, Autrement, 2011.

MITCHELL, Timothy, *Colonizing Egypt*, Cambridge, Cambridge University Press, 1988.

MUDIMBE, Valentin, *The Invention of Africa: Gnosis, Philosophy, and the Order of Knowledge*, Bloomington, Indiana University Press, 1988.

MUNDY, Barbara E., *The Mapping of New Spain: Indigenous Cartography and the Maps of the Relacionès Geogràficas*, Chicago, Chicago University Press, 1996.

MUS, Paul, *L'angle de l'Asie*, Paris, Hermann, 1977.

MYINT-U, Thant, *The Making of Modern Burma*, Cambridge, Cambridge University Press, 2001.

MYOE, Maung Aung, *In the Name of Pauk-Phaw: Myanmar's China Policy Since 1948*, Singapour, Institute of Southeast Asian Studies, 2011.

NGÔ, Đức Thọ, PAPIN, Philippe NGUYỄN Văn Nguyên (eds), *Đồng Khánh địa dư chí*, Hanoi, Institut Hán Nôm; EFEO, 2003.

NGÔ, Quốc Dũng, *Biên-Giới Việt-Trung 1885–2000: Lịch sử thành hình và những tranh chấp*, États-Unis, Dũng Châu, 2005.

NGUYỄN, Thế Anh, *Parcours d'un historien du Viêt Nam. Recueil des articles écrits par Nguyễn Thế Anh*, édité par Philippe Papin, Paris, Les Indes savantes, 2008.

Nguyễn, Thế Anh, "Établissement par le Vietnam de sa frontière dans les confins occidentaux", in Pierre-Bernard Lafont (ed.), *Les frontières du Vietnam*, Paris, L'Harmattan, 1989.

Nguyễn, Thị Hải, *La marche de Cao Bằng*, Paris, Presses d'Inalco, 2018.

Nguyễn, Thi Hanh, *Les conflits frontaliers sino-vietnamiens de 1885 à nos jours*, PhD dissertation, Université Paris 1 Panthéon-Sorbonne, 2006.

Nicolet, Claude, *L'inventaire du monde. Géographie et politique aux origines de l'Empire romain*, Paris, Fayard, 1988.

Niollet, Dominique, *L'épopée des douaniers en Indochine, 1874–1954*, Paris, Éditions Kailash, 1998.

Nordman, Daniel, *Frontières de France. De l'espace au territoire*, Paris, Gallimard, 1998.

Nordman, Daniel, "De quelques catégories de la science géographique: frontière, région et hinterland en Afrique du Nord (XIXe–XXe siècle)", *Annales HSS*, n°5, 1997, p. 969–986.

Nordman, Daniel, *La notion de frontière en Afrique du nord. Mythes et réalités (vers 1830- vers 1912)*, Montpellier III, 1975.

Nordman, Daniel, Raison, Jean-Pierre (eds), *Sciences de l'homme et conquête coloniale. Constitution et usages des sciences humaines en Afrique, XIXe–XXe siècle*, Paris, Presses de l'École normale supérieure, 1980.

Noyes, J. K., *Colonial Space. Spatiality in the Discourse of German South West Africa, 1884–1915*, Chur, Reading, Paris, Philadelphia, Harwood Academic Publishers, 1992.

Nugent, Paul, *Smugglers, secessionists and loyal citizens on the Ghana-Togo frontier. The life of the borderlands since 1914*, Athens/USA/GB, Ohio University Press, 2002.

Nugent, Paul, Asiwaju, Anthony Ijaola, *African boundaries: barriers, conduits and opportunities*, London/NYC, Pinter, 1996.

Numnonda, Thamsook, "The First American Advisers in Thai History", *Journal of the Siam Society*, 62–2, 1974, p. 121–148.

Ó Cadhla, Stiofan, *Civilizing Ireland: Ordnance survey 1824–1842*, Dublin, Portland, Irish Academic Press, 2007.

Okell, J., *A guide to the romanization of Burmese*, London, Luzac; The Royal Asiatic Society of Great Britain and Ireland, 1971.

Owen, Roger, *Lord Cromer. Victorian imperialist, Edwardian proconsul*, Oxford, Oxford University Press, 2004.

Palsky, Gilles, *L'Esprit des Cartes. Approches historiques, sémiologiques et sociologiques en cartographie*, Université Paris XII-Val de Marne, 2003.

Pansini, Valeria, *L'œil du topographe et la science de la guerre. Travail scientifique et perception militaire (1760–1820)*, PhD dissertation, EHESS, Paris, 2002.

Papin, Philippe, "Un laboratoire pour l'étude des transferts", in Hoai Huong Aubert-Nguyen, Michel Espagne (eds), *Le Vietnam. Une histoire de transferts culturels*, Paris, Demopolis, 2015, p. 33–46.

PAPIN, Philippe, *Histoire des territoires de Hà Nội. Quartiers, villages et sociétés urbaines du XIXe au début du XXe siècle*, Paris, Les Indes savantes, 2013.

PAPIN, Philippe, *Histoire de Hanoï*, Paris, Fayard, 2001.

PAPIN, Philippe, *Viêt-Nam. Parcours d'une nation*, Paris, La Documentation française, 1999.

PARK-BARJOT, Rang-Ri, *La Société de construction des Batignolles, des origines à la première guerre mondiale (1846–1914)*, Paris, PUPS, 2005.

PEGG, Richard, *Cartographic traditions in East Asian maps*, Honolulu, University of Hawaii Press, 2015.

PELLETIER, Monique, *Les cartes des Cassini. La science au service de l'État et des provinces*, Paris, CTHS, 2013.

PELLETIER, Philippe, *L'Extrême-Orient. L'invention d'une histoire et d'une géographie*, Paris, Gallimard, 2011.

PELLETIER, Philippe, "La cartographie de l'Asie du Sud-Est et la dénomination de la "Mer du Japon" (Mer de l'Est) aux XVIIe et XVIIIe siècles", *Péninsule*, 54, 2007, p. 58–97.

PELLETIER, Philippe (ed.), *Norao I. Identités territoriales en Asie orientale*, Paris, Les Indes savantes, 2004.

PERDUE, Peter C., *China Marches West*, Harvard, Harvard University Press, 2005.

PHAN, Huy Lê, "Địa bạ cổ ở Việt Nam", in *Tìm về cội nguồn*, Hà Nội, Thế Giới, 1998, vol. 1.

PHASUK, Santanee, STOTT, Philip, *Royal Siamese Maps: War and Trade in Nineteenth Century Thailand*, Bangkok, River Books, 2004.

PICKLES, John, *A History of Spaces. Cartographic Reason, Mapping and the Geo-coded World*, London/New York, Routlegde, 2004.

POISSON, Emmanuel, *Mandarins et subalternes au nord du Viêt Nam. Une bureaucratie à l'épreuve, 1820–1918*, Paris, Maisonneuve & Larose, 2004.

PORTER, Andrew, *Atlas of British Overseas Expansion*, London, Routledge, 1991.

PORTER, Andrew (ed.), *Oxford History of the British Empire. The Nineteenth Century*, Oxford, Oxford University Press, 2001.

PORTER, Bernard, *The Absent-minded Imperialists: Empire, Society, and Culture in Britain*, Oxford, Oxford University Press, 2006.

QUACH-LANGLET, Tam, "La perception des frontières dans l'ancien Vietnam à travers quelques cartes vietnamiennes et occidentales", in Pierre-Bernard LAFONT (ed.), *Les Frontières du Vietnam. Histoire des frontières de la péninsule indochinoise*, Paris, L'Harmattan, 1989, p. 25–62.

RAJ, Kapil., *Relocating Modern Science. Circulation and the Constitution of Knowledge in South Asia and Europe, 1650–1900*, Basingstoke, Palgrave MacMillan, 2007.

RAND, Gavin, "'Martial Races' and 'Imperial Subjects': Violence and Governance in Colonial India, 1857–1914", *European Review of History/Revue européenne d'histoire*, 13–1, 2006, p. 1–20.

Reid, Anthony, *Southeast Asia in the Age of Commerce: 1450–1680*, New Haven; London, Yale University Press, 1988–1993.

Renard, Ronald D., "The Delineation of the Kayah States Frontiers with Thailand: 1809–1894", *Journal of Southeast Asian Studies*, 18, 1987, p. 81–92.

Ritchie, John (ed.), *Australian Dictionary of Biography*, Melbourne, Melbourne University press, 1990, vol.12.

Robequain, Charles, *L'Indochine française*, Paris, Armand Colin, 1935.

Robic, Claire-Marie, "Note sur la notion d'échelle dans la géographie française de la fin du XIXe siècle et du début du XXe siècle", Journée à l'EHESS, *Echelles et territoires*, Paris, 29 avril 2002.

Robinson, Ronald, Gallagher, John, *Africa and the Victorians: The Official Mind of Imperialism*, London, Palgrave Macmillan, 1961.

Rollet de l'Isle, Maurice, *Étude historique sur les ingénieurs hydrographes et le service hydrographique de la marine (1814–1914)*, Paris, Imprimerie nationale, 1951.

Roque, Ricardo, Wagner, Kim A. (eds), *Engaging colonial knowledge: reading European archives in world history*, Basingstoke, Palgrave Macmillan, 2012.

Saada, Emmanuelle, *Les enfants de la colonie: les métis de l'Empire français entre sujétion et citoyenneté*, Paris, Éditions la Découverte, 2007.

Sadan, Mandy, *Being and becoming Kachin. Histories beyond the State in the Borderworlds of Burma*, Oxford, Oxford University press, 2013.

Sadan, Mandy, *A Guide to Colonial Sources on Burma: Ethnic & Minority Histories of Burma in the India Office Records, British Library*, Orchid Press, 2008.

Safier, Neil, "The Confines of the Colony. Boundaries, Ethnographic Landscapes, and Imperial Cartography in Iberoamerica", in Akerman, James R. (ed.), *The Imperial Map, Cartography and the Mastery of Empire*, Chicago, University of Chicago Press, 2009, p. 133–183.

Sahlins, Peter, *Boundaries: The Making of France and Spain in the Pyrenees*, Berkeley, University of California press, 1989.

Said, Edward, *Orientalism*, New York, Pantheon Books, 1978.

Salkin, Geneviève, *Le triple destin de Jules Harmand. Médecin, explorateur, diplomate*, Paris, Economica, 1992.

"Savoirs autochtones et écriture de l'histoire en situation coloniale (XIXe–XXe siècle)", *Outre-Mers*, t. 94, n° 252–253, 2006.

Schaffer, Simon, Roberts, Lissa, Raj, Kapil, Delbourgo, James (eds), *The Brokered World: Go-Betweens and Global Intelligence, 1770–1820*, Sagamore Beach, Science History Publications, 2009.

Schiavon, Martina, *Itinéraires de la précision. Géodésiens, artilleurs, savants et fabricants d'instruments de précision en France, 1870–1930*, Nancy, PUN-Éditions universitaires de Lorraine, 2014.

SCHIAVON, Martina, "Les officiers géodésiens du Service géographique de l'armée et la mesure de l'arc de méridien de Quito (1901–1906)", *Histoire & mesure*, XXI-2, 2006, p. 55–94.

SCHOBER, Juliane, "Colonial Knowledge and Buddhist Education in Burma", in Ian HARRIS (ed.), *Buddhism, Power and Political Order*, Routledge, 2007, p. 52–70.

SCHWARTZBERG, Joseph, "A Nineteenth Century Burmese Map Relating to French Colonial Expansion in Southeast Asia", *Imago Mundi*, 46, 1994, p. 117–127.

SCHWARTZBERG, Joseph, "Southeast Asian Geographical maps", in John Brian HARLEY, David WOODWARD (eds), *History of Cartography*, Chicago, University of Chicago Press, 1992, vol. 2, book 2, p. 741–827.

SCOTT, James C., *The Art of Not Being Governed: An Anarchist History of Upland Southeast Asia*, Yale, Yale University Press, 2009.

SEEKINS, Donald M., *Historical dictionary of Burma (Myanmar)*, Lanham, Scarecrow press, 2006.

SELLIER, Jean, *Atlas des peuples d'Asie méridionale et orientale*, Paris, La Découverte, 2001.

SIBEUD, Emmanuelle, "Post-colonial et Colonial Studies: enjeux et débats", *Revue d'histoire moderne et contemporaine*, 51, n°4 bis, 2004.

SIBEUD, Emmanuelle, *Une science impériale pour l'Afrique? La construction des savoirs africanistes en France, 1878–1930*, Paris, EHESS, 2002.

SIMON, Hélène, *Auguste Pavie, explorateur en Indochine*, Rennes, Éditions Ouest-France, 1997.

SINGARAVÉLOU, Pierre (ed.), *Les empires coloniaux, XIXe–XXe siècle*, Paris, Éd. Points, 2013.

SINGARAVÉLOU, Pierre, *Professer l'Empire. Les « sciences coloniales » en France sous la IIIe République*, Paris, Publications de la Sorbonne, 2011.

SINGARAVÉLOU, Pierre (ed.), *L'Empire des géographes. Géographie, exploration et colonisation, XIXe–XXe siècle*, Paris, Belin, 2008.

SOUBEYRAN, Olivier, "La géographie coloniale. Un élément structurant dans la naissance de l'école française de géographie", in Michel BRUNEAU, Daniel DORY, *Les enjeux de la tropicalité*, Paris, Masson, 1989, p. 82–90.

SPIVAK, Gayatri Chakravorty, "Can the subaltern speak?", in *Marxism and the Interpretation of Culture*, London, Macmillan, 1988, p. 24–28.

STASZAK, Jean-François, "Ethnogéographie et savoirs géographiques: quelques problèmes méthodologiques et épistémologiques", *Bulletin de l'Association de géographes français*, 73–1, 1996, p. 39–54.

STOLER, Ann Laura, *Along the Archival Grain. Epistemic Anxieties and Colonial Common Sense*, Princeton, Princeton University Press, 2009.

SUÁREZ, Thomas, *Early mapping of Southeast Asia*, Singapour, Periplus Editions, 1999.

SUNDERLAND, David, *Managing the British empire: the crown agents, 1833–1914*, London, Royal Historical Society, 2004.

SUREMAIN, Marie-Albane de, "Les cartes ethno-démographiques de l'Afrique occidentale: enjeux d'une construction 1952–1963", *Mappemonde*, 92–4, 2008.

SUREMAIN, Marie-Albane de, "Coutumes, coutumiers", in Sophie DULUCQ, Jean-François KLEIN, Benjamin STORA (eds), *Les mots de la colonisation*, Toulouse, Presses universitaires du Mirail, 2007, p. 37–38.

SUREMAIN, Marie-Albane de, "Ambitions positivistes et savoirs autochtones: les traditions orales dans le Bulletin du CEHSAOF/IFAN", in Sophie DULUCQ, Colette ZYTNICKI (eds), *"Savoirs autochtones, XIXe-XXe siècle", Outre-Mers*, 2006, p. 33–46.

SURUN, Isabelle, TERTRAIS, Hugues, "Introduction – Les continents orphelins?", *Mondes(s)*, n° 3, *Inventions des continents*, juillet 2013, p. 7–27.

SURUN, Isabelle, "L'exploration de l'Afrique au XIXe siècle: une histoire précoloniale au regard des postcolonial studies", *Revue d'histoire du XIXe siècle*, n°32, 2006–1, p. 21–39.

SURUN, Isabelle, *Dévoiler l'Afrique ? Lieux et pratiques de l'exploration (Afrique occidentale, 1780-1880)*, Paris, Editions de la Sorbonne, 2018.

TAGLIACOZZO, Eric, *Secret Trades, porous Borders: Smuggling and States along a Southeast Asian Frontier, 1865–1915*, New Haven, Yale University Press, 2005.

TAILLARD, Christian (ed.), *Norao II. Intégrations régionales en Asie orientale*, Paris, Les Indes savantes, 2004.

TARLING, Nicholas (ed.), *The Cambridge history of Southeast Asia.* Vol. 2, *The nineteenth and twentieth centuries*, Cambridge, Cambridge University Press, 2008.

TENG, Emma, *Taiwan's Imagined Geography. Chinese Colonial Travel Writing and Pictures, 1683–1895*, Harvard, Harvard University Asia Center, 2004.

TAYLOR, Keith. W., *A History of the Vietnamese*, Cambridge, Cambridge University Press, 2013.

TERTRAIS, Hugues, *L'Asie pacifique au XXe siècle*, Paris, Armand Colin, 2015.

TERTRAIS, Hugues (ed.), *Atlas des guerres d'Indochine, 1940–1990. De l'Indochine française à l'ouverture internationale*, Paris, Autrement, 2004.

TESSIER, Olivier, BOURDEAUX, Pascal, *Đà Lạt. Et la carte créa la ville...*, Hanoi, Nhà Xuất Bản Tri Thủc/École Française d'Extrême-Orient, 2013.

THAW KAUNG, Sithu, *Aspects of Myanmar history and culture*, Yangon, Gant-gaw-Myaing Sarpay, 2010.

THOMAS, Frédéric, *Histoire du régime et des services forestiers français en Indochine de 1862 à 1945*, Hanoi, Thế Giới, 1999.

TILLEY, Helen, *Africa as a Living Laboratory: Empire, Development, and the Problem of Scientific Knowledge, 1870–1950*, Chicago, University of Chicago Press, 2011.

TILLION, Germaine, *Il était une fois l'ethnographie*, Paris, Seuil, 2000.

TILLION, Germaine, *L'Algérie aurésienne*, Paris, La Martinère/Perrin, 2001.

TRAGER, Frank, KOENIG, William, *Burmese Sit-tàns 1764–1826: records of rural life and administration*, Tucson, University of Arizona Press, 1979.

TSUBOÏ, Yoshiharu, *L'empire vietnamien face à la France et à la Chine*, Paris, L'Harmattan, 1987.

TULLY, John A., *France on the Mekong. A History of the Protectorate in Cambodia, 1863–1953*, Lanham, University Press of America, 2002.

TURNBULL, David, *Maps are Territories. Science is an Atlas*, Chicago, Chicago University Press, 1993.

TURNER, Alicia, *Saving Buddhism. The Impermanence of Religion in Colonial Burma*, Honolulu, University of Hawaii Press, 2014.

VANDEN BUSSCHE, Eric, *Contested Realms. Colonial Rivalry, Border Demarcation, and State Building* in *Southwest China, 1886–1960*, PhD Dissertation, Stanford University, 2014.

VAN SCHENDEL Willem, "Geographies of knowing, geographies of ignorance: jumping scale in Southeast Asia", *Environment and Planning D: Society and Space*, 20–6, 2002, p. 647–668.

VASQUEZ Jean-Michel, *La cartographie missionnaire en Afrique. Science, religion et conquête (1870–1930)*, Paris, Karthala, 2011.

VENAYRE, Sylvain, *La gloire de l'aventure. Genèse d'une mystique moderne 1850–1940*, Paris, Aubier, 2002.

WALKER, Andrew, "Seditious State-Making in the Mekong Borderlands: The Shan Rebellion of 1902–1904", *Sojourn: Journal of Social Issues in Southeast Asia*, 29–3, 2014, p. 554–590.

WALKER, Andrew, *The Legend of the Golden Boat : Regulation, Trade and Traders in the Borderlands of Laos, Thailand, China, and Burma*, Honolulu, University of Hawaii Press, 1999.

WALLER, Derek J., *The Pundits. British exploration of Tibet and Central Asia*, Lexington, University Press of Kentucky, 1990.

WANG, Jianping, *Concord and Conflict. The Hui Communities of Yunnan Society*, Stockholm, Almqvist & Wiksell, 1996.

WEBB, Adrian, *The Expansion of British Naval Hydrographic Administration, 1808–1829*, Thèse de doctorat soutenue à l'Université d'Exeter, 2010.

WERNER, Michael, ZIMMERMANN, Bénédicte, *De la comparaison à l'histoire croisée*, Paris, Seuil, 2004.

WHITE, Richard, *The Middle Ground: Indians, Empires, and Republics in the Great Lakes region, 1650–1815*, Cambridge, Cambridge University Press, 1991.

WHITMORE, John, "Cartography in Vietnam", in John Brian HARLEY, David WOODWARD (eds), *History of Cartography*, Chicago, University of Chicago Press, 1992, vol. 2, book 2, p. 478–508.

WINICHAKUL, Thongchai, *Siam mapped: a history of the geo-body of a nation*, Honolulu, University of Hawaii press, 1994.

WOODMAN, Dorothy, *The making of Burma*, London, Cresset Press, 1962.

YEE, Cordell, "Taking the world's measure: Chinese maps between observation and text", in John Brian HARLEY, David WOODWARD (eds), *History of Cartography*, Chicago, University of Chicago Press, 1992, vol. 2, book 2, p. 96–127.

ZINK, Michel, DRÈGE, Jean-Pierre (eds), *Paul Pelliot. De l'histoire à la légende*, Paris, Académie des Inscriptions et Belles-Lettres, 2013.

Index

www.ingramcontent.com/pod-product-compliance
Lightning Source LLC
Chambersburg PA
CBHW071325230926
51849CB00002B/321
* 9 7 8 9 0 0 4 4 5 6 2 1 1 *